C0-ALO-939

VEGETATION DYNAMICS & GLOBAL CHANGE

Vegetation Dynamics & Global Change

Edited by
Allen M. Solomon and Herman H. Shugart

Chapman & Hall
New York • London

IIASA

First published in 1993 by
Chapman and Hall
an imprint of
Routledge, Chapman & Hall, Inc.
29 West 35th Street
New York, NY 10001-2299

Published in Great Britain by
Chapman and Hall
2-6 Boundary Row
London SE1 8HN

Printed in the United States of America on acid free paper.

Library of Congress Cataloging in Publication Data

Vegetation dynamics and global change / Allen M. Solomon & Herman H.
Shugart, editors.
p. cm.
Based on a series of conferences held in 1988 and 1989.
Includes bibliographical references and index.
ISBN 0-412-03671-1 (cloth)—ISBN 0-412-03681-9 (paper)
1. Vegetation dynamics—Congresses. 2. Vegetation and climate—Congresses. 3. Forest ecology—Congresses. 4. Climatic changes—Congresses. 5. Vegetation dynamics—Mathematical models—Congresses. 6. Vegetation and climate—Mathematical models—Congresses. 7. Forest ecology—Mathematical models—Congresses.
I. Solomon, Allen M., 1943- . II. Shugart, H. H.
QK910.V42 1992
581.5—dc20 92-35042
CIP

British Library Cataloguing in Publication Data also available.

Contributors

P.J. Burton
University of British Columbia
Department of Forest Science
270-2357 Main Mall, Vancouver, B.C., Canada V6T 1W5

W.H. Chomentowski
Institute for the Study of Earth, Oceans, and Space
University of New Hampshire
Complex Systems Research Center
Durham, NH 03824, USA

B.J. Choudhury
Hydrological Sciences Branch
NASA/Goddard Space Flight Center
Greenbelt, MD 20771, USA

W.P. Cramer
University of Trondheim College of Arts and Sciences
Department of Geography
N-7055 Dragvoll, Norway

W.R. Emanuel
Environmental Sciences Division
Oak Ridge National Laboratory
PO Box X, Oak Ridge, TN 37831-6038, USA

A.D. Friend
Institute of Terrestrial Ecology
Edinburgh Research Station
Bush Estate, Penicuik, Midlothian, Scotland, U.K.

M.R. Fulton
Industrial Engineering Department
Texas A&M University
College Station, TX 77843-3131

G. Grabherr
Institute of Plant Physiology
University of Vienna
Althanstrasse 14, A-1090 Vienna, Austria

J.P. Grime
Unit of Comparative Plant Ecology (NERC)
Department of Animal and Plant Sciences
The University, Sheffield, S10 2TN, UK

S. Kojima
College of Liberal Arts
Toyama University
3190 Gofuku, Toyama City, 930 Japan

Ch. Körner
Botanical Institute
University of Basel
Schoenbeinstrasse 6, CH-4056 Basel, Switzerland

R. Leemans
Global Change Department
National Institute for Public Health and Environmental Protection, RIVM
PO Box 1, 3720 Bilthoven, The Netherlands

Ph. Martin
Commission of the European Communities
Joint Research Center
TP440I-21020
Ispra, Italy

R.A. Monserud
Intermountain Research Station
USDA Forest Service
1221 S. Main St., Moscow, ID 83843, USA

B. Moore III
Institute for the Study of Earth, Oceans, and Space
University of New Hampshire, Complex Systems Research Center
Durham, NH 03824, USA

W.M. Post
Environmental Sciences Division
Oak Ridge National Laboratory
PO Box 2008, Oak Ridge, TN 37831-6335, USA

I.C. Prentice
Department of Plant Ecology
University of Lund
Ö. Vallgatan 14, S-223 61, Lund, Sweden

B.N. Rock
Institute for the Study of Earth, Oceans, and Space
University of New Hampshire, Complex Systems Research Center
Durham, NH 03824, USA

C. Rosenzweig
Columbia University and NASA/Goddard Institute for Space Studies
2880 Broadway, New York, NY 10025, USA

H.H. Shugart
Department of Environmental Sciences, Clark Hall
University of Virginia
Charlottesville, VA 22903, USA

D.L. Skole
Institute for the Study of Earth, Oceans, and Space
University of New Hampshire, Complex Systems Research Center
Durham, NH 03824, USA

T.M. Smith
Department of Environmental Sciences, Clark Hall
University of Virginia
Charlottesville, VA 22903, USA

A.M. Solomon
U.S. Environmental Protection Agency
ERL-C
200 SW 35 Street
Corvallis, OR 97333

F.I. Woodward
Unit of Comparative Plant Ecology
Department of Animal and Plant Sciences
The University, Sheffield S10 2TN, UK

Contents

III Measuring Global Vegetation Change

IV Modeling Global Vegetation Change

Preface

During the summer of 1987, a series of discussions[1] was held at the International Institute for Applied Systems Analysis (IIASA) in Laxenburg, Austria, to plan a study of global vegetation change. The work was aimed at promoting the International Geosphere-Biosphere Programme (IGBP), sponsored by the International Council of Scientific Unions (ICSU), of which IIASA is a member. Our study was designed to provide initial guidance in the choice of approaches, data sets and objectives for constructing global models of the terrestrial biosphere. We hoped to provide substantive and concrete assistance in formulating the working plans of IGBP by involving program planners in the development and application of models which were assembled from available data sets and modeling approaches. Recent acceptance of the "IIASA model" as the starting point for endeavors of the Global Change and Terrestrial Ecosystems Core Project of the IGBP suggests we were successful in that aim.

The objective was implemented by our initiation of a mathematical model of global vegetation, including agriculture, as defined by the forces which control and change vegetation. The model was to illustrate the geographical consequences to vegetation structure and functioning of changing climate and land use, based on plant responses to environmental variables. The completed model was also expected to be useful for examining international environmental policy responses to global change, as well as for studying the validity of IIASA's experimental approaches to environmental policy development.

The project was based on a conference comprised of 75 scientists from 20 nations held at IIASA in April 1988. The meeting brought together the principal and potential contributors to the project. Among model users present were IGBP leaders who were to observe, advise on, and assimilate from, the entire project as it unfolded. Also, scientists were invited who could delineate the ecological relationships and principles relevant to model characteristics, and who could contribute to model construction. The meeting participants defined the function of the global vegetation modeling system and, therefore, how it could be used and by whom. They discussed the salient vegetation responses to climate and soils which must be incorporated, the land use and soils effects and responses to be included, the geographic climate data sets and classifications available, the modeling approaches which were most appropriate for the task, and the nature of the model interfaces with its prospective users.

The meeting results were applied during a summer-long workshop at IIASA in 1988 to develop a strategy for modeling global vegetation change from available data sets and numerical approaches. Twelve ecologists and modelers[2] and six graduate students[3] spent much of the three summer months of 1988 at IIASA. As a consequence of interactions among the participants, the group provided a tremendous range of experiences and approaches and produced an imaginative and well researched modeling strategy.[4]

The strategy which resulted from the April 1988 conference and the contributions by the participants of the 1988 summer workshop was the basis for a second summer workshop in 1989. The 1989 activities were devised to construct components of the global vegetation dynamics model, and were preceded by planning meetings in Charlottesville, Virginia in fall, 1988 and Sofia, Bulgaria in spring, 1989. Along with the established scientists,[5] graduate students[6] in 1989 had been chosen for their interests and capabilities in the summer research tasks.

In addition to efforts by these enthusiastic participants, there were invaluable contributions by many others too numerous to mention. However, we would be remiss if we did not gratefully acknowledge the constant efforts of Marilyn Brandl, the project administrative assistant who provided much of the organizational support for the project, and of Erica Schwartz, whose editing skills at IIASA and at University of Virginia catalyzed the production of the book. Financial resources were provided by IIASA national member organizations through their annual contributions, as well as by a grant from the A.P. Sloan Foundation, and by the U.S. National Science Foundation through Grant INT87-06669 to the American Academy of Arts and Sciences.

The 1988 IIASA Global Vegetation Conference, the 1988 and 1989 IIASA Summer Vegetation Change Workshops, and the IIASA Biosphere Dynamics Project in which these activities were embedded, were tremendously stimulating arenas for approaching the objectives of global change research. Although the project activites were responsible for many substantive scientific products, including the papers which compose this book, they also are and will be the source of considerably more long-term, if not intangible, productivity. The intellectual working relationships among the participants which the project engendered and the research ideas on global change which they provoked will probably be the foundation of productive scientific research for many years. IIASA's most valuable product, enduring scientific working relationships among researchers who would otherwise never have had the opportunity to interact, has been well served by the participants in our research. It is fitting that the IIASA project on biosphere dynamics which sustained our work, presently entertains a different outlook and direction, and one which is also likely to provide a similarly invaluable product. No less is deserved by the national member countries of IIASA which are responsible for the opportunities which we have accepted to examine global biospheric change.

Allen M. Solomon
Corvallis, Oregon

Herman H. Shugart, Jr.
Charlottesville, Virginia

Preface Notes

1. Participants included A. M. Solomon, H. H. Shugart, W. C. Clark (now at Harvard University, Cambridge, MA), R. E. Munn (now at Climate Canada, Downsview, ON), and M. Y. Antonovsky (IIASA).

2. The visiting scientists included Gordon Bonan (now at NCAR, Boulder, CO), William R. Emanuel (Oak Ridge National Laboratory, Oak Ridge, TN), Sandy P. Harrison (now at Lund University, Sweden), Janos P. Hrabovszky (Vienna, Austria), Mikhail D. Korzuhkin (Center for Global Change, Russian Academy of Sciences, Moscow, Russia), I. Colin Prentice (now at Lund University, Sweden), Thomas M. Smith (now at University of Virginia, Charlottesville, VA) and Mikhail M. Ter-Mikhaelian (now at Ontario Ministry of Natural Resources, Sault Ste. Marie, Canada). In addition to the visiting scientists, four IIASA staff members were involved (Allen M. Solomon, Project Leader; Victor O. Targulian, now at the Institute of Geography, Russian Academy of Sciences, Moscow, Russia; Sergei Pitovranov, now at VNIISI, Russian Academy of Sciences, Moscow, Russia; Ferenc Toth, now at University of Economics, Budapest, Hungary).

3. Graduate students in the IIASA Young Scientist's Summer Program (YSSP) included Nedialko Nikolov, now at U.S. Forest Service, Ft. Collins, Colorado; Harry Helmisaari, University of Uppsala, Sweden; Sandra Lavoral, National Research Center, Montpelier, France; Rik Leemans, now at R.I.V.M., the Netherlands; Robin Webb, now at NOAA, Boulder, CO; Alexander Minin, Moscow State University, Moscow, Russia.

4. Prentice, I. C., Webb, R.S., Ter-Mikhaelian, M. T., Solomon, A.M., Smith, T. M., Pitovranov, S. E., Nikolov, N. T., Minin, A. A., Leemans, R., Lavorel, S., Korzukhin, M. D., Hrabovsky, J. P., Helmisaari, H.O., Harrison, S. P., Emanuel, W. R. and Bonan, G. B. (1989). *Developing a Global Vegetation Dynamics Model: Results of an IIASA Summer Workshop*. RR-89-7, International Institute for Applied Systems Analysis, Laxenburg, Austria.

5. The senior scientists participating in the summer 1989 research activities included

Wolfgang Cramer (University of Trondheim, Norway), Sandy P. Harrison, Mikhail D. Korzuhkin, Lev Nedorezov (Siberian Branch, Russian Academy of Sciences, Krasnoyarsk, Siberia), I. Colin Prentice, H. H. Sugart, Jr. (University of Virginia), Thomas M. Smith, and F. Ian Woodward (now at The University, Sheffield, UK). IIASA staff included R. Leemans, Allen M. Solomon, Victor O. Targulian, and Ferenc Toth. Other visitors contributing to the summer workshop were Harry O. Helmisaari and Nedialko Nikolov (both of whom won summer scholarships to IIASA based on their work as YSSP students the previous year) and Erica Schwarz (now at University of Edinburgh, UK).

6. Students included Nadja M. Chebakova (Institute for Forests and Wood, Krasnoyarsk, Siberia), Andrew D. Friend (now University of Edinburgh, Scotland), Mark R. Fulton (now at Texas A&M University, College Station, Texas), Ruiping Gao (Academia Sinica, Shenyang, PRC), Sergei V. Goryachkin (Moscow State University, Moscow, Russia), Philippe Martin (now at Macquarie University, Australia), and Brian E. McLaren (now at Michigan Technological University, Michigan).

PART I

Introduction

1

Global Change

Herman H. Shugart

Introduction

Both scientists and nonscientists are aware that the earth and its ecological systems are dynamic when one considers relatively long time-scales. That there was a major continental ice cap a few tens of thousands of years ago is sufficiently part of the common wisdom that children's cartoons can use a "caveman" motif in the context of an "ice age" without any need for explanation. The occurrence of major extinctions of biotic groups (such as the "dinosaurs") in the past is not difficult for most people to imagine. In general, we are quite willing to believe that, viewed over the eons, the earth has been an extremely dynamic planet.

What is novel—newsworthy to the public and challenging to the scientist—are observations of shorter time-scale, relatively rapid changes in atmospheric and surface features of the earth, and the strong evidence that some of these changes are being induced by human activities. It appears that we are producing measurable changes in major earth systems, but we have relatively little knowledge as to how the earth's systems actually operate. The scientific challenges that attend these topical concerns have inspired a major international research program (the International Geosphere-Biosphere Program—IGBP), which was chartered by the International Council of Scientific Unions in 1986, "to describe and understand the interactive physical, chemical, and biological processes that regulate the total earth system, the unique environment that it provides for life, the changes that are occurring in this system, and the manner in which they are influenced by human activities."

This book treats one of several problems related to global change, namely, the prediction of the response of the global pattern of vegetation to a change in climate. Changes in the composition of the earth's atmosphere (such as the change in the amount of carbon dioxide—CO_2) are, sensu stricta, components of global climate change, as are possible changes in global temperature and precipitation patterns. There are several observations of global-scale change that shape the scientific issues of vegetation dynamics and global change. Three examples are our increasing appreciation of the nature and magnitude of past changes in vegetation in response to climatic change, the measured variation of the atmospheric CO_2 concentration over the past three decades, and the ability to survey the global terrestrial surface by using sensors carried on orbiting platforms. The past vegetation dynamics were discussed recently for boreal regions by Solomon (1992). The other two examples are discussed briefly in the two sections that follow.

Observations of Change in the Atmospheric CO_2 Concentration

In 1957, Keeling (e.g., Keeling 1983) began measuring the concentration of CO_2 in the atmosphere at Mauna Loa, Hawaii. He has continued these observations to this time. The initial observations indicated an annual fluctuation in the amount of CO_2 in the atmosphere over Mauna Loa. Further observations confirmed that the amount of CO_2 in the atmosphere had been increasing exponentially since the initial 1957 measurements. This exponential increase was found to be correlated with the release of CO_2 into the atmosphere from the burning of fossil fuels.

Figure 1.1 shows the pattern of variation in CO_2 measured at four different stations (Barrow, Alaska [BRW]; American Samoa [SMO]; Mauna Loa Hawaii [MLO]; South Pole [SPO]) from 1973 to 1983 (Harris and Bodhaine 1983). While there is a considerable difference in the amount of oscillation in these data, there is a clear tendency for the measured levels of CO_2 to increase regularly throughout this record. This figure can be thought of as illustrating the "breathing of the planet." It shows regular oscillations of the CO_2 concentrations in the Northern Hemisphere over three annual cycles and a much more constant pattern in the Southern Hemisphere (Harris and Bodhaine 1983).

Carbon dioxide is an essential component of plant photosynthesis, and when one sees seasonal or systematic multiple-year change in CO_2 levels, questions immediately arise as to whether these changes might in some way be altering the way plants function (chapter 3). If the functioning of plants is changed in some way, what other changes might ensue if the CO_2 in the atmosphere continues to increase? There is evidence that the number of stomata per unit area on plant leaves has decreased in response to the increase in CO_2 concentration since the industrial revolution (Woodward 1987; see also chapters 4–7). At the different levels of CO_2 in the atmosphere that were thought to prevail in the past, were conditions so different from those today as to confound our ability to interpret historical accounts of the vegetation (chapter 2)?

The important role of biological processes in affecting the atmosphere also arises when considering the level of CO_2 in the atmosphere (chapter 6). When one computes the amount of CO_2 that has been released into the atmosphere

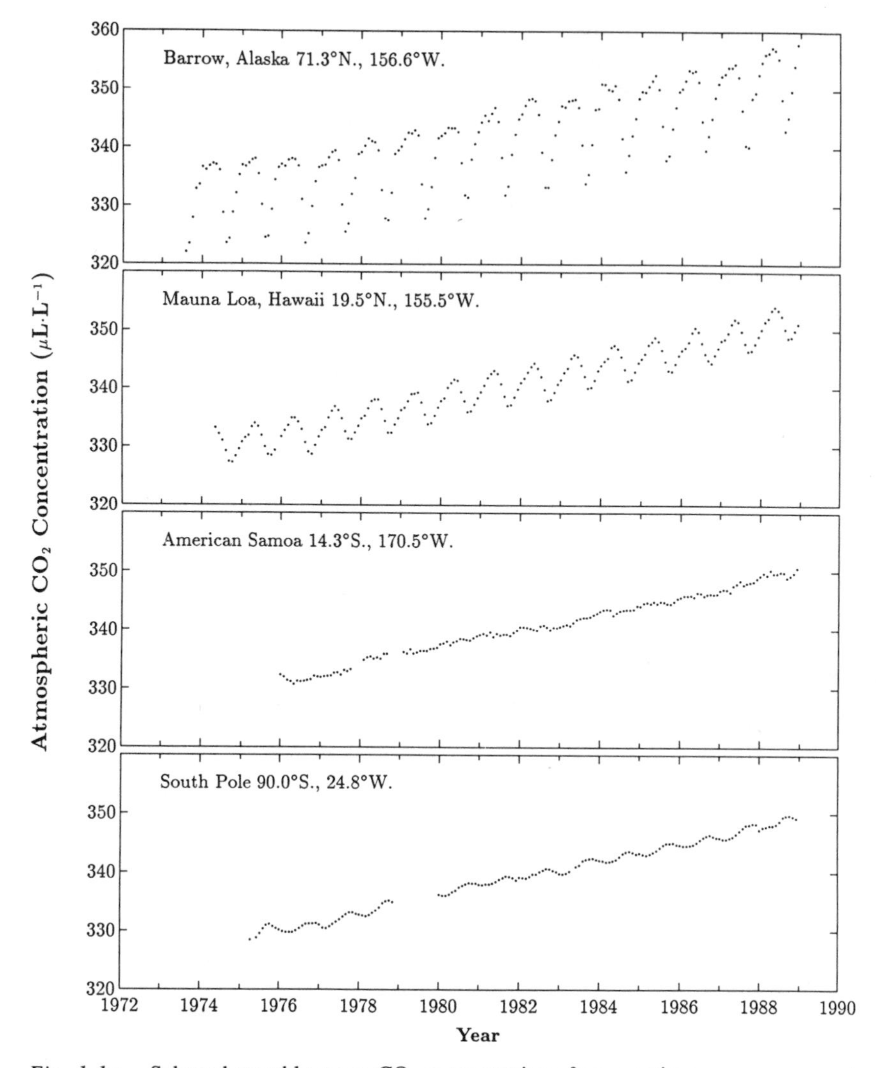

Fig. 1.1a. Selected monthly mean CO_2 concentrations from continuous measurements at National Oceanic and Atmospheric Administration/Geophysical Monitoring for Climate Change (NOAA/GMCC) stations at four locations (Barrow, Alaska [BRW]; American Samoa [SMO]; Mauna Loa Hawaii [MLO]; South Pole [SPO]) from 1973 to 1983 (from Harris and Bodhaine 1983).

through the burning of fossil fuels, subtracts the amount of CO_2 thought to be taken up by the oceans, and further subtracts the amount that appears to remain in the atmosphere (producing the regular increase shown in Fig. 1.1), about half of the CO_2 produced by fossil fuel burning remains unaccounted for. Is this "missing carbon" being taken up by land plants or is our understanding of the processes that transfer CO_2 from the atmosphere to the ocean incomplete?

At the annual time-scale, we would like to know to what degree the terrestrial vegetation is involved in causing the annual oscillations in CO_2 in the Northern Hemisphere (Fig. 1.1, chapters 2, 6 and 7).

An important consequence of an increased ambient level of CO_2 in the atmosphere is related directly to the role of this relatively rare atmospheric constituent in the heat balance of the earth. CO_2 is a "greenhouse gas" that is transparent to visible light but is opaque to long-wave radiation. Since much of the radiation that penetrates the earth's atmosphere is reradiated as long-wave radiation, CO_2 in the atmosphere functions much like the glass in a greenhouse, allowing light to enter and trapping heat (Fig. 1.2a). Clouds absorb long-wave radiation (a "greenhouse" function) but also reflect incoming radiation back to space (a cooling function). Other atmospheric components are also active in the earth's energy balance, notably ozone (O_3), man-made chlorofluorocarbon compounds, and water vapor which act as "greenhouse gases." Some of these other atmospheric components are increasing at rates that exceed the rate for CO_2. An estimate of the possible effects of these atmospheric components based on projected increases from 1980 to 2030 is shown in figure 1.3a (Bolin *et al.* 1986). These computations are based on relatively simple models that do not attempt to take into account the interactive nature of and the feedback effects among the components of the atmosphere (Dickinson 1986).

These feedback effects have been incorporated into massive computer models (general circulation models, or GCMs) in an attempt to understand the global climate system and to assess the effects of changes in the atmosphere such as might occur from the observed change in atmospheric CO_2 (Dickinson 1986). The resultant climate models are complex (Fig. 1.2b) and, at present, even the largest and fastest computers are unable to solve the equations at a fine spatial level (solutions are typically for an earth divided into large blocks ca. 2° latitude by 2° longitude) or with representations of the effects of the oceans which change dynamically as the climate changes.

The GCMs vary with respect to the way that processes important to the earth's climate are represented in a given formulation and in terms of assumptions used to approximate conditions that cannot be simulated in the models directly (e.g., the formation of clouds, the energy exchange between the atmosphere and the ocean, or the formation of ice at sea). Despite these model differences, the models converge in that they predict an increase in the earth's temperature as a consequence of increased atmospheric CO_2. The models show substantial

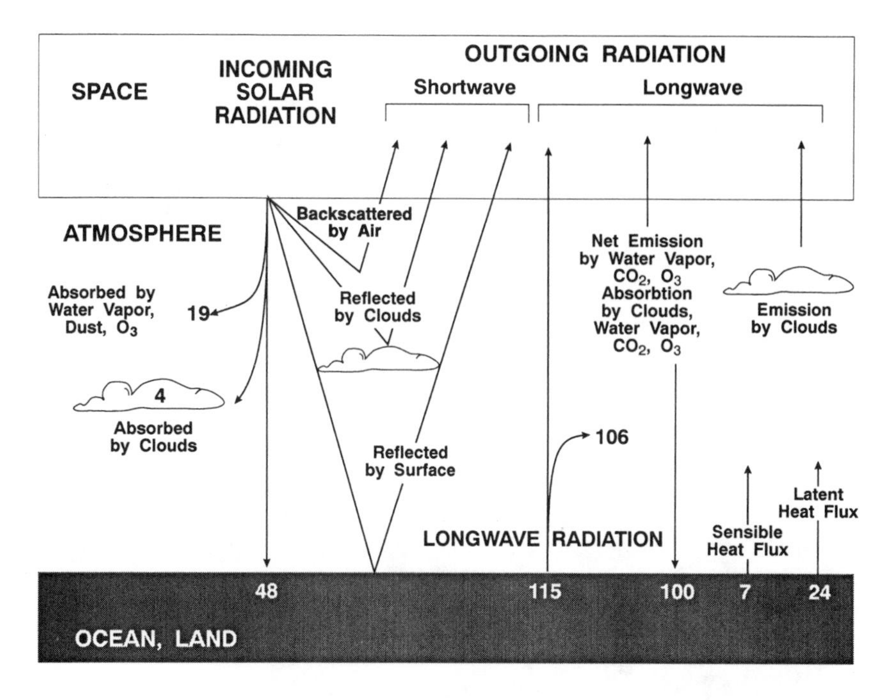

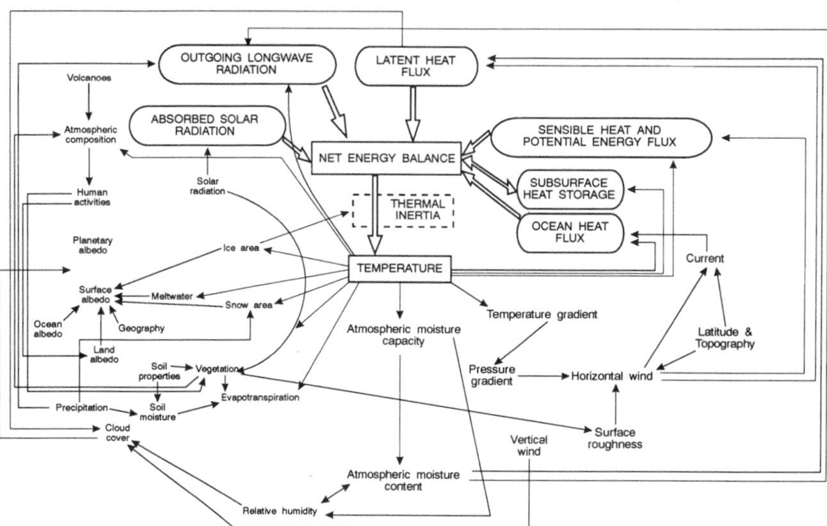

Fig. 1.2.a. Schematic diagram of the global average components of the earth's energy balance (from MacCracken and Luther 1985). *b.* Schematic illustration of the climatic cause-and-effect (feedback) linkages that are typically included in numerical models of the earth's climate system (from MacCracken 1985; based on Roboch 1985).

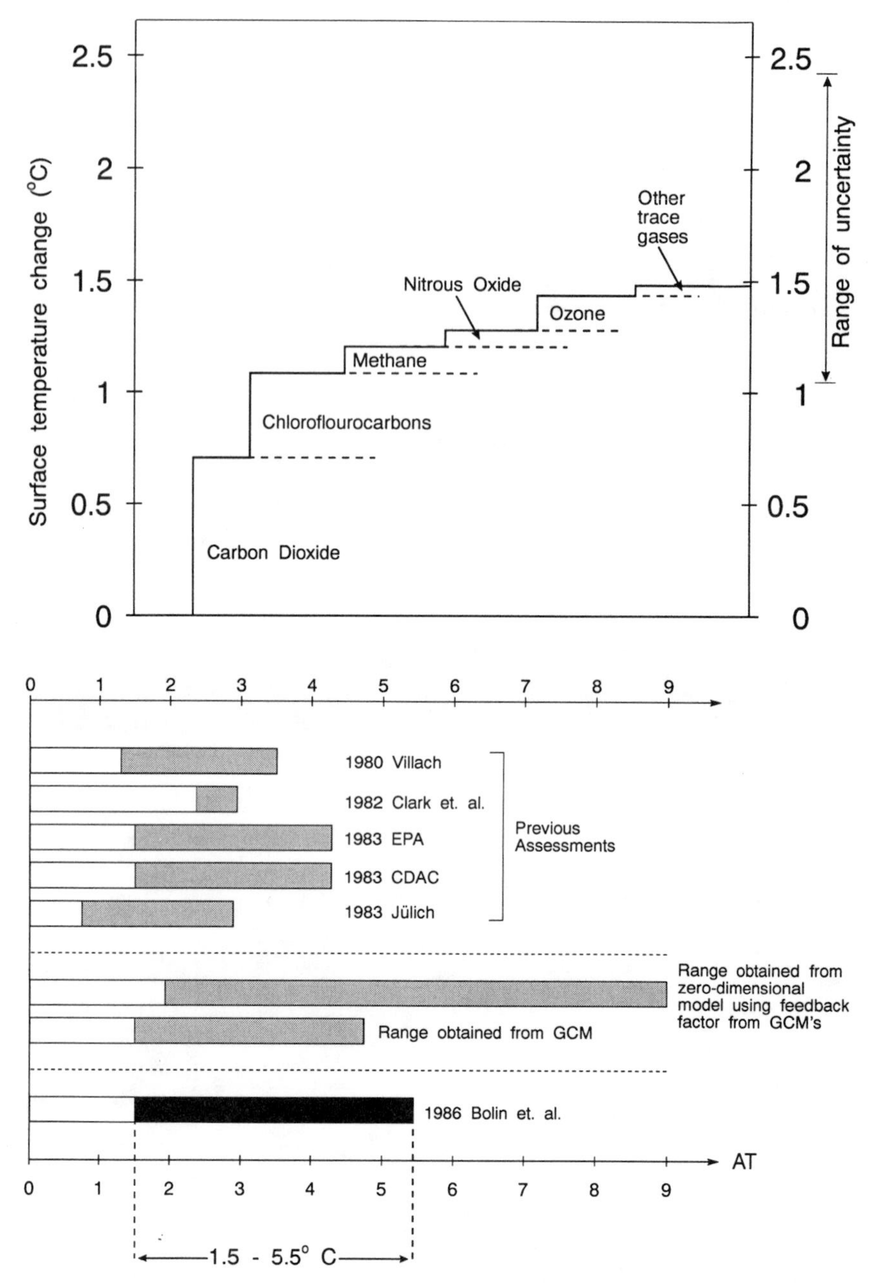

Fig 1.3.a. Cumulative surface temperature warming due to an increase in CO_2 and other trace gases from 1980 to 2030 as computed by a one-dimensional model (from Ramanathan *et al.* 1985). In these calculations the atmospheric concentration of CO_2 is expected to double in this time period. *b.* Estimates of GCM models that include dynamic feedbacks for the global warming expected from a doubling of CO_2 (from Bolin *et al.* 1986).

differences in the patterns of temperature change in space and particularly with regard to precipitation (Dickinson 1986).

There is a considerable level of uncertainty in the predictions of the rates of increase of CO_2 over the coming years, in the reliability of the GCMs and in the effects of the behavior of the oceans under a changed climate. Nevertheless, there is a need for a clearer understanding of the response of vegetation to climate change. This is true both with respect to the effect of climate on the vegetation (chapters 9 and 10) and with respect to the role of the terrestrial biota in the carbon cycle (chapters 6 and 9).

Because of the short time-scale of predicted environmental change (Schneider 1989a & b), the most important issues related to ecosystem response in response to climate-change scenarios should concern transient responses and factors affecting these responses (chapters 3, 4, 7, 10, and 15). These include direct consideration of the magnitude and time-scale of climate change, shifts in the seasonality of temperature and precipitation, and changes in climate-mediated disturbance regimes (Solomon *et al.* 1984; Neilson *et al.* 1989; Smith and Tirpak 1989). Important ecosystem factors that must be considered include the ecosystem state in terms of age structure or successional status of the vegetation (Davis and Botkin 1985; Urban and Shugart 1989; Urban *et al.* 1989) and soil characteristics (Pastor and Post 1988; Bonan *et al.* 1990). These considerations underscore the fundamental concern that the transient responses of ecosystems may be dramatically different from the responses extrapolated from studies of plant species or communities in isolation.

Observations of Vegetation Dynamics from Satellites

The manned space programs and the development of orbiting platforms with sensors capable of monitoring features on the surface of the earth have provided images that have allowed a generation to "see" what the earth looks like when viewed from space. Making the earth visible as a whole has indubitably amplified the interest in global ecological studies much as the invention of the microscope electrified the scientists of another era. Along with the important role of making global studies more tangible, the use of satellites and remote sensing of the earth's surface has provided fundamental observations that fuel a need to better understand global vegetation patterns and dynamics (chapter 8).

For example, Tucker *et al.* (1986a) related the variations in global CO_2 shown in Figure 1.1 to the variation in the "greenness" of the total earth surface. Tucker *et al.* (1986b) used the advanced very high resolution radiometer (AVHRR) that is carried on the National Oceanic and Atmospheric Administration's "weather" satellites (NOAA-6, NOAA-7, and NOAA-8 sun-synchronous, polar-orbiting, operational satellites) to map the greenness of vegetation at the continental and global scale. These same satellites are used to develop the satellite maps that are frequently used as part of the weather-report section of many television evening news programs.

Two of the spectral bands detected by the satellites (one in the visible-red part of the spectrum and one in the near infrared) have been used to map green leaf area or intercepted photosynthetically active radiation (Tucker 1979; Curran 1980; Kumar and Monteith 1982). By analyzing data collected over the past several years by these satellites, estimates have been obtained of the rates of clearing of tropical rainforest (Tucker *et al.* 1986b), of the types of vegetation cover in Africa (Tucker *et al.* 1985), and of the world distribution of green leaf biomass (Tucker *et al.* 1986a).

In this latest application, the periodic variation in the amount of green leaf biomass (as indicated by the NOAA-7 satellite AVHRR sensor) was determined for the same period as is shown in the data of Harris and Bodhaine (1983; see Fig. 1.1). The resultant pattern of "greenness" varying over 3 years and with latitude is shown in Figure 1.4. One notes a striking similarity in the pattern of Figure 1.4 and that seen in Figure 1.1a for global CO_2 dynamics. Tucker *et al.* (1986a) state in their conclusions that "Our analysis demonstrates the measurable link between atmospheric CO_2 drawdowns and terrestrial NDVI [*greenness index*, ed.] dynamics and suggests that there may be quantitative relationships between multi-temporal satellite data and atmospheric CO_2 drawdowns." This is indeed the case, and serves as but one example of the scientific challenges being produced by the observational capacity provided by satellites (chapter 9).

Historical Background for Global Ecological Studies

Many of the basic concepts used in global ecology originate in the works of Clements (1916; an emphasis on dynamic interactions), Gleason (1926; the importance of species attributes in dynamic systems), Tansley (1935; the ecosystem concept), and Watt (1947, the relationship between internal dynamics and spatial patterns). The theories that were developed by these early ecologists proved difficult to apply in a formal mathematical fashion to the complex natural systems for which they were intended. The eventual development of mathematical models based on these concepts has clearly been catalyzed by the increased availability of computers (chapter 12). Some of the chapters that follow will discuss the application of several different models to problems in global ecology with the overall intent of providing an impression of the issues, capabilities, and challenges involved in understanding vegetation dynamics and climatic change.

Pattern and Process in Ecological Studies

A fundamental theme in the biological sciences that is echoed in cellular biology, genetics, morphology, population biology, and ecology involves the relationship between geometrical structures of living things and the processes that attend these

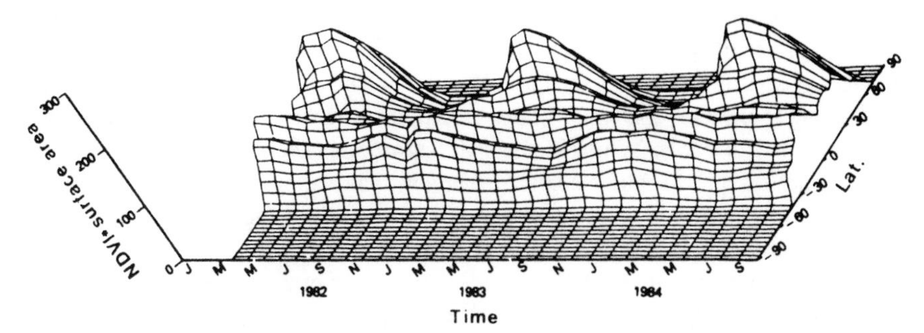

Fig. 1.4. A weighted index based on the visible-red and the near-infrared channels of the NOAA-7 satellite and indicative of green leaf biomass plotted against time and latitude zone. The seasonal effects seen in the northern latitudes in the dynamics of CO_2 are also seen in the northern latitudes of this graph of "greenness" dynamics. Note also the relatively constant values for the equatorial latitudes and the influence of deserts in depressing greenness in the 20° to °N latitude zone (from Tucker *et al.* 1986a).

structures. The organization of tissues is intimately related to their physiological function. It is important to the function of the DNA molecule that it is a spiral helix. The adaptive implications of the morphology of plants and animals is central to both taxonomy and evolutionary biology.

Pattern and Process

In biology, these themes are variously described as "structure and function" or "pattern and process." Depending upon the examples chosen, there may be an emphasis on the manner in which processes influence pattern. For example, what changes in the morphology of the vertebrate limb develop in the evolution of flight? Or, how do the seeds of plants vary between arid and moist environments? In other examples, the emphasis is on the pattern or structure modifying processes. Pattern and process are linked in a biological yin and yang in which each causes and is caused by the other.

The relationship between form and function, or pattern and process, is a classic ecological theme (Lindeman 1942; Watt 1947; Whittaker and Levin 1977). Bormann and Likens (1979a) pointed out the effects of changes in forest structure on processes such as productivity and nutrient cycling. Many ecologists recognize that pattern and process are mutually causal, with changes in ecosystem processes causing changes in pattern and modifications in ecosystem pattern changing processes. Nonetheless, it is difficult to investigate directly the feedback between pattern and process.

In global ecological studies, what is of particular importance is a knowledge of the degree of dominance of particular causal factors at particular scales. The knowledge of which factors are important at a given scale is also involved in the determination of the "rules" for deciding what should be included in the formulation of a given model. That different phenomena may be invoked when developing models of analogous phenomena at different scales is to a degree responsible for what is categorized as the "art" (as opposed to the science) of ecological modeling (chapters 4–7 and 12). While the determination of importance of processes at a given time- or space-scale is central in model formulation and evaluation, it is a consideration that is neither trivial nor unique to the developers of computer models. For example, in the case of "hierarchy theory" (Allen and Starr 1982; Allen and Hoekstra 1984; O'Neill *et al.* 1986; Urban *et al.* 1987), one sees a focus on expressing relevant mathematical developments in a manner that can provide insight into the ways ecosystems are structured at different scales (chapters 11, 13, and 14).

The categorization of controlling factors important at different space- and time-scales in particular ecosystems has been the topic of several reviews (Delcourt *et al.* 1983; Pickett and White 1985). Historically, this interest is evident in A.S. Watt's (1925) early work on beech forests and elaborated in his now-classic paper on pattern and process in plant communities (Watt 1947). These themes have been reiterated by several subsequent ecologists (Tansley 1935; Whittaker 1953; Bormann and Likens 1979a,b).

The factors governing vegetation structure and ecosystem processes vary considerably within and among biomes. In mesic forests, a frequent constraint is the availability of light. As a forest environment tends from mesic to xeric or nutrient-poor conditions, the effective constraint shifts from above- to below-ground factors (Webb *et al.* 1978; Tilman 1988; Smith and Hudson 1989). Under still drier conditions, forest changes to grassland in which the principal constraint is below-ground, suggesting patterns in the influence of environmental constraints in structuring ecosystems across broad environmental gradients (chapters 11, 13, and 14).

Pastor and Post (1986) used the LINKAGES model to evaluate the principal constraints in forests through time and among trees of different stature. The principal constraint changed from below-to above-ground as trees grew and the canopy closed. The principal constraint also varied with tree size and time in the simulated succession. The dominant individuals were limited by below-ground constraints, while understory individuals shifted from below- to above-ground constraints as they were overtopped. This result cautions against oversimplifying ecosystems in terms of "the primary constraint" and also suggests parallels in patterns of constraint through time and over spatial gradients.

Spatial Scale

A consideration at the basis of the relationship between pattern and process is the intrinsic *scale* of phenomena. This theme is fundamental to modern science. Phenomenological scale can be considered in spatial or temporal dimensions. In either case, scale refers to the range with respect to a fundamental dimension (usually in time or space) associated with a pattern or process of interest. For example, at the time- and space-scales considered in quantum mechanics, the effects of gravity are sufficiently small to be ignored. At the scales typically considered in astrophysical studies, the gravitational effects are a paramount consideration and cannot be ignored.

As is the case with pattern and process, the concept of scale is well developed in the biological sciences. In field studies in ecology, a frequent manifestation of scale involves the determination of the appropriate size of a sample quadrat to survey a given sort of vegetation. A typical question in this regard might be, "Is it more accurate (or more convenient) to survey the variety of trees in a forest by determining the number of trees of different species in a random set of sample areas of 0.1 ha or 1.0 ha?" The appropriate scale of sample quadrats is influenced by the spatial distribution of the tree species so that what might appear to be a straightforward sampling question is actually involved with understanding a fundamental property of a forest.

One of the most intellectually stimulating aspects of global-scale ecological studies is a consequence of the interaction of environmental scientists (oceanographers, meteorologists, climatologists, etc.) with a strong orientation toward physics. This has simultaneously served to create an interest in phenomenological scale in the ecological sciences and to present problems in understanding phenomena at space- and time-scales that are relatively nontraditional to the ecologists. For example, for an atmospheric process to exert a significant successional effect, it must impact the ecosystem in a manner that directly allows for the alteration of local community composition. Some meteorological processes (windfalls, lightning strikes) appear as prominent and persistent. Others, such as differential mortality of trees of different species resulting from unusual temperatures, are much more subtle. Fujita (1981) provides strong evidence for five relatively discrete scales of atmospheric motion (Fig. 1.5), some of which can influence ecological succession. Fujita's five scales are each separated by approximately two orders of magnitude in spatial extent, with the largest, or "A" scale, corresponding to the planetary and synoptic circulations. It is generally the "E" ("mesoscale") circulations which come to mind when one thinks of the direct effects of the atmospheric circulation on forests. The determination of how changes in the frequencies of tropical cyclones (Neumann *et al.* 1981) and differences in the spatial patterns of the paths of such cyclones (Hayden 1981; Shapiro 1982) might be manifested as a change in a forest region is related to considering the forest at an appropriate scale (ca. regions on the order of 10^2 to 10^3 km on a side).

The problem of understanding spatial scale arises in interfacing ecology with physical dynamics of the Earth's surface (chapters 7 and 11), and it is also an important feature of global ecological studies as such. For example, in chapter 4 the problem of understanding the gas, water, and heat exchange of the surface of a leaf (a problem typically considered at space-scales of a few square centime-

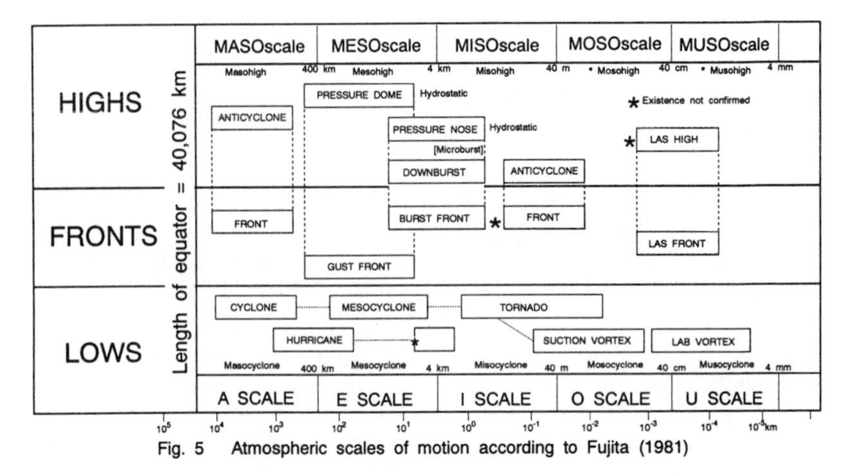

Fig. 1.5. Atmospheric scales of motion according to Fujita (1981).

ters or less) at the scale of a surface several kilometers on a side is discussed in some detail. This is but one of a large number of scale-related problems that have been engendered by the interest in global ecological studies.

Temporal Scale

There are also scale-related responses in time. Temporal scale can be discussed in an example case by considering the response of dynamic systems to periodic variations in important external variables. There is a tendency for many dynamic systems to display what is termed by engineers as "band-limited" behavior. In such systems, the dynamics response of the system to very-high-frequency and very-low-frequency periodic variation is almost zero, and only in an intermediate range of frequencies of variation does the system actually respond. One biological example is the human ear, which does not respond to extremely high- or low-frequency sounds (which we consequently do not hear), but does respond to an intermediate frequency of sound waves (in the audible range). The ear of a dog has a different response, and a dog hears a different range of sound waves from a human. The response of ecological systems to different periodicities of variation in time may also in many cases be band-limited, with a range of frequencies of external factors to which the system is responsive bordered by frequencies either too high or too low to produce a response.

The important consequence of band-limited behavior in ecological systems involves the identification of the typical periodicities in the external factors that are important to the ecological system. As will be discussed in the chapters that follow, the range of important frequencies in external variables as well as the particular external variables that are important depends to a great degree on the nature of the ecological processes under consideration.

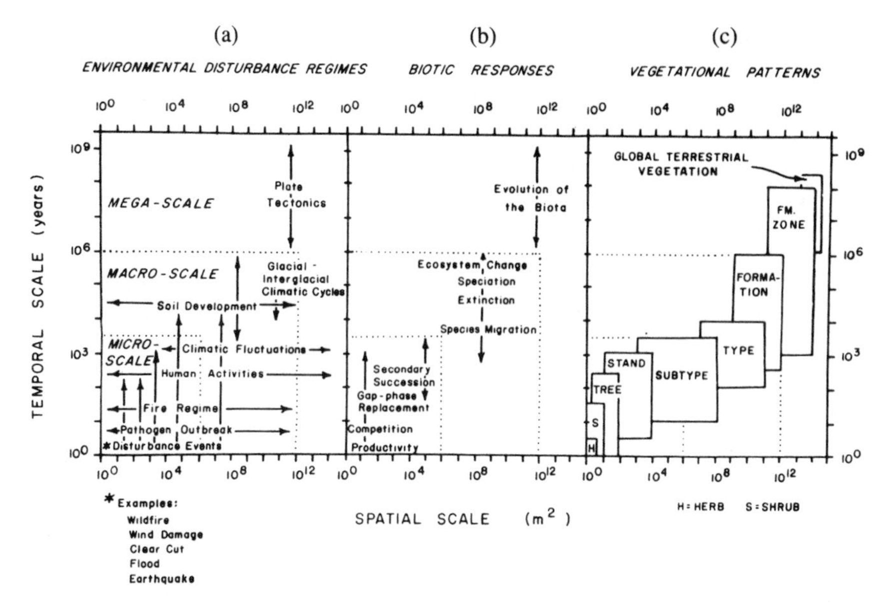

Fig. 1.6.a. Biotic disturbance regimes viewed in the context of the space/time domain. The scale of a particular process shown in the figure reflects the sample intervals typically required to observe the process. *b.* Space/time domain of representative biological processes. *c.* Space/time domain of ecological patterns (from Delcourt *et al.* 1983).

Scale, Pattern, and Process

Figure 1.6 (from Delcourt *et al.* 1983) illustrates the relationship between the scale of several important external environmental conditions (disturbances, Fig. 1.6a) with the scale of important biological processes (Fig. 1.6b). The interaction of these disturbances and processes produces ecological patterns (Fig. 1.6c). The fundamental concept of ecological scale is that the processes that are most important in producing ecological patterns change as a direct function of the temporal and spatial domain. The importance of a particular process when predicting the response of an ecological system to change is related to the time- and space-scales of interest.

The Ecosystem Concept

In a classic paper written by Tansley in 1935, the term "ecosystem" was first defined as an arbitrary system with respect to both its spatial extent and the phenomena considered. This can be seen in the first use of the term by Tansley:

> The more fundamental conception is, as it seems to me, the whole *system* (in the sense of physics), including not only the organism-complex, but also the whole complex of physical factors forming what we call the environment—the

> habitat factors in the widest sense. Though the organisms may claim our primary interest, when we are trying to think fundamentally we cannot separate them from their special environment, with which they form one physical system.
>
> It is the systems so formed which, from the point of view of the ecologist, are the basic units of nature on the earth. . . . These *ecosystems,* as we may call them, are of the most various kinds and sizes. They form one category of the multitudinous physical systems of the universe, which range from the universe as a whole down to the atom.

In this, the first use of the word "ecosystem" in the English language, Tansley stressed that ecosystems are of "various kinds and sizes." This relative arbitrariness and abstraction was viewed by Tansley as a necessary step to the formulation of an ecological science that was on a par with physics and other more established sciences. The value of the ecosystem concept has been proven in the 50 years that have ensued since Tansley coined the term.

If the three panels shown in Figure 1.6 are separated and stacked one on top of the next, then an ecosystem, as defined by Tansley, can be depicted as the set of interacting external variables, ecological processes, and patterns all with equivalent space-time domains. An ecosystem may be large or small with respect to either temporal or spatial scales but there should be an equivalency in the time and space domains in the patterns and processes considered.

Organization of the Other Chapters

Global ecological studies are in what appear to be the early stages of their development and for this reason this book reports the approaches being taken on several fronts. The coming decade should greatly enrich our understanding of the global dynamics of vegetation and considerably increase the examples that could be provided in a book such as this one. It is our hope that this present effort will provide the reader with sufficient background to enjoy the scientific excitement that should attend the future developments in global ecology and to appreciate the scientific challenges of today.

Each chapter that follows has the relatively straightforward intention of providing background on fundamental mechanisms that appear to be important in understanding the global response of the vegetative surface of the earth. The modus operandi of most of the chapters is to use mathematical models (usually designed for implementation on a computer) as a method of exploring the consequences of a given process in producing global patterns. Modeling is playing an ever-increasing role in the development of ecological theory at several scales, from understanding the mechanisms of carbon fixation (Farquahar and Sharkey 1982; Farquhar and von Caemmerer 1982) and plant water balance (Cowan 1982, 1986), to scaling of physiological processes to whole plant function (Reynolds *et al.* 1986), to exploring how ecosystem processes of carbon and nitrogen cycling operate at continental to global scales (Emanuel *et al.* 1984, 1985). Of particular importance is the role of modeling in exploring phenomena which occur at spatial and temporal scales at which extensive direct observation and experimentation are prohibitive, if not impossible. Recent examples include the role of spatial and temporal variation in competition in ecosystem functioning (Wu *et al.* 1985; Sharpe *et al.* 1985, 1986; Walker and Sharpe 1989), extrapolation of the processes of carbon fixation and water balance to the landscape scale to enable the linking of ecosystem models with remotely sensed data (Running and Coughlan 1988), and exploration of the implications of the evolution of plant adaptations to varying environmental conditions for current patterns of ecosystem structure across environmental gradients (Tilman 1988).

What we would hope is that, in reading this book, the reader will obtain an impression of the relations among pattern, process, and scale that have been briefly discussed in this chapter. For example, one will find that the phenomena that dominate the surface dynamics of a leaf (and that have global implications) will only be considered as an average at the level of an individual plant and will be embedded in model assumptions at the scale of a small patch of forest. The larger intention in developing this book is to provide an impression of the phenomenological scale of different mechanisms of importance.

References

Allen, T.F.H. and Hoekstra, T.W. (1984). Nested and non-nested hierarchies: a significant distinction for ecological systems. In *Proceedings of the Society for General Systems Research. I. Systems Methodologies and Isomorphies,* ed. A.W. Smith, pp. 175–80. Lewiston: Intersystems Publications, Courts Library Service.

Allen, T.F.H. and Starr, T.B. (1982). *Hierarchy: Perspectives for Ecological Complexity.* Chicago, Illinois: University of Chicago Press.

Bolin, B., Döös, B.R., Jäger, J. and Warrick, R.A. ed. (1986). *The Greenhouse Effect, Climatic Change, and Ecosystems, (SCOPE 29).* Chichester: John Wiley & Sons.

Bonan, G.B., Shugart, H.H. and Urban, D.L. (1990). The sensitivity of some high-latitude boreal forests to climatic parameters. *Climate Change,* **16,** 9–29.

Bormann, F.H. and Likens, G.E. (1979a). *Pattern and Process in a Forested Ecosystem.* New York: Springer-Verlag.

Bormann, F.H. and Likens, G.E. (1979b). Catastrophic disturbance and the steady state in northern hardwood forests. *American Scientist,* **67,** 660–69.

Clements, F.E. (1916). *Plant Succession: An Analysis of the Development of Vegetation.* Carnegie Institute Publication 242. Washington, D.C.

Cowan, I.R. (1982). Regulation of water use in relation to carbon gain in higher plants. In *Physiological Plant Ecology Encyclopedia of Plant Physiology (NS), Vol. 12B,* ed., O.L. Lange, P.S. Noble, C.B. Osmond and H. Ziegler, pp. 549–87. Berlin: Springer.

Cowan, I.R. (1986). Economics of carbon fixation in higher plants. In *On the Economy of Plant Form and Function,* ed. I.J. Givinish, pp. 133–70. Cambridge: Cambridge University Press.

Curran, P.J. (1980). Multispectral remote sensing of vegetation amount. *Progress in Physical Geography,* **4,** 315–41.

Davis, M.B. and Botkin, D.B. (1985). Sensitivity of cool-temperate forests and their fossil pollen record to rapid temperature change. *Quarternary Research,* **23,** 327–40.

Delcourt, H.R., Delcourt, P.A. and Webb, T. (1983). Dynamic plant ecology: The spectrum of vegetational change in space and time. *Quarternary Science Reviews,* **1,** 153–75.

Dickinson, R.E. (1986). How will climate change? In *The Greenhouse Effect, Climatic Change, and Ecosystems. (SCOPE 29),* ed. B. Bolin, B.R. Doos, J. Jager and R.A. Warwick, pp. 206–70. Chichester: John Wiley & Sons.

Emanuel, W.R., Shugart, H.H., and Stevenson, M.P. (1985). Climatic change and the broad-scale distribution of terrestrial ecosystem complexes. *Climatic Change,* **7,** 29–43.

Emanuel, W.R., Killough, G.G., Post, W.M. and Shugart, H.H. (1984). Modeling terrestrial ecosystems and the global carbon cycle with shifts in carbon storage capacity by land use change. *Ecology,* **65,** 970–83.

Farquhar, G.D. and Sharkey, T.D. (1982). Stomatal conductance and photosynthesis. *Annual Review of Plant Physiology,* **33,** 317–45.

Farquhar, G.D. and von Caemmerer, S. (1982). Modeling of photosynthetic response to environmental conditions. In *Physiological Plant Ecology, Encyclopedia of Plant Physiology (NS), Vol. 1, 12B,* ed. O.L. Lange, P.S. Noble, C.B. Osmond, and H. Ziegler, pp. 549–87. Berlin: Springer.

Fujita, T.T. (1981). Tornadoes and downbursts in the context of generalized planetary scales. *Journal of Atmospheric Science,* **38,** 1511–34.

Gleason, H.A. (1926). The individualistic concept of the plant association. *Bulletin of the Torrey Botanical Club,* **57,** 7–26.

Harris, J.M. and Bodhaine, B.A., eds. (1983). *Summary Report 1982, Geophysical Monitoring for Climatic Change.* Environmental Research Laboratories/NOAA, U.S. Department of Commerce, Washington, D.C.

Hayden, B.P. (1981). Secular variation in Atlantic Coast extratropical cyclones. *Monthly Weather Review,* **109,** 159–67.

International Council of Scientific Unions. (1986). *The International Geosphere-Biosphere Program: A Study of Global Change.* Report No. 1, Final Report of the Ad Hoc Planning Group. ICSU 21st General Assembly, Bern, Switzerland.

Keeling, C.D. (1983). The global carbon cycle: What we know and could know from atmospheric, biospheric and oceanic observations. In *Proceedings of the CO_2 Research Conference: Carbon Dioxide, Science, and Consensus,* pp. II.3-II62, DOE CONF-820970, NTIS, Springfield, Virginia.

Kumar, M. and Montieth, J.L. (1982). Remote sensing of plant growth, In *Plants and the Daylight Spectrum,* ed. H. Smith, pp. 133–44. London: Academic Press.

Lindeman, R.L. (1942). The trophic-dynamic aspect of ecology. *Ecology,* **23,** 399–418.

MacCracken, M.C. and Luther, F.M. (1985). Projecting the Climatic Effects of Increasing Carbon Dioxide (DOE/ER-0237). Washington, D.C.: Department of Energy.

Neilson, R.P., King, G.A., DeVelice, R.L., Lenihan, J., Marks, D., Dolph, J. Campbell, W. and Glick, G. (1989). Sensitivity of Ecological Landscapes and Regions to Global Climatic Change. EPA/600/3-89/073, NTIS No. PB90 120 072/AS. In preparation as monograph in The Ecology of Complex Systems Series, by T.F.H. Allen and D.W. Roberts (Eds.), New York: Columbia University Press.

Neumann, C.J., Cry, G.W., Caso, E.L. and Jarvinen, B.R. (1981). *Tropical Cyclones of the North Atlantic Ocean, 1871–1980*. NOAA, Asheville, North Carolina.

O'Neill, R.V., DeAngelis, D.L., Waide, J.B. and Allen, T.F.H. (1986). *A Hierarchical Concept of the Ecosystem*. Princeton, New Jersey: Princeton University Press.

Pastor, J. and Post, W.M. (1988). Response of northern forests to CO_2 induced climate change. *Nature,* **334,** 55–8.

Pastor, J. and Post, W.M. (1986). Influence of climate, soil moisture and succession on forest carbon and nitrogen cycles. *Biogeochemistry,* **2,** 3–28.

Pickett, S.T.A. and White, P.S. ed. (1985). *The Ecology of Natural Disturbance and Patch Dynamics*. New York: Academic Press.

Ramanathan, V., Singh, H.B., Cicerone, R.J. and Kiehl, J.T. (1985). Trace gas trends and their potential role in climate change. *Journal of Geophysical Research,* **90,** 5547–66.

Reynolds, J.F., Bachelet, D., Leadley, P. and Moorhead, D. (1986). Response of vegetation to carbon dioxide. Assessing the effects of elevated carbon dioxide on plants: Toward the development of a generic plant growth model. Progress Report 023 to U.S. Dept. of Energy.

Robock, A. (1985). An updated climate feedback diagram. *Bulletin of the American Meteorological Society,* **66,** 786–7.

Running, S.W. and Coughlan, J.C. (1988). A general model of forest ecosystem processes for regional applications. I. Hydrological balance, canopy gas exchange and primary production processes. *Ecological Modelling,* **42,** 125–54.

Schneider, S.H. (1989a). The greenhouse effect: Science and policy. *Science,* **243,** 771–81.

Schneider, S.H. (1989b). Global warming: Is it real and should it be part of a global change program? In *Global Change and Our Common Future,* ed. R.S. DeFries and T.F. Malone, pp. 209–19. Washington, D.C.: Committee on Global Change, National Research Council, National Academy Press.

Shapiro, L.J. (1982). Hurricane climatic fluctuations. Part 1: Patterns and cycles. *Monthly Weather Review,* **110,** 1007–23.

Sharpe, P.J.H., Walker, J., Penridge, L.K. and Wu, H. (1985). A physiologically based continuous-time Markov approach to plant growth modeling in semi-arid woodlands. *Ecological Modeling,* **29,** 189–213.

Sharpe, P.J.H., Walker, J., Penridge, L.K., Wu, H. and Rykiel, E.J. (1986). Spatial considerations in physiological models of tree growth. *Tree Physiology,* **2,** 403–21.

Smith, T.M. and Huston, M. (1989). A theory of the spatial and temporal dynamics of plant communities. *Vegetatio,* **83,** 49–69.

Smith, J. and Tirpak, D. (1989). *The Potential Effects of Global Climate Change on the United States*. U.S. Environmental Protection Agency, Washington, D.C.

Solomon, A.M. (1992). The nature and distribution of past, present and future boreal forests: Lessons for a research and modeling agenda. pp. 291–307. In *A Systems Analysis of the Global Boreal Forest,* ed. H.H. Shugart, R. Leemans and G.R. Bonan, New York: Cambridge University Press.

Solomon, A.M., Tharp, M.L., West, D.C., Taylor, G.E., Webb, J.M. and Trimble, J.C. (1984). Response of unmanaged forests to CO_2-induced climate change: Available information, initial tests and data requirements. U.S. Dept. of Energy, Washington, D.C.

Tansley, A.G. (1935). The use and abuse of vegetational concepts and terms. *Ecology* **16,** 284–307.

Tilman, D. (1988). *Plant Strategies and the Dynamics and Structure of Plant Communities*. Princeton: Princeton University Press.

Tucker, C.J. (1979). Red and infrared linear combinations for monitoring vegetation. *Remote Sensing Environment,* **8,** 127–50.

Tucker, C.J., Fung, I.Y., Keeling, C.D. and Gammon, R.H. (1986a). Relationship between atmospheric CO_2 variations and a satellite derived vegetation index. *Nature,* **319,** 195–9.

Tucker, C.J., Townshend, R.G. and Goff, T.E. (1985). African land cover classification using satellite data. *Science,* **227,** 369–74.

Tucker, C.J., Townshend, J.R.G., Goff, T.E. and Holben, B.N. (1986b). Continental and global scale remote sensing of land cover. pp. 221–241. In *The Changing Carbon Cycle: A Global Analysis,* ed. J.R. Trabalka and D.E. Reichle, New York: Springer-Verlag.

Urban, D.L., O'Neill, R.V. and Shugart, H.H. (1987). Landscape Ecology. *Bioscience,* **37,** 119–27.

Urban, D.L. and Shugart, H.H. (1989). Forest response to climatic change: A simulation study for southeastern forests. In *The Potential Effects of Global Climate Change on the United States,* eds. J.B. Smith and D.A. Tirpak, pp. 3-1 to 3-45. EPA-230-05-89-054, U.S. Environmental Protection Agency, Washington, D.C.

Urban, D.L., Shugart, H.H. and Smith, T.M. (1989). Forest response to environmental change: A factorial model. In, Forests of the World: Diversity and Dynamics (abstracts), ed. E. Sjogren. *Studies in Plant Ecology,* **18,** 47–9.

Walker, J., Sharpe, P.J.H., Penridge, L.K. and Will, H. (1989). Ecological field theory: The concept of field tests. *Vegetatio* **83,** 81–95.

Watt, A.S. (1925). On the ecology of British beechwoods with special reference to their regeneration. Part 2, sections II and III. The development of the beech communities on the Sussex Downs. *Journal of Ecology,* **13,** 27–73.

Watt, A.S. (1947). Pattern and process in the plant community. *Journal of Ecology*, **35,** 1–22.

Webb, W., Szarek, S., Lauenroth, W.K., Kinerson, R. and Smith, M. (1978). Primary productivity and water use in native forest, grassland, and desert ecosystems. *Ecology*, **59,** 1239–47.

Whittaker, R.H. (1953). A consideration of climax theory: The climax as a population and a pattern. *Ecological Monographs*, **23,** 41–78.

Whittaker, R.H. and Levin, S.I. (1977). The role of mosaic phenomena in natural communities. *Theoretical Population Biology*, **12,** 117–39.

Woodward, F.I. (1987). Stomatal numbers are sensitive to increases in CO_2 from pre-industrial levels. *Nature*, **327,** 617–18.

Wu, Hsin-I, Sharpe, P.J.H., Walker, J. and Penridge, L.K. (1985). Ecological field theory: A spatial analysis of resource interference among plants. *Ecological Modeling*, **29,** 215–43.

PART II

Biotic Responses to Global Environmental Change

2

Biospheric Implications of Global Environmental Change

Allen M. Solomon and Wolfgang Cramer

A currently popular question is whether the global warming, measured during the past century, and especially in the decade of the 1980s, is the warming expected to occur because of human-produced (anthropogenic) increases of atmospheric greenhouse gases (GHGs). These include not only CO_2, but also methane, ozone, water vapor, and other radiatively active gases. The answer to this question has great political significance, serving as a basis for potential governmental actions.

However, the answer to the question may be of little functional consequence. A fundamental fact cannot be circumvented: GHGs will warm the atmosphere if their concentration continues to increase (Schneider 1989). Atmospheric physicists may disagree on how much warming can be expected, or on whether resultant increases in the intensity of the hydrological cycle will reduce or enhance the warming, but they do not disagree on the relationship between concentrations of greenhouse gases and temperatures in the lower atmosphere; the basis for expecting a general warming with increasing greenhouse gases is grounded in the laws of physics and is not in dispute. However, as we discuss later, the expectation of an identifiably singular warming is quite unlikely; the magnitude and timing of warming will vary from season to season, year to year, and place to place. In fact, some places may become cooler as a result of global warming (Washington 1989).

In addition to the obvious inevitability of global-scale warming in response to increasing GHG concentrations, it is clear that the warming which will result is likely to be quite similar to the warming which has already been measured over the past 100 or 125 years, during which GHGs have also increased (Wigley and Jones 1981). Hence, if the warming already measured is not the warming which will occur because of GHG, the similarity of the two warmings makes the question moot.

Another fact which cannot be evaded is that we will never be able to provide rigorous scientific proof that atmospheric warming has occurred as a result of GHG concentrations. The scientific method requires replication of both the experimental treatment (increasing GHGs) and of the control system (no change in GHGs), and we have only one world to study. We may compare global temporal sequences, for example, the world under glacial conditions 18 000 years ago with the world today, but these comparisons are informative, not definitive, as tests of scientific hypotheses. As a result, the controversy over the presence or absence of anthropogenic warming cannot be resolved. One will always be able to find scientists (including us) who will state unequivocally that GHG-induced warming has not been proven scientifically, and one will also be able to find scientists (again, including us) willing to state their unequivocal belief that GHG-induced warming is occurring now.

Instead of basing mitigative actions on the availability of a scientifically rigorous answer to the question, "Is it here yet?", we can suggest potential political responses must be based on the nature of the global environmental changes which must follow increasing GHG concentrations, and on the potential biospheric responses to those changes. This information should allow us to define the most important cause-and-effect properties of the global system to include the hypotheses which we will test. Testable hypotheses provide the scientific basis for learning and applying the needed knowledge. Hypotheses are tested when formal models of the hypothesis are compared with data describing the system our model should mimic. The tested mathematical models, then, can simulate future conditions, providing the illustrations of effects required for informed decision-making. The remainder of this chapter examines the nature of the known global environmental changes now underway which are relevant to biospheric functioning, the biospheric responses of particular interest with regard to the survival of the earth's biotic resources, and the most appropriate methods available for testing and implementing global biospheric models.

Changing Atmospheric Carbon Dioxide

The most definitive and politically significant environmental data set relevant to understanding global change consists of the records of atmospheric CO_2 measured from 1958 until the present at Mauna Loa Observatory, Hawaii (Keeling *et al.*, 1982). Begun during the International Geophysical Year, the familiar record illustrated in Figure 2.1 irrefutably establishes the presence of a significant upward trend in atmospheric CO_2 concentrations. The average atmospheric CO_2 concentration in 1958, 315 ppm (parts per million), had increased in very consistent annual steps to 352 ppm by 1989 (Keeling *et al.* 1989). During the past 200 years, the atmospheric concentration of CO_2 has increased by 25%, from about 280 ppm before A.D. 1750 (Friedli, *et al.*, 1986).

Increases of similar proportion of other GHGs, especially methane, nitrous

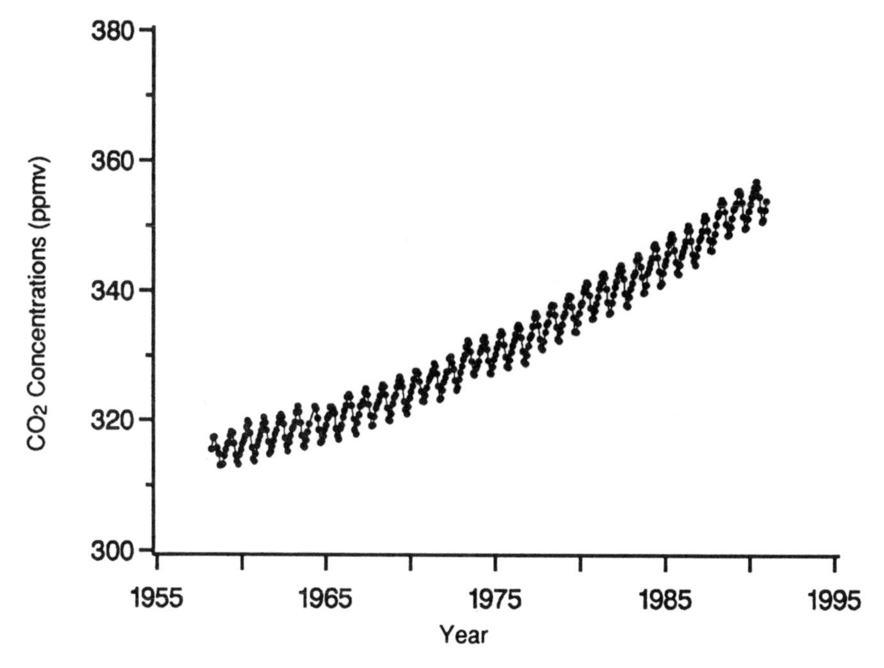

Fig. 2.1. Monthly atmospheric carbon dioxide concentrations measured at Mauna Loa, Hawaii, March 1958 to January, 1991 (from Boden, et al., 1992).

oxide, and chlorofluorocarbons, also have been detected; together they probably affect the atmosphere about as much as CO_2 does alone. However, the increase in CO_2 is the most thoroughly documented, and in addition, CO_2 at present and predicted concentrations is hypothesized to have direct "fertilization" effects on plant growth. Therefore, the focus of the following discussion is on CO_2, although the reader must be aware that it does not constitute the entire GHG problem.

Future CO_2 Concentrations

Despite the unequivocal trend, the Mauna Loa data generate many questions. One logical question is what the curve will look like in the future. If one merely projects the annual increments of CO_2 measured during the period from 1977 to 1989 (1.592 ppm/year), when the Mauna Loa curve approximated a straight line (Figure 2.1), the atmospheric concentration of CO_2 would reach 600 ppm (double that in the year 1900) in about 155 years. However, the CO_2 increase clearly is not linear but rather forms a geometric progression, increasing by 2–4% each year. Hence, the doubling of CO_2 concentration will probably occur much sooner. Considering the nature of fossil fuel use and the uptake of CO_2 by plants and the oceans, Schneider (1989) suggests that atmospheric CO_2 concentrations will reach 600 ppm between the years 2030 and 2080, an estimate with which other knowledgeable scientists concur (e.g., Trabalka *et al.* 1985; Jäger 1988).

Annual Oscillation in CO_2 Concentration

In addition to the upward trend of average CO_2 concentration, the Mauna Loa record demonstrates a clear annual oscillation in CO_2 concentration (Figure 2.1). Monthly concentrations rise slowly through the winter, peaking in midspring (all 32 annual peaks have been measured in May), and then fall quickly to a minimum in early fall (September or October). The range between maximum and minimum (the amplitude) is not constant and has increased during the period of the record. For example, for the first half of the Mauna Loa data set the amplitude averaged 5.36 ppm, and for the second half, it averaged 5.80 ppm. The source of this oscillation and of the increasing amplitude has also generated controversy.

Source of the Annual Oscillation

The source of the annual oscillation in CO_2 concentration measured at Mauna Loa can be inferred from examining annual oscillations measured elsewhere. Figure 2.2 presents 4 years of monthly values of CO_2 concentrations at 15 sites which form a latitudinal transect from 76° N to 75° S (Gammon *et al.* 1985). This diagram has been referred to as the flying carpet of atmospheric CO_2. As Shugart (this volume) points out, it illustrates the earth breathing. Carbon dioxide is absorbed by the earth, or inhaled, during the growing season, with the greatest atmospheric drawdown being in the Northern Hemisphere and at high latitudes. In the Southern Hemisphere, the annual oscillation is much less intense. The oscillation is also out of phase with that in the north, because the Southern Hemisphere growing season occurs during Northern Hemisphere winter. The growing-season drawdown of CO_2 implicates the global biota as its source, while the terrestrial portion of the biota is implicated because the annual CO_2 oscillation is most extreme at the latitudes which possess the greatest amount of land (or, conversely, the least amount of ocean). All green plants take up CO_2 through the process of photosynthesis during the growing season; some is released again through respiration, and some is used to construct plant tissues, removing that portion of CO_2 from the atmospheric pool. In Figure 2.2, the cumulative characteristics of photosynthetic and respiratory activities of all the plants on earth are illustrated.

Low CO_2 Concentrations in the Annual Oscillation

The idea that the land biota control the minimum atmospheric CO_2 concentration is supported by differences between the annual oscillations at high latitudes and those at lower latitudes, such as Mauna Loa (20° N). The CO_2 drawdown is a rapid event at the highest latitudes (e.g., Point Barrow, Alaska, at 71° N),

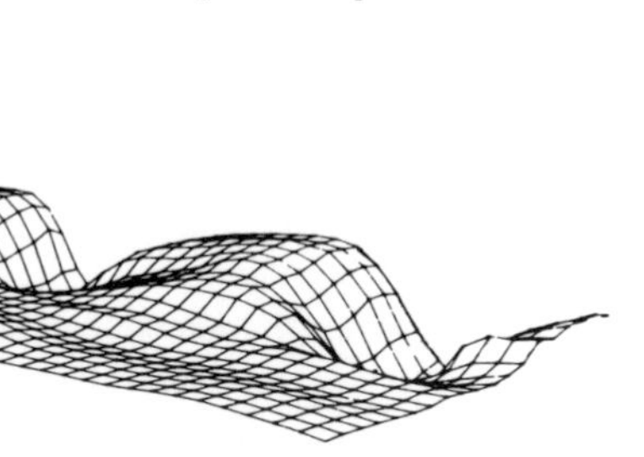

Fig. 2.2. Three dimensional representation of the monthly variation in CO_2 at different latitudes over a three year period (from Gammon, et al., 1985).

much like the onset of the high-latitude growing season. Atmospheric CO_2 concentrations remain high through June, but by August atmospheric concentrations reach the lowest monthly values of the year (Boden *et al.* 1991; 23). After September, atmospheric concentrations begin to rise again. The coincidence of the rapid drawdown of CO_2 and the wide annual amplitude (generally over 14 ppm) with the short, intensely productive growing season of the high latitudes strongly suggests a cause-and-effect relationship. Differences between growing-season length and intensity at Point Barrow and those at lower latitudes also parallel differences in the geography of the annual CO_2 cycle.

High CO_2 Concentrations in the Annual Oscillation

The earth slowly exhales during the remainder of the year. At high latitudes, atmospheric CO_2 concentrations rise rapidly after September, approaching near-maximum values by December. During the 17 years of Point Barrow record tabled by Boden *et al.* (1991), the month of maximum concentrations varied considerably; maximum concentrations occurred most frequently in May (7 times), followed by April (5 times), March (3 times), June (once) and February–January (tie, once).

This emission of CO_2 to the atmosphere involves at least two important sources. The decomposition of plant tissues, releasing CO_2, is one great source of the exhalation. Included here are not only the breakdown of leaves and branches which vegetation drops by the end of every growing season but also the release of carbon in seasonal wildfires (January to March in the tropics of the Northern Hemisphere; July to September in the Southern Hemisphere; Crutzen and Andreae 1990) and the breakdown of organic carbon compounds because of forest clearing (Woodwell *et al.* 1978) or farming (Mann 1986). These processes are active throughout the growing season and continue in the absence of photosynthesis at the end of the growing season. CO_2 release from living plants does cease each year because respiration is a temperature-dependent process, like carbon fixation in photosynthesis.

A second important source of CO_2 is the burning of fossil fuels, which reaches its annual peak in January and February. The wide variability in the month of maximum atmospheric concentration of CO_2 is paralleled by the wide variability in onset and intensity of winter conditions at higher latitudes. Although no one has yet attempted to correlate the individual variations in weather with variations in CO_2-related biotic and human responses, the relationship is likely to be documented in the near future.

Increasing Amplitude: Indicator of Enhanced Terrestrial Carbon Storage?

There is less controversy over what constitutes the sources of atmospheric CO_2 drawdown and liberation than there is over the cause of the increasing seasonal amplitude in the Mauna Loa data. Many oceanographers (e.g., Broecker *et al.* 1979; Tans *et al.* 1990, and others cited therein) believe that the increasing amplitude results from increasing uptake of CO_2 by terrestrial vegetation during the growing season, without a corresponding increase in the release of carbon through respiration. They point to their calculations of the annual amount of carbon which must be taken up by oceans and the amount of carbon known to enter the atmosphere from annual burning of fossil fuels. As Shugart (this volume) points out, these sources and sinks suggest that considerably more CO_2 should be in the atmosphere each year than is actually measured there.

This *missing fraction* is thought to be taken up and stored by terrestrial vegetation through the process of *carbon fertilization*. Photosynthesis requires CO_2 along with water and nutrients in order to fix atmospheric carbon. If CO_2 is present in less-than-optimal concentrations, it should limit growth, just as does less-than-optimal soil moisture or nutrients. Thus, if increased atmospheric CO_2 concentrations decrease the constraint on growth, each plant or plant community should be able to fix and store more carbon than it could at lower concentrations. This negative-feedback system is often cited as a control on the maximum atmospheric CO_2 concentrations which can be reached because the greater is the amount of CO_2 which enters the atmosphere, the greater the amount that will removed by the terrestrial biosphere.

A parallel to this view is the high probability that the terrestrial biosphere was a net sink for carbon during the transition from the last full-glacial, 18 000 years ago, to the postglacial 8 000 to 10 000 years ago (Grove 1984; Solomon and Tharp 1985; Adams *et al.* 1990), despite calculations suggesting that the terrestrial biosphere stored about as much carbon during full-glacial time as during the most recent preindustrial time (Prentice and Fung 1990). This increase in carbon storage during the full-glacial to postglacial transition represents a different process from the carbon fertilization suspected of being responsible for the increase in terrestrial carbon storage which is needed to balance global carbon-cycle models. Nevertheless, it is important to note that glacial-to-interglacial temperature increases were probably paralleled not only by atmospheric CO_2 increases but also by increases in carbon storage in terrestrial ecosystems. The most important caveat in assuming that the glacial-postglacial carbon sequestration could repeat itself in future warming from CO_2 is that there may be little resemblance between the carbon storage processes involved in warming a world which is covered by ice sheets, and possesses a weak hydrological cycle and a vegetation starving for CO_2, and those carbon processes involved in warming a world which has little glacial ice, an intense hydrological cycle, and a vegetation which, if limited by CO_2, is becoming less so every year.

Increasing Amplitude: Indicator of Enhanced Terrestrial Carbon Release?

In contrast to the view that the terrestrial biosphere is taking up increasing amounts of carbon, many ecologists believe that the terrestrial biota is releasing increasing amounts of CO_2 and that this may account for the increased amplitude in the seasonal CO_2 oscillation (e.g., Woodwell *et al.* 1978; Solomon and West 1985; Houghton and Woodwell 1989). Several processes could account for increased carbon release. For example, about half of the carbon which was stored in terrestrial ecosystems before the industrial age has recently reentered the atmosphere following fires, logging, and other land-clearance actions by humans (Olson *et al.* 1983; Post *et al.* 1990). The terrestrial biosphere is calculated to be a current net source of CO_2 of a magnitude between one-tenth and one-half of that released annually by fossil fuel burning (Woodwell *et al.* 1983). Crutzen and Andreae (1990) point out that the tropical deforestation which generates most of the terrestrial source is indeed increasing.

A supporting idea can be derived from the record of temperature and CO_2 in air bubbles in the polar ice caps (e.g., Barnola *et al.* 1987). Throughout the past 160 000 years, the variations in CO_2 match those in the record of temperature (Fig. 2.3). Low temperatures have been accompanied by low atmospheric CO_2 and high temperatures match high atmospheric CO_2. Although the data are not adequate to determine whether temperature change preceded or followed changes in CO_2, many scientists have assumed the more obvious process dominates and have concluded that CO_2 has forced temperature change (e.g., Berner 1990). Certainly the concentration of CO_2 in the atmosphere can control temperature (CO_2 accounts for 50% of the variance in temperature during glacial to interglacial warming, according to Lorius *et al.* 1990).

Houghton and Woodwell (1989) point out, however, that the temperature changes reflected in the ice cores of the past 160 000 years (Fig. 2.3) could have forced the prehistoric CO_2 changes instead of the reverse. As temperature increases, the rate of respiration would increase more than that of photosynthesis, generating a net release of CO_2 (Larcher 1983). Soil respiration may be essentially decoupled from the above-ground photosynthesis-respiration carbon cycling. Soil

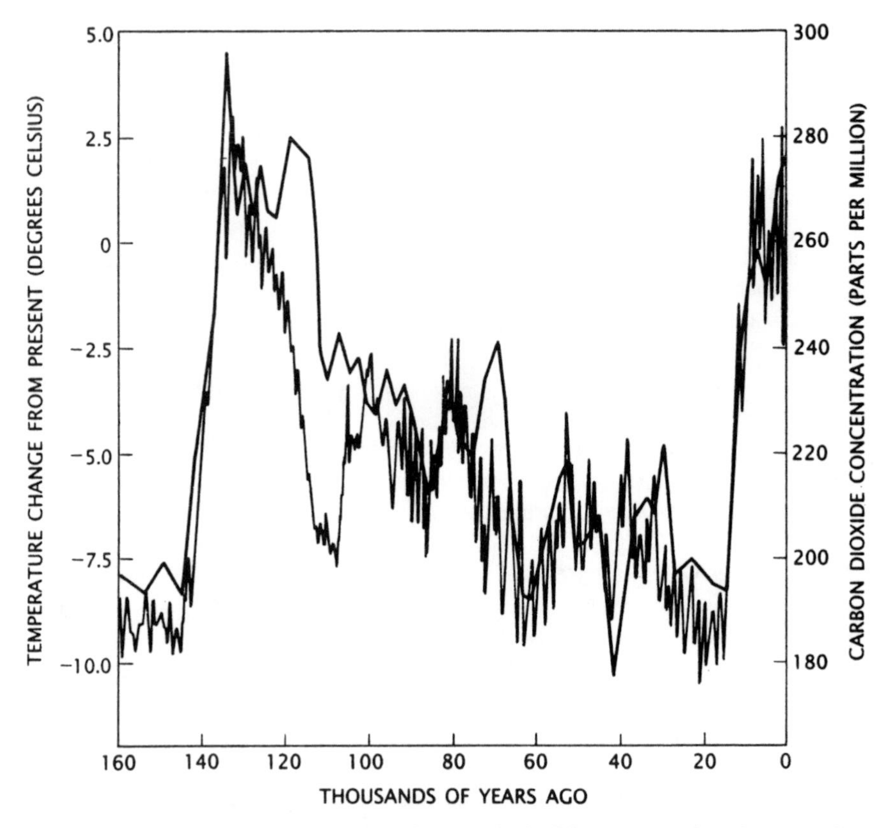

Fig. 2.3. Variations in temperature and atmospheric CO_2 concentrations in Antarctica during the last 160 000 years (from Schneider 1989). Jagged line represents temperature.

respiration may increase exponentially with increasing temperature while operating on organic substrates stored in soils over hundreds and thousands of years (e.g., Billings *et al.* 1982). If this is true, then greater warmth predicted to occur from increasing atmospheric CO_2 could be part of a positive feedback system whereby greater temperature leads to greater release of CO_2, in turn enhancing temperature. The close and direct relationship between CO_2 and temperature over 160,000 years makes it difficult to justify the view that increasing CO_2 will not increase release of carbon to the atmosphere, as the "carbon fertilization" hypothesis requires.

The logical extension of a positive feedback system in global carbon cycling has been examined as having the potential to generate a "runaway greenhouse" (complete evaporation of the oceans). Kasting and Ackerman (1986) conclude that the dynamics of the water vapor–CO_2 system are immune to this effect (i.e., form a negative feedback system), even at atmospheric CO_2 concentrations two orders of magnitude greater than those contemplated from burning all the fossil carbon on earth over several hundred years (e.g., Sundquist 1986). Because the model was aimed at the early life-free earth, they did not examine the concentration of atmospheric CO_2 in the modern world at which biospheric inputs might become unimportant. However, it is clear that an upper limit to any biota-driven positive feedback system must exist.

A mechanism to control long-term changes in temperature, independent of CO_2 concentrations, exists in the form of the widely accepted astronomical theory of glaciation (Imbrie and imbrie 1980). When gradual cyclic shifts in the earth's orbit force global-scale changes in temperature, a resulting atmospheric CO_2 release or depletion could appear to be almost concurrent, based on available time and chemistry measurement techniques. Unfortunately, there is no evidence to indicate that warming actually increases the rate of release of CO_2 from soils to a significantly greater extent than photosynthesis sequesters carbon in new biomass. Indeed, plant respiration and litter decomposition have decreased under enhanced CO_2 in some experimental situations, without adequate explanation (Mooney *et al.* 1991). Also, it appears doubtful that much carbon can be sequestered in the above-ground (Solomon 1986) or below-ground (e.g., Schlesinger 1990) portions of the terrestrial biosphere (with the exception of the potential carbon fertilization process described below) over the short time frame of decades to centuries during which anthropogenic CO_2 is expected to affect terrestrial ecosystems.

Is the amplitude of the seasonal oscillation in atmospheric CO_2 increasing because the terrestrial biosphere is a net storer of carbon or because it is a net producer of carbon? The foregoing discussion does not provide an answer, but instead only indicates that the current information is too sparse and ambiguous to allow prediction of the role of the terrestrial biosphere in amplifying or ameliorating future atmospheric CO_2 concentrations.

Responses to Changing Atmospheric CO_2

As one may have gleaned, the record of changing atmospheric CO_2 concentration raises great concern because of its potential for changing the climatic and biotic status quo, rather than because of any inherent concern with atmospheric chemistry per se. These responses require examination, particularly those relevant to process and pattern in the terrestrial biosphere. Two direct results of increasing atmospheric CO_2 which emerge from the foregoing discussion are of concern: responses by plant growth and responses by the climate system. In addition, several secondary responses are of particular interest. These include the response of the terrestrial biosphere to changing climate, especially in terms of the global carbon-cycling phenomena discussed above, but also including changing biodiversity and effects on endangered species, the ability of species to migrate to newly available niches, and so on. Also of concern is the interaction of an increasingly large human population with lands presently too cool to farm but which may undergo clearance and agriculture in a warmer world.

Ecosystem Response to Increasing Atmospheric CO_2

The expected change in atmospheric CO_2 concentration during the 21st century may directly affect plant growth. The overriding issue is whether effects on plant growth will be translated into effects on properties of global carbon sources and sinks. We should be most concerned about forests, which are likely to be the most important carbon sources and sinks. Although they cover only about 35% of the earth's surface, they produce 65% of the earth's annual net carbon fixation (Whittaker and Likens 1973) and store over 80% of the earth's terrestrial organic carbon (Olson *et al.* 1983). Enhanced atmospheric CO_2 concentrations have obvious effects in speeding the life cycle of annual plants grown in greenhouses. Lemon (1983), Strain and Cure (1985), and Bazzaz (1990) provide detailed analyses of the effects of enhanced CO_2 on photosynthesis, respiration, growth, and development of plants in greenhouse experiments. However, little information has been published on long-term growth responses by perennials, especially trees. Tree seedlings in greenhouse and growth-chamber experiments have exhibited increased growth rates and biomass accumulation and have increased their water use efficiency (amount of carbon fixed per unit of water) (Lemon 1983; Oechel and Strain 1985; Körner, this volume).

As yet, no research data indicate that mature trees growing in forests will be capable of taking advantage of the increases in dry-matter production and drought tolerance found in greenhouse herbs and woody seedlings. Plants acclimate (cease to respond) to raised CO_2 concentration after several days or months (Oechel and Strain 1985; Brown and Higginbotham 1986; Kramer and Sionit 1987), indicating that caution is needed in defining long-term ecological implications on the basis of short-term experiments with high CO_2 concentrations. Indeed, the opposite response (growth loss) could actually occur with enhanced atmospheric CO_2.

Field studies which were developed to measure changes in tree growth in response to acidic precipitation and gaseous air pollutants revealed that annual tree growth has declined (Johnson 1983; McLaughlin *et al.* 1983; Plochmann 1984; Nilsson and Duinker 1987), despite increases in global CO_2 of 25% since about 1850. Even the growth increases at very high altitudes in the White Mountains of California (LaMarche *et al.* 1984), which were claimed to parallel CO_2 increases, are ambiguous at best. For example, similar growth increases in the San Bernardino Mountains to the south (Jacoby 1986) have now been attributed largely to precipitation anomalies (Jacoby, pers. comm. 1988), and in the Sierra Nevada a short distance to the west (Graumlich 1991), to climate variables not measured near the White Mountains trees. Also, the timing of enhanced tree growth at these temperature-limited growth sites coincides as closely with the warming of the past century as with the CO_2 increases. In contrast to these results, high-latitude trees in Canada, Alaska, and northern Europe seem quite uniformly to have undergone growth declines in the past 30–50 years (Schweingruber 1992) despite measurements to the contrary in Finnish forests (Harri *et al.* 1984; Hari and Arovaara 1988).

If carbon fertilization is occurring in forest ecosystems and it cannot be detected above ground, then it must be sought below ground. Initial experiments examining root growth of tree seedlings in chambers containing high CO_2 (Norby *et al.* 1984, 1986, a,b) indicate not only greater growth of roots than shoots, but also a greater mobilization in the soil solution of nutrients necessary for growth, particularly when experimental plants were under substantial water and nutrient limitations. Current understanding of the role of the terrestrial biosphere in the global carbon cycle could be fundamentally changed if ecosystems are found to commonly possess this ability to enhance nitrogen- and water-use efficiency under stress while a major portion of annual net productivity is shunted to grow more and larger roots. More recent studies on long-term growth of root-perennials in salt marshes (Drake 1989; Mooney *et al.* 1991) produced similar results in the absence of water and nutrient limitations. Again, there is no evidence to confirm or deny that these processes are occurring in the field. If they are occurring, they will be detected in the real world only with great difficulty because below-ground processes are notoriously inconvenient to handle and manipulate in the field.

Plant communities will respond to changes in climate and CO_2 "fertilization," if at all, as a function of changing competitive advantages among species. Körner (this volume) presents compelling arguments for producing a hierarchy of species responses, based on ecological functional properties (including below-ground properties) of classes of species. This approach negates the simplistic view that all species would benefit from increased CO_2 or that responses are so unpredictable that all species must be measured before any projection can be made (e.g., Strain 1987). In fact, growth advantages conferred on one species must result in growth losses in less-competitive species in a complex, but predictable, way. The fundamental problem is that there is presently no body of ecological theory which can successfully predict the outcomes of competition at any specified time or place when it occurs under increasing effects of carbon fertilization. Indeed, one cannot even provide an accurate scientific estimate of whether forest stands, for example, will be positively or negatively affected, although opinions range from expecting strong positive growth anomalies in all trees (e.g., Goudriaan 1986) to strong negative anomalies (e.g., Solomon 1988).

Climatic Responses to Increasing Atmospheric CO_2 and Other Greenhouse Gases

The global climate change predicted for the next century is expected to be spatially and seasonally nonuniform. Consider one scenario of the warming projected for a doubling of CO_2, shown in Figure 2.4 (Manabe and Stouffer 1980), which

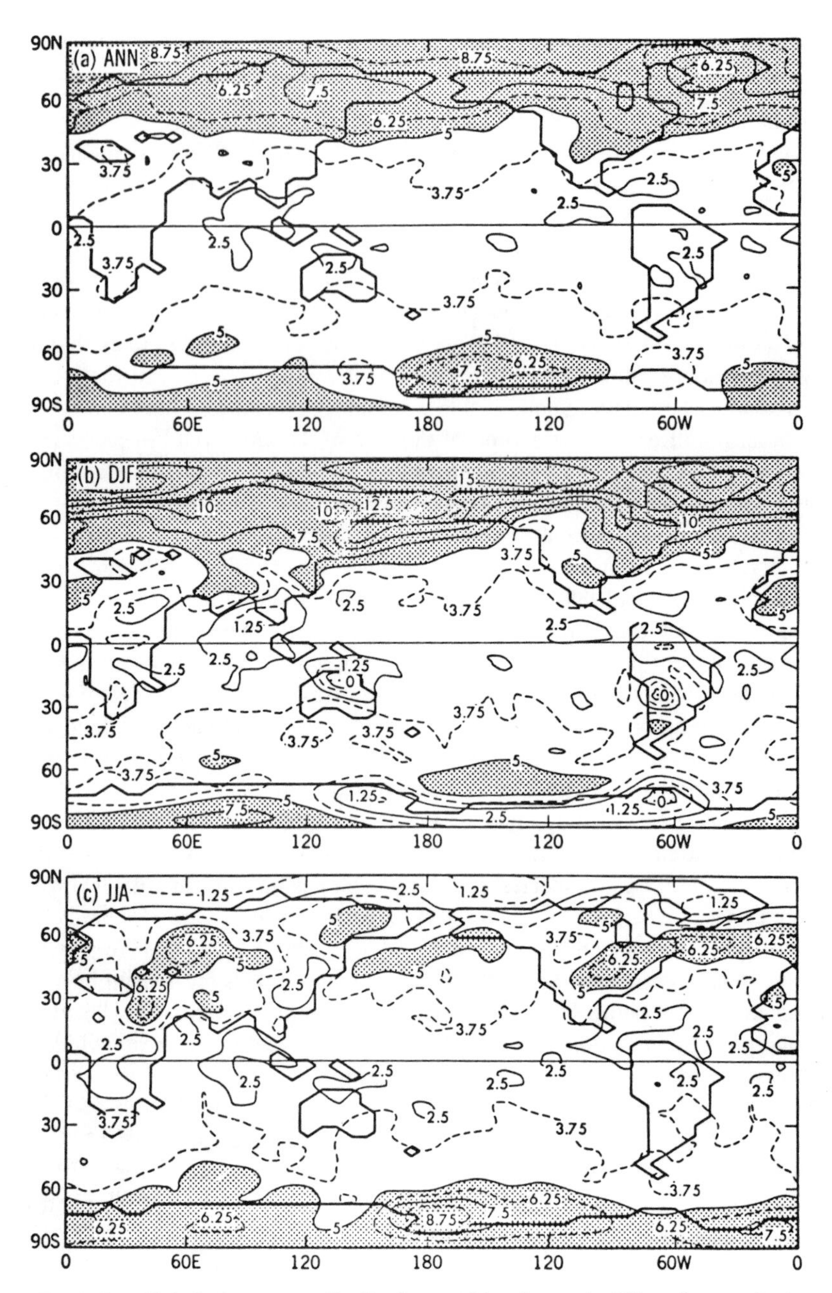

Fig. 2.4. Global temperature distribution resulting from a doubling of atmospheric CO_2 concentrations, projected by a general circulation model (GCM) of the atmosphere. a. Annual average temperature. (b) December–January–February (Northern Hemisphere winter) temperature. c. June–July–August (Northern Hemisphere summer) temperature (from Manabe and Stouffer 1980).

results from simulations of the earth's climate by a general circulation model (GCM) of the atmosphere. Although it differs in detail from other scenarios, as each does from the others (e.g., Schlesinger and Michell 1985; Cess *et al.* 1989), the trends of geographic and seasonal variation produced by all the GCMs are similar.

The warming becomes more intense with increasing latitude (Figure 2.4a–c) and is more intense in the winters (Figure 2.4b) than in summers (Figure 2.4c). While summer is expected to warm the least, it is expected to warm the quickest, as winter air temperatures are modulated by ocean surface temperatures which change more slowly than air temperatures (Thompson and Schneider 1982). The patterns portrayed in Figure 2.4 do not necessarily indicate that climate *effects* are greatest at highest latitudes. For example, actual year-to-year (interannual) variation is also greatest at highest latitudes and least at lower latitudes. As a result high-latitude plant and animal species are adapted to wide variation and may be largely unaffected. Also, the signal-to-noise ratio describing the relationship between systematic global warming and nondirectional temperature variation is predicted to be greatest in neither high nor low latitudes but rather in midlatitudes (Wigley and Jones 1981).

It is important to note that the magnitude of predicted warming not only varies from place to place and season to season but may in some cases even be expressed as cooling. Recent simulations with the NCAR (National Center for Atmospheric Research) coupled atmosphere-ocean GCM produced a pattern of warming over much of the globe with strong winter cooling in northern Scandinavia and Russia. The simulated Gulf Stream, which warms much of northern Europe in its present path along the Norwegian coast, migrates up the western coast of Greenland (Washington 1990).

The trend of greater warmth in winter than in summer would greatly increase the length of the growing season without much apparent increase in average growing season temperature. Plant communities of the highest latitudes, which encounter long summer days with critically short growing seasons and which are most sensitive to warmth, may be expected to be most affected by enhanced warmth. Extension of the growing season can have deleterious effects as well as the obvious advantages to productivity. Warming of the atmosphere early in the season, when the soil moisture may still be locked up in snow, can cause needle loss in conifers, as photosynthesis is induced in the absence of the required free soil moisture. Earlier thaw of soils and evapotranspiration, in the absence of enhanced precipitation, could reduce overall growth.

The sensitivity of plants and plant communities to interannual variations is probably inversely related to latitude, with those high-latitude plants which experience the greatest climatic variation being the most plastic in their environmental requirements, and hence most insensitive to weather and climate extremes. However, the trend of warming with greater latitude is still significant because plants and plant communities are increasingly sensitive to warmth per se with increasing latitudes such that warmth is the primary growth-limiting factor at highest latitudes. In contrast, there is seemingly always enough warmth for routine growth at low latitudes, and hence the amount and seasonal distribution of precipitation are of critical importance there.

A few generalizations about precipitation in a warmer world can be made. The hydrological cycle is expected to increase in intensity, resulting in increased precipitation and increased evapotranspiration. There is no indication yet that the ratio of precipitation to evapotranspiration will change. Midcontinental areas could be provided with less precipitation and experience more frequent droughts (Manabe and Wetherald 1987) and coastal regions could experience enhanced moisture (rain, fog, and so on). This pattern is reflected in derived soil moisture distributions generated by GCMs (Fig. 2.5). Unfortunately, more precise estimates are unavailable. Precipitation scenarios from GCMs are much less reliable than temperature scenarios, primarily because processes which determine precipitation generally ensue at smaller spatial scales than are accommodated in most GCMs. This poor reliability is illustrated by the weak agreement among different models (Schlesinger and Mitchell 1985).

The rate of warming is also expected to be nonuniform. A widely quoted scenario for increases in globally averaged temperature (Fig. 2.6) was developed by global change experts at Villach, Austria, in the fall of 1987 (Jäger 1988). The gradual warming of the past century was expected to accelerate greatly toward the end of the 20th century. This prediction was made before the years of 1987, 1988, 1990 and 1991, which became the warmest years recorded during the last century (Kerr 1991, 1992). The rapid increase in temperature, shown

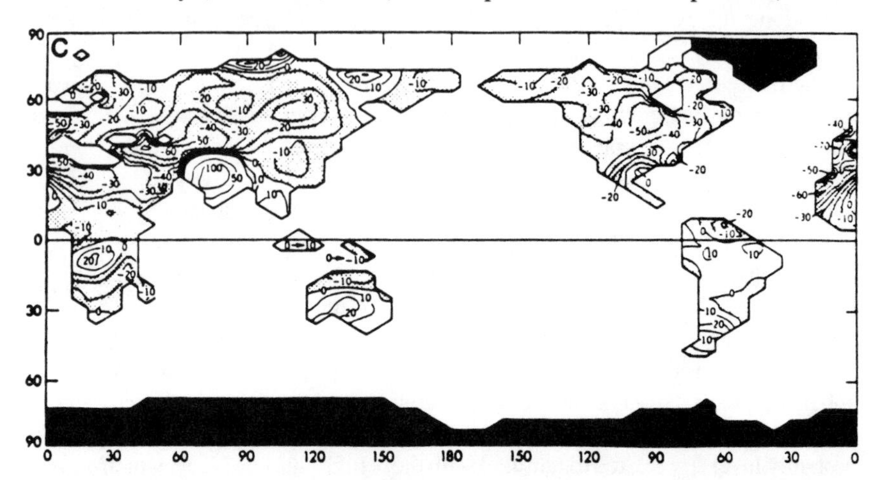

Fig. 2.5. Global soil moisture distribution as a percentage of model control soil moisture in (Northern Hemisphere) summer resulting from a doubling of atmospheric CO_2 concentrations, projected by a GCM of the atmosphere (from Manabe and Wetherald 1987).

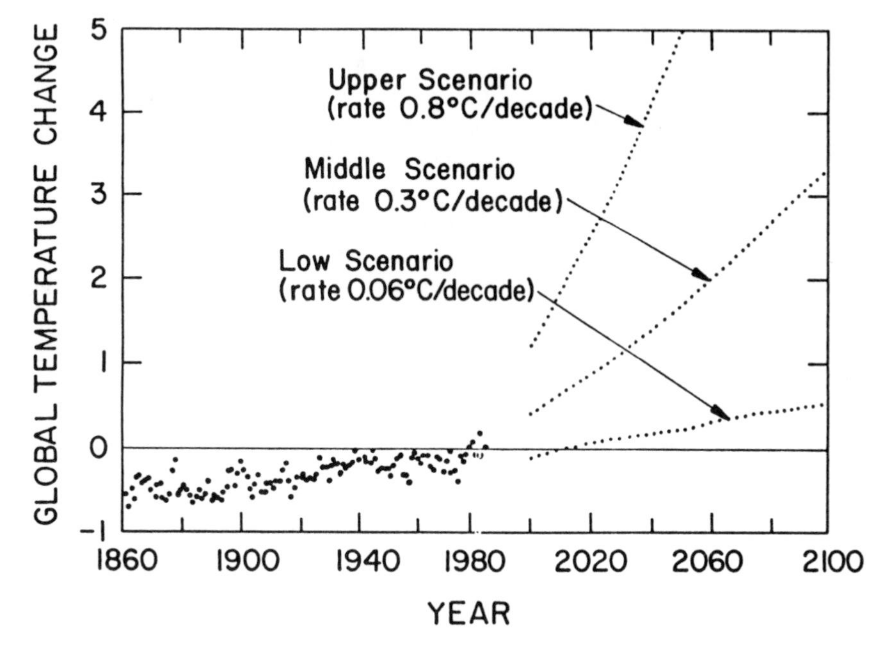

Fig. 2.6. Temperature change measured from 1860 to 1985 and projected to the year 2100, in response to continued emissions of atmospheric greenhouse gases (GHGs). Values are plotted as differences from 1985 values. Each curve includes the time lags in the climatic response resulting from the ocean's heat capacity. The middle curve is based on continuing present trends of GHGs except for chlorofluorocarbons (CFCs). The upper curve reflects accelerated GHG emissions and high climate sensitivity predicted by some GCMs. The lower curve reflects the result of radically curtailed GHG emissions and relatively low climate sensitivity (from Jäger 1988).

after the 1970s in Figure 2.6, was induced when the climate response began to overtake the climate forcing by overcoming lags in the climate system caused particularly by slowing warming ocean surfaces (Thompson and Schneider 1982). The rapid temperature change may allow the warming to emerge from the inherent variability of temperature within about a decade (Hansen 1989; Schneider 1989).

Ecosystem Responses to Climate

Ecosystems are expected to respond in fundamental ways to the changes in climate. These have been reviewed in detail (Solomon and West 1985; Shugart *et al.* 1986; Overpeck and Bartlein 1989), and they provide the impetus for most of the chapters which follow in this book. Brief consideration of ecosystem response suggests several important areas of concern.

The natural limits and composition of the globe's primary producers, plants, may change significantly as a result of the future climate changes implied by results from GCMs. This expectation is based on the assumption that climate determines the boundaries of plant species and ecosystems and that current geographic distributions of plant populations and ecosystems are more or less in balance with current distributions of the controlling climate variables. Rapid climate change would result in redistribution of climate properties, decoupling ecosystem distributions from climatic ones, and rendering extant populations and ecosystems incompatible with the conditions under which they would be growing. Restoration of the climate-vegetation coupling would require dieback and replacement of most plant communities and ecosystems (Solomon 1986). Although stress-related tree death may require only a few years (Waring 1987), many decades may be required to replace trees, even in environments not undergoing chronic climate change.

Current concerns surrounding this destruction of extant ecosystems focuses on preservation of biodiversity and endangered species and on preserving the genetic resources of the earth. The biota receiving most of the present concern have been reduced by the actions of man: harvesting forests, replacing natural biotic communities with agriculture, emitting dangerous pollutants, and introducing species from other regions (Peters 1988). The issue for the future is that biota will also be threatened by changing extremes in climate and chemical content of the troposphere.

Characteristically, endangered species are specialized and inflexible with respect to their environmental requirements. They cover very little geographic area and possess little of the genetic variability required to adapt to new conditions. If the unique set of circumstances in which they can thrive is disrupted or disappears, the species also disappears. Thus, endangered species are confined to very few locations or habitats and, by definition, are present in very low numbers. Threatened biotic communities, like threatened species, similarly possess unique requirements and thus are also uncommon (Peters and Darling 1985).

Current management approaches for preserving endangered species or rare and specialized communities almost invariably include setting aside isolated reserves where the endangered biota can live undisturbed. As a result, endangered species and biotic reserves are limited to a few isolated point locations, making them particularly vulnerable to a significant climatic change. The weather conditions that they require may shift space only slightly and still depart completely from the limited area in which the species or communities grow. Recent analyses indicate that although about 30% of the earth's vegetation would be reclassified to new categories under the climate of a CO_2 doubling, these changes would encompass 50% of the earth's nature reserves (Leemans 1991).

Ecosystem destruction also implies great changes in forests and forest products. Direct choices by society will determine hardwood forest product availability during the next few decades. Currently, forest destruction is rapid in the tropics, where much of the global hardwood growing stock is found. Thus, the availability of hardwood products may not depend much on climatic effects of global environmental change.

However, human activities which produce changes in climate and atmospheric chemistry may indirectly define the future of softwood growing stock. The circumpolar boral forest in the Northern Hemisphere contains three-fourths of the softwood growing stock of the world (U.S.D.A. Forest Service 1982). Climates that occur nowhere today in the boreal forests may soon displace the temperature and precipitation regimes under which the northern ecosystems currently exist. A potential result is massive forest dieback (Solomon 1986), at first in southern boreal forests where growing stock is most accessible to the forest industry. Similar stresses, with widespread tree mortality, are possible in more temperate regions as well.

For both biodiversity and forest products concerns, the speed of warming rather than its magnitude may be the most crucial characteristic of global change for defining resultant effects on ecosystems. During the alternation of ice age and interglacial climates during the past two million years, temperature changes were more intense than any expected to result from a doubling of atmospheric CO_2 and other greenhouse gases. For example, temperatures increased by 5°C in midlatitudes of eastern North America and Europe at the end of the last glacial, 10 000 to 15 000 years ago, and by 6°C at the end of the previous glacial, 125 000 to 130 000 years ago (Fig. 2 of Webb 1986), with little loss of species diversity (e.g., Adams and Woodward 1989).

However, the apparent speed of the warming events has been relatively slow during these geologic time periods (e.g., 0.1–0.2°C per century; Fig. 2 of Webb 1986). In contrast, the rate of temperature change, even of the moderate scenario developed by the scientists at Villach for the 21st century, was 15–30 times as great (3°C per century, see Fig. 2.6). A moderate warming of 2.4–3.0°C (summer) and 3.6–4.2°C (winter) per century in midlatitudes was suggested by the Villach group (Jäger, 1988; Tab. 1). Latitudinal temperature gradients of 0.5°C/100 km in summer and 1.5°C/100 km in winter (midlatitude North America; Moran, 1972; Cohen 1973) will induce summer isotherms to shift northward 120–150 km per century and winter isotherms to shift northward 540–630 km per century.

Migration Response to Rapid Warming

Seeds are carried from parent plants by animals or the wind, frequently tens and occasionally hundreds of meters away. If they are carried beyond the boundaries of the plant population and if conditions at the sites of seed deposition favor seed germination and growth, the seeds will establish new subpopulations which then may reach reproductive maturity and also issue far-traveling seeds. Over many decades and centuries of repeated establishment and transport, a plant population may spread a great distance from its starting point. The ecological significance of this migration process in terms of rapid climate change is that it requires long time periods, particularly among the slowest-growing species, forest trees.

Thus, the discrepancy between known migration rates of trees and the expected displacement rate of temperature isotherms causes considerable concern among ecologists. The maximum migration rates calculated in eastern North America are 10–15 km per century for trees possessing animal-transported seeds (zoophilous) and 20–30 km per century for trees possessing wind-transported (anemophilous) seeds (Solomon *et al.* 1984). These rates of displacement are consistent with those measured at 10–35 km per century, and 20–40 km per century, respectively, from pollen which accumulated in eastern North America during the past 20 000 years (Davis 1981; Tab. 10.2). Gear and Huntley (1991) measured migration responses of 35–80 km per century by the anemophilous Scots pine in northern Scotland to a "rapid" midpostglacial climate change.

Although these tree migration rates are an order of magnitude less than expected rates of isotherm displacement (120–630 km per century, described above), they may be optimistic estimates for conditions expected during the 21st century. Both the calculated and measured tree migration rates assume that barriers to seed transport do not exist and that sites for establishment of new trees are readily available. Davis *et al.* (1986) have shown that tree migration from the Lower to the Upper Peninsula of Michigan was delayed by as much as a thousand years because of the barrier presented by lakes Huron and Michigan, which narrowed to as little as 10 or 20 km at the Straits of Mackinac during the Nipissing Great Lakes stage, 4 000–6 000 years ago (Larsen 1987). The agricultural landscapes of the temperate regions also present wide barriers to seed transport and contain few sites on which migrating species can become established.

Life-cycle Response to Rapid Warming

Another potential difficulty presented by rapid rates of climate change involves survival of species growing in place. Plants require a certain length of time to complete their life cycles; trees require many years. Life cycles consist of several stages: the establishment of seedlings, the subsequent growth of established seedlings to reproductive maturity, and the production of healthy seeds which finally become available for establishment as seedlings. Each stage requires different environmental conditions. Seedlings of eastern hemlock (*Tsuga canadensis*) require moderate but constant soil moisture during several successive growing seasons while mature trees have less need for constant soil moisture than for nutrient-rich soils. (See Fig. 2.7). The requirements of separate life stages are not normally a concern because conditions necessary to complete one stage usually occur in the same places as conditions needed for other stages. Thus, the soil disturbance and sunlight required by jack pine (*Pinus banksiana*) seedlings (Fig. 2.7) will be present in the same region as the specific ranges of sunlight and warmth that mature jack pine trees require. However, the differences in the

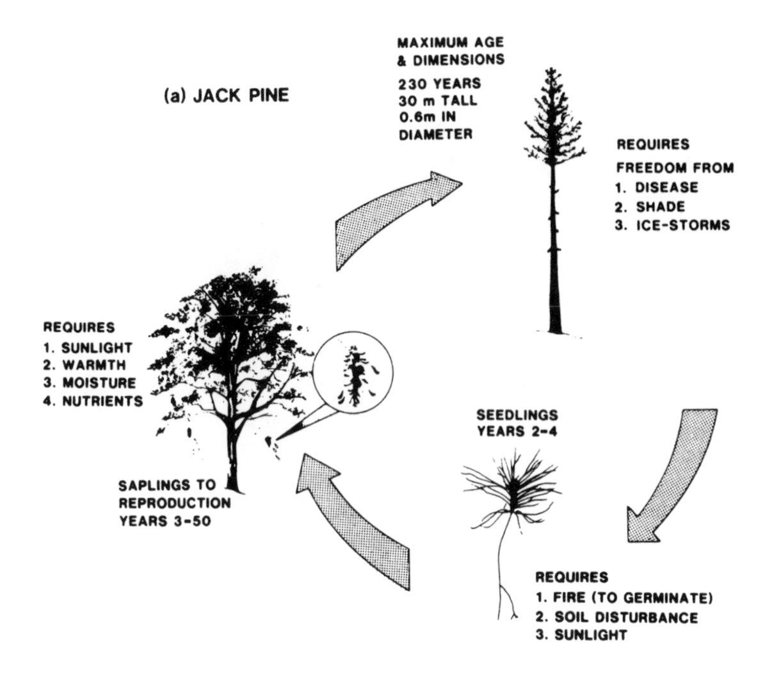

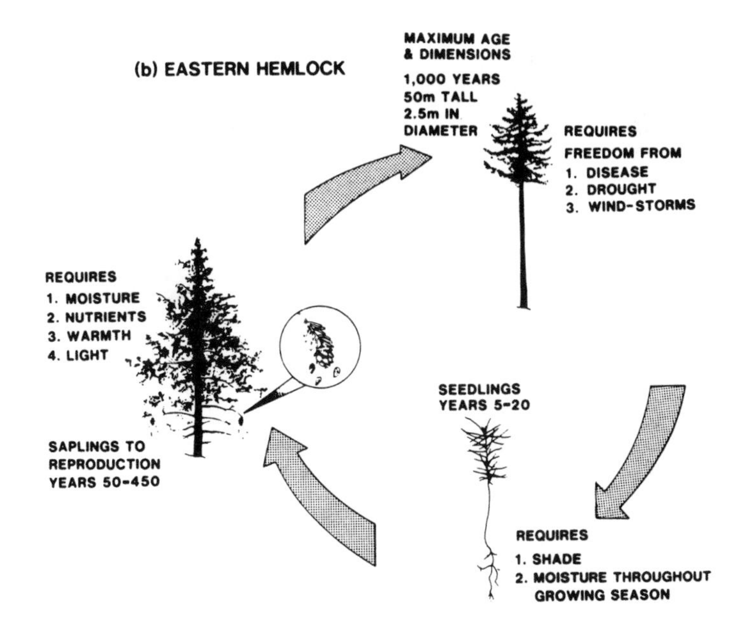

Fig. 2.7. Life-cycle stages of two temperate North American tree species and the critical environmental variables required for continued vigor in each stage (from Solomon and West 1985).

predicted rate of displacement of summer and winter isotherms described above suggest that life-cycle stages could become uncoupled if, for example, one stage is limited by winter warmth and another stage by summer growing seasons.

Perhaps a more probable difficulty arising from the rapid rate of the expected climate change is the intrinsic timing of tree life cycles. The most rapidly maturing trees, represented in Figure 2.7 by the jack pine, require a minimum of 5 or 10 years to reach reproductive maturity with longer periods needed in the closed-canopy forests which store the majority of the earth's terrestrial carbon. Slow-growing species such as eastern hemlock require 50–450 years to reach reproductive maturity after seeds germinate. The wide range of values given for the duration of a life cycle in each species is derived from environmental differences from place to place within plant populations; jack pines under stress from lack of water or nutrients grow much more slowly than when these variables are not limiting. Hemlocks grow much more slowly in the shade of closed forests, although the shade is required to meet their needs of continuous but moderate soil moisture.

The question of concern here is whether the life cycle of individual trees can be completed before changing climate at their growth sites eliminates the conditions required for successful growth in one or more life stages (Solomon and Leemans, 1990). The climate associated with a doubling of CO_2, possible within 50–100 years, could kill trees before they reach reproductive maturity. Even if reproductive maturity is reached, the seeds the trees produce may be unsuitable. Trees whose seedlings can survive today belong to mature trees which changing climate may prohibit from growing there in 50–100 years. Mature trees which can grow there in 50 or 100 years may be derived from seedlings which cannot survive climatic conditions which prevail today.

A spontaneous response by vegetation must occur, possibly involving the loss of slow-growing species and the selection of species which complete their life cycles very quickly—that is, early successional trees and shrubs which may be the species which also migrate and colonize most quickly. In addition, the ability to reach reproductive maturity in the shortest amount of time requires that the successful tree species grow in the sun—that is, in the absence of closed-canopy forests, again suggesting early successional species. In essence, this suggests the chronic loss of today's dense, closed-canopy forests which contain great carbon storage capacity in tropical, temperate, and boreal regions. A succeeding open landscape of scattered trees which form woodlands or savannas would also contain less than half the carbon storage capacity (e.g., Olson *et al.* 1983). The stresses induced by a chronically warming climate or by atmospheric pollutant loads, soil acidification, and so on could reduce growth rates of individual trees further, increasing still further the chances that life cycles would be incomplete, whatever species were present.

This is the point in the logic train on predicted environmental change and ecological response at which the uncertainties begin to overwhelm the available knowledge. For example, if direct fertilization by enhanced CO_2 concentrations does occur, its most probable effect would be to ameliorate the stress-related slowing of growth, thereby allowing trees to reach reproductive maturity more quickly than they would otherwise and possibly canceling the negative effects of climate stress. It is important to note that there is absolutely no evidence to suggest that life cycles of perennial species such as trees could be shortened by increased atmospheric CO_2 concentrations, although the phenomenon is well known in annual plants.

Summary

The foregoing discussion examines the nature of current and future global environmental changes and describes the primary concerns and uncertainties in the corresponding responses by the terrestrial biosphere. The discussion focuses on defining the reasons why the terrestrial biosphere must be studied and the questions that the studies must answer. Clearly, the terrestrial biosphere is the primary modulator of global change and therefore must be understood if future global change is to be accurately predicted.

The single phenomenon of greatest significance is the indisputable trend in atmospheric CO_2 concentration measured at Mauna Loa Observatory since 1958. The record indicates a geometrically increasing average annual CO_2 concentration which may reach twice the amount present in 1900 by the year 2030. In addition, the CO_2 record demonstrates a seasonal oscillation—increasing in the summer, decreasing in winter—and a subtle-but-persistent increase in the annual amplitude of this oscillation. The Mauna Loa data led to the measurement of atmospheric CO_2 concentrations in other regions of the world, revealing a systematic global pattern in which the annual oscillation is most pronounced at high northern latitudes and in the Northern Hemisphere. In the Southern Hemisphere, annual oscillations in atmospheric CO_2 are much less intense and the seasonal cycle is out of phase with that in the Northern Hemisphere.

Analysis of these CO_2 records implicates the terrestrial biosphere as the primary driver of the annual cycle. The increasing amplitude, however, has produced the greatest amount of controversy. Many oceanographers believe that the amplitude is increasing because the terrestrial biosphere is increasingly storing more carbon than it is releasing. They point to data and models which indicate that oceans can absorb less carbon than is leaving the atmosphere each year. They believe the process responsible is carbon fertilization, the increase in growth rates which plants are known to undergo when they are fumigated with CO_2 while growing in pots in greenhouses. The view that biospheric carbon can increase during warming is supported by recent research which indicates that during the full-glacial to postglacial warming of 10,000–15,000 years ago, carbon simultaneously increased in the atmosphere and was sequestered in the terrestrial biosphere because of a net replacement of ecosystems containing low amounts of carbon by ecosystems with large amounts of carbon.

A contradictory view is expressed by some terrestrial ecologists who consider that credible evidence exists to suggest the opposite cause of the increasing amplitude: they believe that carbon is being released in steadily greater amounts than carbon is being sequestered each year. They point to the significant amounts of carbon being released from land clearance and forest burning, which make the terrestrial biosphere a net source of CO_2. They also mention the importance of soil respiration in liberating CO_2, a process which will increase with warming, especially in the highest latitudes. Their view is supported by parallel records of CO_2 and temperature embedded in polar ice caps. Data of the past 160,000 years indicate a close and direct relationship between CO_2 and temperature such that increasing temperature may be the cause of increasing CO_2 through its effect of enhancing soil respiration. If the oceanographers are correct, and the terrestrial biosphere is a continuously increasing carbon sink, measurements of tree growth are demonstrating that the missing carbon is not in stems and branches of trees. Instead, collection of fundamental data and creation of accurate environmental models may eventually show that the carbon is present in below-ground parts (roots), a subsystem which is very difficult to study over the decadal time scales necessary for documenting increasing carbon storage in natural systems.

In addition to the direct effects of CO_2 on vegetation, we are also concerned about phenomena derived from indirect effects—that is, from relevant properties of climate change. The projected climate changes of most interest include world-wide warming of increasing magnitude with latitude, and with summer warming occurring earlier than that in winter (although the warming in winter will eventually be of greater magnitude than warming in summer). Because of this latter feature, growing-season length is likely to increase much more than will average annual temperature, particularly at higher latitudes.

Precipitation, which should increase along coasts and decrease in continental regions, and soil moisture, which may decline in midcontinental regions, are both much less predictable than temperature. Processes generating precipitation are only coarsely represented in climate models, while soil moisture cannot be accurately predicted because of the absence of good precipitation and evaporation projections. The rate of warming is expected to be rapid compared with any other warming in the past 100 000 years. It is also expected to accelerate in the coming decades, such that within this decade, GHG-induced warming should identifiably emerge from the natural variation in global temperature.

The nature of plant population and ecosystem responses to these climate shifts is not specific enough to provide concrete or accurate predictions of ecological behavior. For example, increasing growing-season length could increase the growth and survival of species, especially at high latitudes. On the other hand, the ecotypes (ecological varieties) which are currently competing successfully in each population will be subject to a temperature and moisture climate in which they do not presently grow. Consequently, they may undergo moderate to severe dieback, requiring many decades to regrow.

Of greatest vulnerability are endangered species. By definition, they are species of few, isolated populations, having limited adaptability and ability to compete with other species. The isolated populations, frequently growing in reserves which are similarly vulnerable, can become climatically inappropriate quite rapidly. Less vulnerable, but of equal concern, is the world's softwood growing stock, three-fourths of which is concentrated in boreal forests. Dieback in these forests could severely stress the global forest-products industry.

The rate rather than magnitude of any climate change appears to have the potential to generate the greatest long-term impact. Plants have survived greater climate changes in the past two million years than those projected for the next century, but they appear to have encountered only rates of change which were slow enough to allow species to migrate and populations to adapt. Migration rates on modern landscapes are likely to be much slower than those of prehistoric time, but even the latter are an order of magnitude too slow to allow species to keep pace with the changing geography of climate. In addition, the longest-lived tree species will be vulnerable to extinction if they are unable to complete their life cycles (seed establishment, growth to reproductive maturity, seed production) before conditions at any one location become so hostile that mortality results.

None of these ecological consequences of global environmental change will necessarily occur. Species and individuals have proven to be much more plastic in their requirements and responses than a systems approach to ecological research would allow. Yet, the systems approach is the only scientific technique available for examining the complex and often-undocumented relationships among organisms and between organisms and their changing environment. It is to this goal that the remaining chapters of this book are directed.

References

Adams, J. M., Faure, H., Faure-Denard, L., McGlade, J. M. and Woodward, F. I. (1990). Increases in terrestrial carbon storage from the last glacial maximum to the present. *Nature,* **348,** 711–14.

Adams, J. M. and Woodward, F. I. (1989). Patterns in tree species richness as a test of the glacial extinction hypothesis. *Nature,* **339,** 699–701.

Barnola, J. M., Raynaud, D., Korotkevich, Y. S. and Lorius, C. (1987). Vostok ice core provides 160 000-year record of atmospheric CO_2. *Nature,* **329,** 408–14.

Bazzaz, F. A. (1990). The response of natural ecosystems to the rising global CO_2 levels. *Annual Review of Ecology & Systematics,* **21,** 167–96.

Berner, R. A., (1990). Atmospheric carbon dioxide levels over Phanerozoic time. *Science,* **249,** 1382–86.

Billings, W. D., Luken, J. O., Mortensen, D. A., and Peterson, K. M. (1982). Arctic tundra: A source or sink for atmospheric carbon dioxide in a changing environment? *Oecologia,* **53,** 7–11.

Boden, T. A., Sepanski, F.J. and Stoss, F.W. (1990). *Trends '91: A Compendium of Data on Global Change. ORNL/CDIAC-46.* Oak Ridge, TN: Oak Ridge National Laboratory.

Broecker, W. S., Takasashi, T., Simpson, J. J. and Peng, T.-H. Fate of fossil fuel carbon dioxide and the global carbon budget. *Science,* **206,** 409–17.

Brown, K. and Higginbotham, K. O. (1986). Effects of carbon dioxide enrichment and nitrogen supply on growth of boreal tree seedlings. *Tree Physiology,* **2,** 223–32.

Cess, R. D., Potter, G. L., Blanchet, J. P., Boer, G. J., Ghan, S. J., Kiehl, J. T., Le Treut, H., Li, Z.-X, Liang, X.-Z., Mitchell, J. F. B., Morcrette, J.-J, Randall, D. A., Riches, M. R., Roeckner, E., Schlese, U., Slingo, A., Taylor, K. E., Washington, W. M., Wetherald, R. T., and Yagai, I. (1989). Interpretation of cloud-climate feedback as produced by 14 atmospheric general circulation models. *Science,* **245,** 513–16.

Cohen, S. B. (1973). (1973). *Oxford World Atlas.* New York: Oxford University Press.

Crutzen, P. J. and Andreae, M. O. (1990). Biomass burning in the Tropics: Impact on atmospheric chemistry and biogeochemical cycles. *Science,* **250,** 1669–78.

Davis, M. B. (1981). Quaternary history and the stability of forest communities. In *Forest Succession: Concepts and Application,* ed. D. C. West, H. H. Shugart, and D. B. Botkin, pp. 132–53. New York: Springer-Verlag.

Davis, M. B., Woods, K. D., Webb, S. L. and Futyma, D. (1986). Dispersal versus climate: Expansion of *Fagus* and *Tsuga* into the Upper Great Lakes region. *Vegetatio,* **67,** 93–103.

Drake, B. G. (1989). Effects of Elevated Carbon Dioxide on Chesapeake Bay Wetlands. IV. Ecosystem and Whole Plant Responses, April–November, 1988. Prog. Report No. 051, Research, Atmospheric and Climate Research Division, U.S. Department of Energy, Washington, D.C.

Friedli, H., Loetscher, H., Oeschger, H., Siegenthaler, U., and Stauffer, B. (1986). Ice core record of $^{13}C^{12}C$ ratio of atmospheric CO_2 in the past two centuries. *Nature,* **324,** 237–38.

Gammon, R. H., Sundquist, E. T. and Fraser, P. J. (1985). History of carbon dioxide in The atmosphere. In *Atmospheric Carbon Dioxide and the Global Carbon Cycle,* ed. J. R. Trabalka, pp. 25–62. DOE/ER-0239, Washington, D.C. U.S. Dept. Energy.

Gear, A. J. and Huntley, B. (1991). Rapid changes in the range limits of scots pine 4 000 years ago. *Science,* **251,** 544–47.

Goudriaan, J. (1986). Simulation of ecosystem response to rising CO_2, with special attention to interfacing with the atmosphere. In *Climate-Vegetation Interactions,* ed. C. Rosenzweig and R. Dickinson, pp. 49–53. UCAR Report OIES-2, NASA, Office for Interdisciplinary Earth Studies, P.O. Box 3000, Boulder, Colorado.

Graumlich, L. J. (1991). Subalpine tree growth, climate, and increasing CO_2: An assessment of recent growth trends. *Ecology,* **72,** 1–11.

Grove, A. T. (1984). Changing climate, changing biomass, and changing atmospheric CO_2. *Progress in Biometeorology,* **3,** 5–10.

Hansen, J. E. (1989). A time to cry wolf. Letter submitted to *New York Times,* July, 1989 (pers. comm.).

Hari, P., Arovaara, H., Raunemaa, T. and Hautojaervi, A. (1984). Forest growth and energy production, a method for detecting trends in growth potential of trees. *Canadian Journal Forest Research,* **14,** 437–40.

Hari, P. and Arovaara, H. (1988). Detecting CO_2 induced enhancement on the radial increment of trees: Evidence from northern timber line. *Scandinavian Journal Forest Research,* **3,** 67–74.

Houghton, R. A. and Woodwell, G. M. (1989). Global climatic change. *Scientific American,* **260,** 18–26.

Imbrie, J. and Imbrie, J. Z. (1980). Modeling the climatic response to orbital variations. *Science,* **207,** 943–53.

Jacoby, G. C. (1986). Long-term temperature trends and a positive departure from the climate-growth response since the 1950s in high elevation lodgepole pine from California. In *Climate-Vegetation Interactions,* ed. C. Rosenzweig and R. Dickinson, pp. 81–83. Rept. OIES-2, UCAR, Boulder, Colorado.

Jäger, J. W. (1988). *Developing Policies for Responding to Climatic Change.* WCIP-1, WMO/TD- No. 255, W.M.O., Zurich, Switzerland.

Johnson, A. H. (1983). Red spruce decline in the northeastern U.S.: Hypotheses regarding the role of acid rain. *Journal of Pollution Control Association,* **33,** 1049–54.

Kasting, J. F. and Ackerman, T. P. (1986). Climatic consequences of very high carbon dioxide levels in the earth's early atmosphere. *Science,* **234,** 1383–85.

Keeling, C. D., Bacastow, R. B., and Whorf, T. P. (1982). Measurements of the concentration of carbon dioxide at Mauna Loa Observatory, Hawaii. *Carbon Dioxide Review: 1982,* ed. W. C. Clark, pp. 377–85. NY: Oxford University Press.

Keeling, C. D., Bacastow, R. B., Carter, A. F., Piper, S. C., Whorf, T. P., Heimann, M., Mook, W. G., and Roeloffzen, H. (1989). A three-dimensional model of atmospheric CO_2 transport based on observed winds: 1. Analysis of observational data. In *Aspects of Climate Variability in the Pacific and the Western Americas.* ed. D. H. Peterson, *Geophysical Monographs,* **55,** 165–235. Amer. Geophys. Union, Washington D.C.

Kerr, R. A. (1991). Global temperature hits record again. *Science,* **251,** 274.

Kerr, R. A. (1992). 1991: Warmth, chill may follow. *Science,* 255, 281.

Kramer, P. J., and Sionit, N. (1987). Effects of increasing carbon dioxide concentration on the physiology and growth of forest trees. In *The Greenhouse Effect, Climate Change, and U.S. Forests,* ed. W. E. Shands and J. S. Hoffman, pp. 219–46. Washington D.C.: Conservation Foundation.

LaMarche, V. C., Graybill, D. A., Fritts, H. C. and Rose, M. R. (1984). Increasing atmospheric carbon dioxide: Tree ring evidence for growth enhancement in natural vegetation. *Science,* **225,** 1019–121.

Larcher, W. (1983). *Physiological Plant Ecology.* 2nd ed. NY: Springer-Verlag.

Larsen, C. E. (1987). *Geological History of Glacial Lake Algonquin and the Upper Great Lakes*. Reston, Virginia: U.S. Geol. Surv. Bull. 1801, U.S.G.S.

Leemans, R. (1991). Ecological and Agricultural aspects of global change. In: *Environmental Implications of Global Change,* J. P. Pernetta, ed. I.U.C.N. Goland, Switzerland, pp. 21–38.

Lemon, E. R. (1983). *CO_2* and Plants: The Response of Plants to Rising Levels of Carbon Dioxide. Boulder, Colorado: Westview Press.

Lorius, C., Jouzel, J., Raynaud, D., Hansen, J. and Le Treut, H. (1990). The ice-core record: Climate sensitivity and future greenhouse warming. *Nature,* **347,** 139–45.

Manabe, S. and Stouffer, R. J. (1980). Sensitivity of global climate model to an increase of CO_2 concentration in the atmosphere. *Journal of Geophysical Research,* **85,** 5529–54.

Manabe, S. and Wetherald, R. T. (1987). Reduction in summer soil wetness induced by an increase in atmospheric carbon dioxide. *Science,* **232,** 626–28.

Mann, L. K. (1986). Changes in soil carbon storage after cultivation. *Soil Science,* **142,** 279–288.

McLaughlin, S. B., Blasing, T. J., Mann, L. K., and Duvick, D. N. (1983). Effects of acid rain and gaseous pollutants on forest productivity: A region scale approach. *Journal of the Air Pollution Control Association,* **33,** 1042–49.

Mooney, H. A., Drake, B. G., Luxmoore, R. J., Oechel, W. C., and Pitelka, L. F. (1991). Predicting ecosystem responses to elevated CO_2 concentrations. *BioScience,* **41,** 96–104.

Moran, J. M. (1972). *An analysis of periglacial climatic indicators of late glacial time in North America*. Ph.D. thesis. University Wisconsin, Madison.

Nilsson, S. and Duinker, P. N. (1987). Extent of forest decline in Europe: A synthesis of survey results. *Environment,* **29,** 4–9, 30–31.

Norby, R. J., Luxmoore, R. J., O'Neill, E. G. and Weller, D. G. (1984). Plant responses to elevated atmospheric CO_2 with emphasis on belowground processes. ORNL/TM-9426, Oak Ridge National Laboratory, Oak Ridge, Tennessee.

Norby, R. J., O'Neill, E. G. and Luxmoore, R. J. (1986a). Effects of atmospheric CO_2 enrichment on the growth and mineral nutrition of *Quercus alba* seedlings in nutrient-poor soil. *Plant Physiology,* **82,** 83–9.

Norby, R. J., Pastor, J. and Melillo, J. M. (1986b). Carbon-nitrogen interactions in CO_2-enriched white oak: Physiological and long-term perspectives. *Tree Physiology,* **2,** 233–41.

Oechel, W. C., and Strain, B. R. (1985). Native species responses to increased atmospheric carbon dioxide concentration. In *Direct Effects of Increasing Carbon Dioxide on Vegetation,* ed. B. R. Strain and J. D. Cure, pp. 118–54. DOE/ER-0238. U.S. Dept. Energy, Washington D.C.

Olson, J. S., Watts, J. A. and Allison, L. J. (1983). *Carbon in Live Vegetation of Major World Ecosystems*. ORNL-5862 Oak Ridge National Laboratory, Oak Ridge, Tennessee.

Overpeck, J. T. and Bartlein, P. J. (1989). Assessing the response of vegetation to future climate change. Ecological response surfaces and paleoecological model validation. In *The Potential Effects of Global Climate Change on the United States: Appendix D—Forests,* ed. J. B. Smith and D. A. Tirpak, pp. 1-1 to 1-32. U.S. Env. Protection Agency, Washington D.C.

Peters, R. L. (1988). Overview of conservation implications of the greenhouse effect. Conference on Consequences of the Greenhouse Effect for Biological Diversity. World Wildlife Fund, Washington D.C. (Abstract).

Peters, R. L. and Darling, J. D. S. (1985). The greenhouse effect and nature reserves. *BioScience,* **35,** 707–17.

Plochmann, R. (1984). Air pollution and the dying forests of Europe. *American Forester,* **90,** 17–21, 56.

Post, W. M., Peng, T.-H., Emanuel, W. R., King, A. W., Dale, V. H., and DeAngelis, D. L. (1990). The global carbon cycle. *American Scientist,* **78,** 310–26.

Prentice, K. C. and Fung, I. Y. (1990). The sensitivity of terrestrial carbon storage to climate change. *Nature,* **346,** 48–51.

Schlesinger, M. E. and Mitchell, J. F. B. (1985). Model projections of the equilibrium climate response to increased carbon dioxide. In *Projecting the Climatic Effects of Increasing Carbon Dioxide,* ed. M. C. MacCracken and F. M. Luther, pp. 81–147. DOE/ER-0237, U.S. Dept. of Energy, Washington D.C.

Schlesinger, W. H. (1990). Evidence from chronosequence studies for a low carbon-storage potential of soils. *Nature,* **348,** 232–34.

Schneider, S. H. (1989). The changing climate. *Scientific American,* **261,** 38–47.

Schweingruber, F. H. (1992). Spatial hemispheric reconstructions of summer temperatures. *Proc. Symp. Boreal Forests: State, Dynamics and Anthropogenic Influence,* U.S.S.R. State Comm. on Forests, Moscow, (in press).

Shugart, H. H., Antonovsky, M. Y., Jarvis, P. G. and Sandford, A. P. (1986). CO_2, climatic change and forest ecosystems. In *The Greenhouse Effect, Climatic Change, and Ecosystems,* ed. B. Bolin, B. R. Doos, J. Jaeger, and R. A. Warrick, pp. 475–521. New York: John Wiley.

Solomon, A. M. (1986). Transient response of forests to CO_2-induced climate change: Simulation modeling experiments in eastern North America. *Oecologia,* **68,** 567–79.

Solomon, A. M. (1988). Ecosystem theory required to identify future forest responses to changing CO_2 and climate. In *Ecodynamics: Contributions to Theoretical Ecology,* ed. W. Wolff, C.-J. Soeder, and F. R. Drepper, pp. 258–274, Berlin: Springer-Verlag.

Solomon, A. M. and Leemans, R. (1990). Climatic change and landscape ecological response: Issues and analysis. In *Landscape Ecological Impact of Climatic Change,* ed. M. M. Boer and R. S. de Groot, pp. 293–311. Amsterdam: JOS Press.

Solomon, A. M. and Tharp, M. L. (1985). Simulation experiments with late Quaternary carbon storage in mid-latitude forest communities. In *The Carbon Cycle and Atmospheric CO_2: Natural Variations Archean to Present*. ed. E. T. Sundquist and W. S. Broecker, pp. 235–50. Geophys. Monogr. 32, Amer. Geophys. Union, Washington D.C.

Solomon, A. M., Tharp, M. L., West, D. C., Taylor, G. E., Webb, J. W. and Trimble, J. L. (1984). *Response of Unmanaged Forests to CO_2-Induced Climate Change: Available Information, Initial Tests, and Data Requirements*. TR009, DOE/NBB-0053, U.S. Dept. of Energy, Washington D.C.

Solomon, A. M. and West, D. C. (1985). Potential responses of forests to CO_2-induced climate change. In *Characterization of Information Requirements for Studies of CO_2 Effects: Water Resources, Agriculture, Fisheries, Forests and Human Health,* ed. M. R. White, pp. 145–69. DOE/ER-0236, U.S. Dept. Energy, Washington D.C.

Strain, B. R. (1987). Direct effects of increasing atmospheric CO_2 on plants and ecosystems. *Tree,* **2,** 18–21.

Strain, B. R., and Cure, J. D. ed. (1985). *Direct Effects of Increasing Carbon Dioxide on Vegetation*. DOE/ER-0238. U.S. Dept. Energy, Washington, D.C.

Sundquist, E. T. (1986). *Geologic analogs: Their value and limitations in carbon dioxide research*. In *The Changing Carbon Cycle: A Global Analysis,* ed. J. R. Trabalka and D. E. Reichle, pp. 371–402. New York: Springer-Verlag.

Tans, P. P., Fung, I. Y. and Takahashi, T. (1990). Observational constraints on the global atmospheric CO_2 budget. *Science,* **247,** 1431–38.

Thompson, S. L. and Schneider, S. H. (1982). Carbon dioxide and climate: The importance of realistic geography in estimating the transient temperature response. *Science,* **217,** 1031–33.

Trabalka, J. R., Edmonds, J. A., Reilly, J. M., Gardner, R. H. and Vorhees, L. D. (1985). Human alterations of the global carbon cycle and the projected future. In *Atmospheric Carbon Dioxide and the Global Carbon Cycle,* ed. J. R. Trabalka, pp. 247–87. DOE/ER-0239, U.S. Dept. of Energy, Washington D.C.

U.S.D.A. Forest Service. (1982) *An Analysis of the Timber Situation in the U.S. 1952–2030. For. Res. Rept. No. 23,* U.S. Dept. Agr., Washington D.C.

Waring, R. H. (1987). Characteristics of trees predisposed to die. *BioScience,* **37,** 569–74.

Washington, W. (1990). Where's the Heat? The ocean may be the missing sink. *Nat. Hist.* **90** (3), 66–72.

Webb, T. III. (1986). Is vegetation in equilibrium with climate? How to interpret late-Quaternary pollen data. *Vegetatio,* **67,** 75–91.

Whittaker, R. H. and Likens, G. E. (1973). Carbon in the biota. In Woodwell, G. M. and E. V. Pecan, eds., *Carbon in the Biosphere,* CONF-720510, U.S. Atomic Energy Comm., Washington D.C.

Wigley, T. M. L., and Jones, P. D. (1981). Detecting CO_2-induced climate change. *Nature,* **292,** 205–208.

Woodwell, G. M., Whittaker, R. H., Reiners, W. A., Likens, G. E., Delwiche, C. C. and Botkin, D. B. (1978). The biota and the World carbon budget. *Science,* **199,** 141–46.

Woodwell, G. M., Hobbie, J. E., Houghton, R. A., Melillo, J. M., Moore, B., Peterson, B. J. and Shaver, G. R. (1983). Global deforestation: Contribution to atmospheric carbon dioxide. *Science,* **222,** 1081–86.

3

CO_2 Fertilization: The Great Uncertainty in Future Vegetation Development

Christian Körner

Introduction

Even the most conservative estimates of the future development of the global CO_2 level forsee a doubling of this essential plant "food" in the next century. The investigation of plant responses to these enhanced supplies of CO_2 is perhaps one of the greatest challenges in ecology. In the words of B.R. Strain, "few endeavors could be more important than the study of the ecological and evolutionary effects of another century of complex, but unidirectional, environmental changes affecting the entire globe" (Strain 1987). The steady increase of atmospheric CO_2 is perhaps the most significant of these global changes.

A large body of information exists on plant CO_2 responses and excellent reviews have been published. (See Strain and Cure 1985.) I do not intend to rereview the literature on the subject. Rather I will summarize established responses and try to pinpoint some of the problems associated with predicting large-scale vegetation responses.

By the end of 1987 approximately 1000 articles and book chapters had been published on plant-CO_2 interactions, out of approximately 10 000 papers on CO_2 phenomena. The information contained in these publications has been qualified and ranked with respect to the level of integration from cellular or leaf level to ecosystem level, based on the bibliography by Strain and Cure (1986). A graphical presentation is shown in Figure 3.1.

The survey is based on a random sample of 339 out of 1032 references listed by Strain and Cure (1986). Since the census was completed the number of papers has accumulated to approximately 1150, 13% of which have been rated as irrelevant for the current purpose (e.g., pure meteorological studies, experiments with CO_2 in anaerobic conditions, soil CO_2, etc.), leading us to the present status of about 1000 papers. About 72% provide original data (Tab. 3.1), although

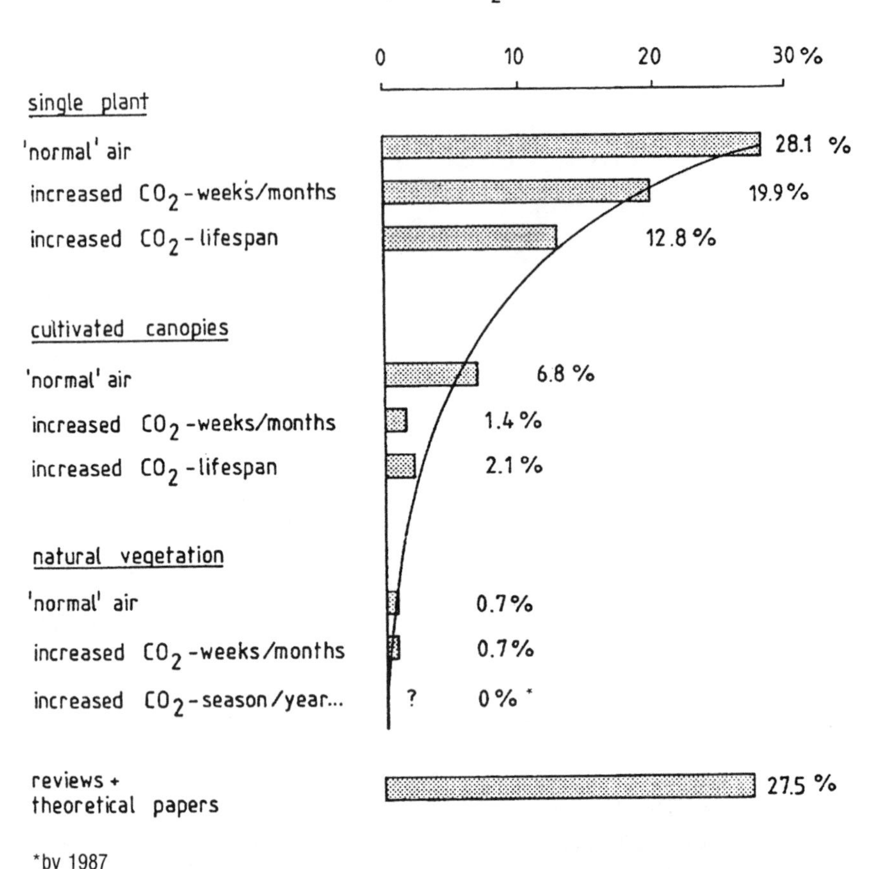

Fig. 3.1. Prior research on plant CO_2 responses ranked by the level of structural complexity and the duration of experimental exposure to increased CO_2. The relative frequency of references was estimated from abstracts of ca. 1000 papers listed by Strain and Cure (1986). Since this survey was completed, several papers on *in situ* responses have been published covering arctic tundra (Oechel and co-workers) and salt marshes (Drake and co-workers).

many repeat presentations of the same experiment. About half of these "original" papers deal with the physiology of short-term CO_2 responses of plants grown in "normal" air (e.g., CO_2 dependence of photosynthesis). The remaining publications (about 370) treat responses of plants grown in altered CO_2 regimes for a certain period or their full life span. More than half of these (ca. 220) deal with effects of periodic increases of CO_2 level (for several weeks or certain growth phases). About 128 papers provide responses to lifetime exposure of contrasting

Table 3.1. Statistical data on CO_2-plant interaction literature.[a]

Code	Category	%
R	**Reviews**, minireviews, crop-grower advice, scientific public-relations articles	20.0%
M	**Models**, theoretical papers	7.5%
	Individual Plant Studies (60.8%)	
P1	Initial physiological responses to suddenly altered CO_2 levels in plants grown in "normal" air (e.g., gas exchange studies, minutes to hours of exposure)	28.1%
P2	Responses in plants grown under manipulated CO_2 levels for several weeks or certain life phases (e.g., photosynthetic acclimation studies)	19.9%
P3	Accumulative responses of plants grown for their whole life under manipulated CO_2 conditions (largely harvest experiments)	12.8%
	Plants Cultivated in Stands (10.3%)	
C1	Short-term aspects of CO_2 exchange in canopies developed in normal air (agrometeorological considerations of CO_2 fluxes, etc.)	6.8%
C2	Canopies maintained for several weeks under manipulated CO_2 regimes	1.4%
C3	Canopies exposed to altered CO_2 for whole plant life (harvest experiments)	2.1%
	Natural Vegetation (1.4%)	
N1	Short-term responses to altered CO_2 in canopies grown in "normal" air	0.7%
N2	Responses during and after several weeks of growth in manipulated CO_2 regime	0.7%
N3	Full-season or extended CO_2 treatments	0.0%[b]

[a]All relevant papers (1000 at present), were categorized and the percentages that resulted are shown.

[b]The census was completed before the tundra and salt marsh papers cited in the text appeared.

CO_2 regimes (mostly biomass or fruit yields of annual glasshouse crops). About 43 papers address CO_2 responses of plants grown in canopies rather than in isolation. With few exceptions these are "artificial" stands, i.e., monocultures or two-species mixtures. No forest responses are reported. Only a handful of these papers deal with natural vegetation for a full growing season.

With respect to a global modeling objective, it is also interesting to evaluate the species that have been investigated. According to the keyword register of Strain and Cure's bibliography, about 100 wild vascular plant species are mentioned in the indexed publications. The majority of these species grew in "normal" air (P1 in Tab. 3.1) and were exposed to higher CO_2 levels only during gas-exchange measurements. Most publications refer to glasshouse-grown crop species (number of publications): tomato (100), soybean (67), corn (54), wheat (38), cucumber (23), sunflower (22), roses (20), lettuce (21), cotton (19), carnation (19), pea (16), sugar beet (15). These crop-plant species appear in 414 original papers. About 50 other cultivated nontree species have been studied in one to six papers each (on average two to three publications per species). In total, papers on cultivated species comprise more than two-thirds of all original papers. Among the remaining papers, investigations of agricultural weeds and forest-tree seedlings form major groups.

It is quite evident that modeling global vegetation change in response to CO_2

on the basis of this extremely unbalanced source of information will be a difficult task. Some pitfalls arising from adopting information from plant-level responses (P-level, Tab. 3.1) for modeling purposes at the ecosystem level (N-level, Tab. 3.1) will be addressed in later sections.

Established Trends of Plant Responses

Detailed reviews of plant responses to CO_2 have been written or edited by Williams (1978), Bolin *et al.* (1979), Lemon (1983), Kimball and Idso (1983), Wittwer (1984), Solomon and West (1985), Strain and Cure (1985), Eamus and Jarvis (1989), Bazzaz (1990), and others. The trends elaborated from published experiments can be summarized as follows:

1. The rate of photosynthesis in C3 plants increases immediately following exposure to CO_2 levels greater than present ambient.
2. This initial response is often reduced or may even disappear under long-term exposure to increased CO_2 levels, indicating a homeostatic trend of carbon fixation (Fig. 3.2).
3. Yet, biomass yield and fruit production under controlled conditions increased in almost all experiments.
4. Increased CO_2 levels reduce stomatal conductance and the rate of transpiration. Quantitative estimates are on the order of 34% of water saved for a doubling of CO_2.
5. Whether water consumption on a ground-area basis is affected depends upon the extent to which parallel changes in leaf area index occur.
6. The content of nonstructural carbohydrates generally increases under high CO_2, while the concentration of mineral nutrients is reduced.
7. Consequently, the food quality of leaf tissue declines, leading to increased per capita requirements of biomass by herbivores.
8. Dark respiration tends to increase (although there are also reports of decreasing dark respiration rates), while photorespiration decreases, but long-term effects are uncertain.
9. Sink size is the major determinant of maintenance of a higher photosynthetic rate under elevated CO_2.
10. Dry-matter allocation patterns change, and root/shoot ratios tend to increase, but exceptions do exist.
11. C3 plants, i.e., the majority of wild and cultivated species, show a more pronounced initial response than C4 plants like corn, sugarcane, and sorghum. This will alter competitive strength, a possibly significant aspect in weed control.

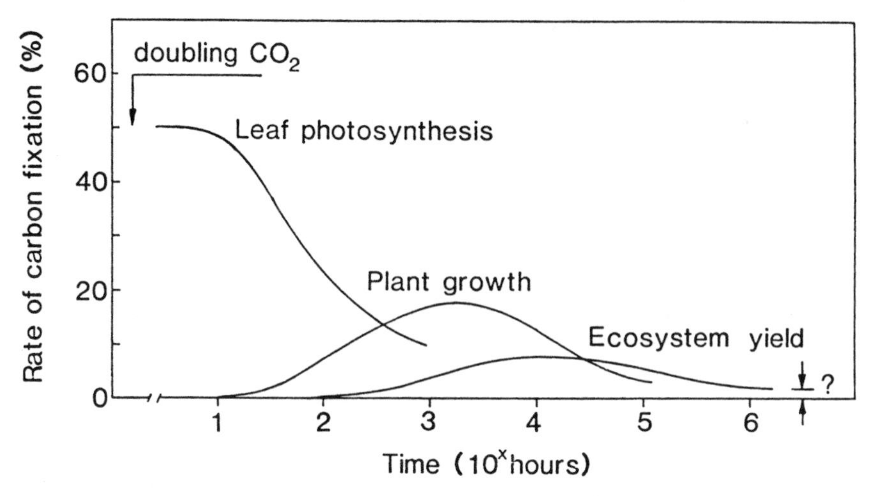

Fig. 3.2. Schematic response of carbon gain to increased CO_2 concentration at different levels of complexity of the carbon-fixing system. For simplicity, an initial doubling of the CO_2 concentration and a nonseasonal climate were assumed, and responses are plotted on an exponential time-scale.

12. Plants with nitrogen-fixing symbionts tend to benefit more from enhanced CO_2 supplies than others.
13. Positive responses are enhanced by increased temperature and light supply.
14. Increased leaf area production leading to a higher leaf area index (LAI) counteracts the stimulation of leaf photosynthesis. The greater the LAI, the smaller the gain per unit leaf area, due to mutual shading of leaves.
15. Increased leaf turnover rate reduces the potential returns of photosynthetic yield per leaf.
16. Increased CO_2 accelerates development and may thus reduce the efficiency of biomass or fruit production as a result of a reduced seasonal growth duration.

Evidently, strong initial photosynthetic responses to experimentally increased CO_2 are diminished by a multitude of negative feedback processes that come into play as the high CO_2 level prevails. This is best evidenced at the leaf level. Many experiments showed that exposure to elevated CO_2 reduces photosynthetic capacity, when measured at the original low CO_2 level, and that these reduced specific rates correlate with reductions in ribulose biphosphate carboxylase activity (e.g., Downton *et al.* 1980; von Caemmerer and Farquhar 1984; DeLucia, *et al.* 1985; Yokota and Canvin 1986).

A problem central to the evaluation of CO_2 effects, which has been widely ignored in the literature, is the distinction between *relative* and *absolute* effects. Much of the controversy about the significance of CO_2 effects under certain growth constraints (such as shade, nutrient shortage, or drought) would not exist were it not for this distinction. When "greater CO_2 responses" under stress conditions are reported, this is usually only true on a relative scale. A good example is photosynthesis in deep shade. Due to a reduction in the minimum quantum demand for net CO_2 uptake (light compensation point) under higher CO_2 levels, net photosynthesis measured at the light level of the original light compensation point, but at increased CO_2 concentration, is increased infinitely (any value divided by zero), but the absolute gain may be only 1% compared to that measured under full sunlight. Yet, such small relative changes may be significant for the survival of a particular plant. Water shortage may enhance the relative effects of rising CO_2 on growth when compared with nonstressed controls. (See reviews by Morison 1985; and Strain and Cure 1986). However, the absolute growth increment under moist conditions is likely to be many times larger. Also, nutrient impoverishment does not preclude a positive response to rising CO_2 (e.g., Strain 1985; Norby *et al.* 1986). Significant absolute increases of biomass production will, however, depend upon optimization of nutrient and light regimes, conditions more likely to occur in vigorously growing agricultural crops than in natural vegetation (e.g., Goudriaan and Ajtay 1979).

In addition, there is a scaling problem. For instance, a decrease in water consumption per unit leaf area under enhanced CO_2 may be offset by increased LAI on a land area basis (Acock and Allen 1985). Transitionally, the plant water balance and landscape water balance may be improved, but once this leads to increased land coverage by vegetation (as estimated, for example, by Idso and Quinn 1983 for Arizona) the original degree of drought stress and hydrological strain, though at higher biomass per area, will be reestablished. Another example of such scaling effects is provided by Nijs *et al.* (1989), who found that the canopy light compensation point of photosynthesis in a grass sward increased under high CO_2 (due to greater self-shading), the opposite effect to that found at the leaf level.

When all the hundreds of experiments with isolated plants grown under artificial conditions with increased CO_2 are considered, one wonders what happens to these plants in the field? What happens when competition for space, light, mineral nutrients, water, and other resources is accounted for and when species of contrasting life cycles, allocation patterns, and morphology form a mixed-plant canopy? What if dynamics of recycling processes, microbial and soil responses to increased litter production, and animal interactions are considered as well? A number of researchers successfully analyzed the interference of two or three variables (e.g., CO_2, water and nutrients; or CO_2, water, and competition between one species and another species; or competition among several different species under contrasting CO_2 levels) under controlled conditions (e.g., Carter and Peterson 1983; Bazzaz and Carlson 1984; Wray and Strain 1986; Reekie and Bazzaz 1989). Although these experiments represent a first step, they nevertheless resemble great oversimplifications compared to the real world. Thus, if we choose not to wait and see how the real world will behave, the use of models is imperative (Solomon 1988).

However, to make reliable predictions, modelers need data from experiments conducted at higher levels of organization. There is no way today, nor will there be in the near future, to link elementary physiological responses of plants and other organisms to the long-term behavior of an ecosystem. Despite extremely complex and logical abstractions, the classical SUBGROW model, developed in California in the 1970s (Fick *et al.* 1975), could not even quantify the sugar yield of a uniform field of sugar beet based on physiological relationships alone. The reason for this limitation of such functional models was not computational but rather the lack of empirical data and unknown cybernetic interrelations within and between plants—a situation that has not changed much since then. If global vegetation changes are to be predicted, the long-term responses of plant populations or communities of species need to be known. Figure 3.1 shows that the most promising level of response analysis for a prognostic model of natural vegetation, the community level, has been largely ignored.

Responses Measured in Nature

Initial results of studies on CO_2 responses in natural plant communities confirm that leaf responses do not necessarily translate into community and ecosystem responses. The first community-based, long-term field investigations of the effect of doubling CO_2 are those carried out in arctic tundra communities by Oechel and his group (e.g., Tissue and Oechel 1987). After 2 years of CO_2 fumigation, the vegetation showed no measurable yield response in contrast to all the high-yield greenhouse experiments with crops that showed dramatic changes. These field measurements in Alaska confirmed results of phytotron experiments with excavated “microcosms” of tundra vegetation by Billings *et al.* (1984), which showed that increased CO_2 has little or no effect on net ecosystem carbon balance over a simulated 3-month arctic summer period. They concluded that CO_2 is not a limiting factor in ecosystem production in the tundra. In contrast, salt marsh communities of much simpler structure, with greater nutrient availability and higher temperatures, did show increased biomass production under enhanced CO_2 supply in the field (Curtis *et al.* 1989).

Except for a short-term gas-exchange analysis (Wong and Dunin 1987), larger trees or parts of forests have not been exposed to artificially elevated CO_2. Evidence from tree-ring analysis is still conflicting. For low elevations no significant change has been observed so far (see Solomon 1988), further supporting the view that natural vegetation may be at best weakly responsive to higher atmospheric CO_2 concentrations. In contrast, at high elevation, consistent increases in ring width were detected in *Pinus aristata* (La Marche *et al.* 1984). However, the effect on ring width of parallel shifts in weather at the growth site (warming, increased precipitation) could not be separated from hypothesized direct effects of CO_2. The problem most confounding tree-ring analysis, but hardly ever addressed, is the increased atmospheric nitrogen input into ecosystems in the Northern Hemisphere in recent decades. Only the monitoring of a global network of reference sites, each with contrasting inputs of nitrogen compounds, will permit a separation of these two atmospheric signals.

Stomatal density has been shown to decrease when plants grow under high CO_2 in controlled environments. This is an anatomical expression of homeostasis of photosynthetic gas exchange. Consequently one would expect that leaves of plants growing today would develop fewer stomata per unit leaf area than those which grew before the rise in CO_2. In accord with this hypothesis, Woodward (1987) detected a decrease in stomatal density of herbarium samples of eight tree species collected during the past 200 years in southern England. Apparently, herbaceous plants do not yet exhibit such a differentiation. Data from over a hundred species from middle European low- and high-altitude sites indicate no change (Körner 1988). The major determinant of stomatal density in plants is the local light climate, difficult to reconstruct in historical sampling sequences. The confounding effects of plant density (e.g., age of trees) and air pollution (or dust, or dust-induced haze) on stomatal density will be difficult to separate from pure CO_2 effects.

Responses Measured in Managed Communities

Agricultural and horticultural ecosystems are much simpler, being mostly monospecific, synchronized, fertilized, and embedded in a homogeneous abiotic and biotic environmental matrix that is managed and yield-optimized. Downton *et al.*(1987) provide a particularly good example of the complex interrelations between yield and CO_2 supply. This study is one of the few investigations of mature woody plants (in this case young orange trees). The study provides experimental evidence for most of the responses listed above and illustrates increases in fresh fruit yield of 35% in trees grown under 800 μl/l CO_2, despite homeostatic responses in photosynthesis. Similar estimates exist for many other crop plants of the C3 type. Kimball and Idso (1983) compiled published responses for 430 crop plants that were exposed to various elevated levels of CO_2 (mostly more than 600 μl/l) and found that the most frequently observed increase in marketable yield was 20%.

If only those results that were obtained under a clearly defined CO_2 concentration are considered, each μl/l increase in CO_2 resulted in an average increase in yield of 0.1% compared to a control at 330 μl/l, i.e., an increase by 33% for a change from 330 to 660 μl/l. Since the current increase in CO_2 averages around 1.65 μl/l per annum, the corresponding yield increase should be around 0.165%

per year, provided other resources do not become limiting. Acock and Allen (1985) and Cure (1985) provide excellent reviews of the current knowledge of crop responses.

Wild Plants: The Overwhelming Influence of Morphotype

Since a large scientific investment is made worldwide in plant CO_2-response studies, there is a great need to coordinate these efforts and focus at the right level of organismic function. One of the most important components in plant performance is the heritable morphology and physiognomic plasticity of species. The literature on plant CO_2 responses largely ignores this facet. There is a long way between a photosynthetic light- or CO_2-response curve of a leaf and the response of a canopy; leaf area index is merely one of the relevant components (e.g., Eamus and Jarvis 1989). For the individual competitor, there are many other features that ultimately determine success. Structural features such as leaf angle, leaf size, bud position, internode length, branching patterns, bud dominance hierarchy, adult plant size, and so on can be more significant than the photosynthetic behavior of a sunlit leaf. The dominant taxa of final successional stages in forests tend to have a very reduced photosynthetic capacity compared with the pioneers (e.g., Bazzaz 1979). This was recently confirmed in an elegant investigation in hedgerow communities of central Europe by Küppers (1985). Qualities completely different from those investigated by gas-exchange analysis may determine whether a species and its physiological constitution will become dominant in a natural plant community. Increased CO_2 levels may or may not influence a species' overall performance, depending upon such structural features. Models based on photosynthetic base lines and which do not account for these dynamic structural aspects are likely to fail.

In order to visualize the problem, consider a very simple natural plant community: a meadow consisting of only one type of life form and a small number of species. Each of these species shows a different short-term response of leaf photosynthesis to CO_2 partial pressure (Fig. 3.3). If we assume that these physiological characteristics are preserved while ambient CO_2 levels increase by only 100 μl/l, rate increases can be approximated from these plots for each species. The increase varies from +50% in *Erigeron acre* at the left of the diagram to almost zero in *Primula elatior* at the right, with largely different potentials for CO_2 increases toward saturation concentrations in all species.

By no means do these data reflect proportional yield increases. Two examples illustrate that yield control is largely at the morphogenetic and phenologic level in such a community. *Primula elatior* may translate its marginal photosynthetic profit of increased CO_2 into real growth increment, since it precedes most of the other species in spring, utilizing high light and nutrient levels. A species like *Polygonum bistorta*, with a relatively low responsiveness but large flat leaves,

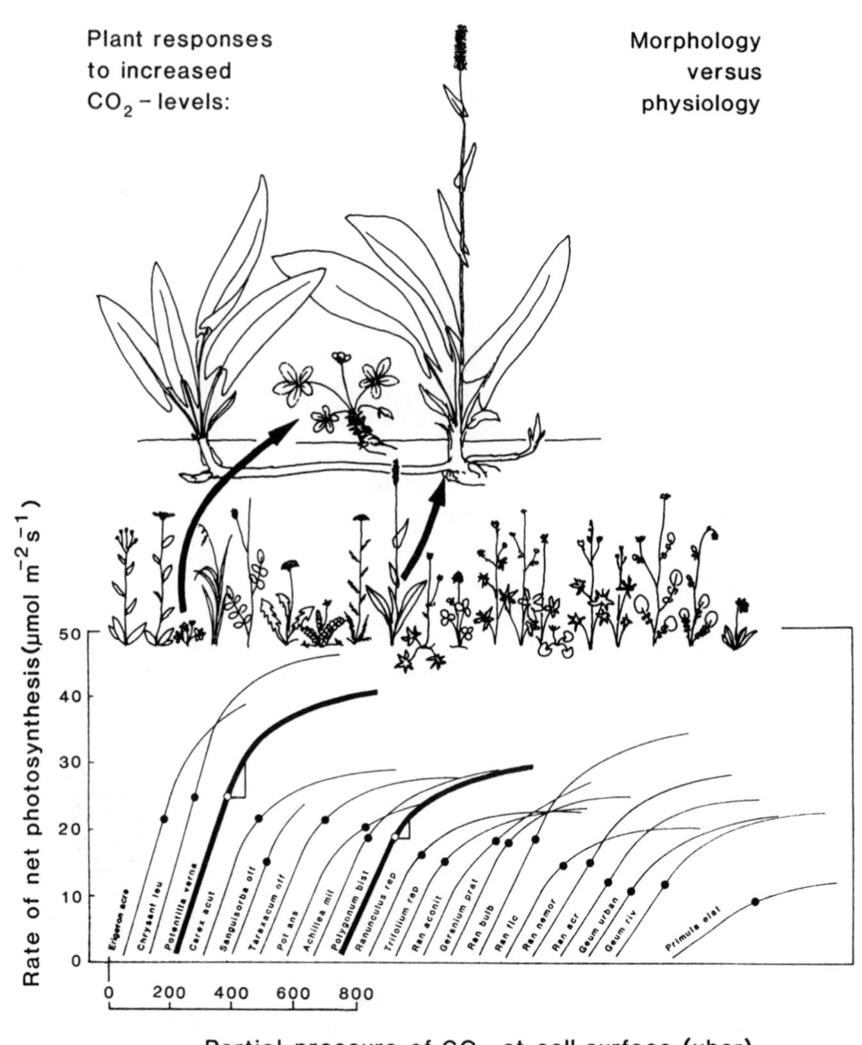

Fig. 3.3. CO_2 responses of photosynthesis in meadow plants. Triangles indicate the potential increase in photosynthesis for a 100-ppm increase in atmospheric CO_2 concentration (ignoring acclimative adjustment). The uppermost illustration represents an example in which the greater photosynthetic responsiveness of one species (e.g., *Potentilla* sp.) will not translate into greater biomass yields because a less-responsive species (e.g., *Polygonum* sp.) has a more competitive shoot morphology (from Körner 1991a, reproduced with permission of Blackwell Scientific Publications, Edinburgh.)

will outcompete more efficient, but narrow-leaved plants like *Erigeron* and *Potentilla* that suddenly receive more shade. Thus, stature and leaf size reduce or even reverse effects of isolated photosynthetic behavior.

This example should emphasize the need for experiments at higher levels of integration. Inevitably these will be statistical approaches including their inherent heuristic weaknesses. Physiological baselines at the leaf level indicate potential short-term responses that generally will not translate into ecosystem responses in a predictable manner and extent. Hence, Strain and Bazzaz (1983) called for an ecological approach to the analysis of CO_2 responses of vegetation, which is repeated here.

Response Hierarchies—A First Approximation

On the basis of present knowledge, hierarchy scenarios can be developed that may assist both in selecting research areas and in developing modeling strategies. Coarse CO_2-response scales (biomass response larger than/smaller than) may permit semiquantitative approximations. I would suggest that CO_2 response units (CRUs) range from zero response (responsiveness factor 1.0) to maximum possible response (open-end scale, but with responsiveness factors realistically not larger than perhaps 2). For practical reasons, the maximum biomass response to a doubled CO_2 concentration could be assumed to be +50%, i.e., a responsiveness factor of 1.5 CRU. Agricultural crops of C3 plants under optimal growth conditions average around 1.33 (Kimball and Idso 1983), while C4 plants will range close to 1.06 CRU (Cure 1985). Some desert communities may attain a relative profit close to the approximated increase in water-use efficiency estimated for crops for a doubling in CO_2 concentration (+34%; Cure 1985), yielding a potential factor near 1.3 CRU, whereas arctic tundra communities rate close to 1.0 CRU. The following attempts at ranking responses from least (left) to greatest (right) relative profit from enhanced CO_2 still contain a great number of speculative elements:

Individual Plant Properties

-Evergreen $<$ deciduous woody $<$ perennial $<$ annual herbaceous

-Old tree $<$ young tree $<$ seedling

-Late successional $<$ early successional species

-Small fruit sink $<$ large fruit sink species

-No special storage organs $<$ special below-ground storage organs

-Horizontal leaf position $<$ vertical leaf position

-Broad-leafed $<$ narrow-leafed

-C4 plant $<$ C3 plant

-Non-nitrogen fixing < nitrogen fixing
-Nonmycorrhizal < mycorrhizal

Vegetation/Ecosystem Properties

-Forest < grassland
-Woody crop < herbaceous C3 crop
-Not seriously water limited (A) < water limited (B)
  -If A: tundra < boreal < other
  -If B: other < Mediterranean < semiarid < arid
-Low altitude < high altitude
-Submersed aquatic < emergent aquatic < terrestrial
-Nonsaline < saline
-Nutrient deficient < non-nutrient deficient
-Cold < moderate < warm

Needless to say, few of these sequences are based on a sound experimental background and numerous redundancies are included in this list. Many more such sequences could perhaps be developed; an extension of this hierarchy approach is published elsewhere (Körner 1991). To illustrate the possible advantage of such mental experiments, let us ask which plant would most likely exhibit the greatest response to rising CO_2. Following the above sequences it should be a terrestrial, herbaceous crop plant with great sink size, which uses the C3 pathway, is a legume or is otherwise fertilized, has a low leaf area index, and is grown in warm temperatures at high altitude. This could be an Andean Mountain potato crop, cultivated in wide rows and under good soil conditions. The reader is invited to test other combinations.

Future Research

The Enclosure Approach

Systematic comparative investigations designed in a manner similar to those of the International Biological Program are required to fill the above CO_2 response hierarchy with real data for short- and medium-term responses. Strain and Bazzaz (1983) have provided a simple environmental gradient matrix that could serve as a guideline. The program should be strictly community- or ecosystem-oriented and standardized with respect to plant treatment, timing, and type of responses analyzed.

A priority catalog of measurements needs to be developed. Priority should be given to growth analysis, fractionation of biomass production within and between species, canopy leaf area changes, water consumption, effects on soil-, plant-, and litter-mineral pool sizes and qualities, and the associated microbial and herbivore responses to changed "food quality." With respect to the microclimatic changes by enclosures, minimum disturbance approaches need to be adopted that use open-top or free-air CO_2 enrichment methodology (Drake *et al.* 1985)

The Historical Approach

Another possibility for establishing plant and vegetation responses to global increases of CO_2 is the comparative analysis of plants that grew in preindustrial and modern climates. Monitoring structural or chemical properties of plant tissues from periods of contrasting CO_2 levels may reveal trends of change which may be of possible predictive value for the future. Of these, three have been adopted in the past: the analysis of tree rings, the analysis of stomatal density, and the analysis of carbon isotope composition.

Carbon isotope composition in earlier preserved carbohydrates and in those presently formed could reveal changes in water-use efficiency and photosynthetic behavior. Farquhar (1980) has suggested analyzing the isotope composition in tree rings as an indicator of CO_2 effects. In short-term exposure of herbaceous plants to high CO_2, no significant changes have been found. Perhaps the reason is that the ratio between internal and external CO_2 partial pressure is independent of the external level in the ranges relevant here. However, this holds only if no acclimatization to the altered CO_2 level has taken place. So far neither tree-ring data (Freyer and Francey 1981) nor fossil carbon from periods of higher CO_2 concentration (Bach 1985) have revealed significant changes if the decline of atmospheric 13C due to biomass and fossil fuel burning is accounted for (ca. 2% reduction). More tree-ring studies and "calibrations" of the isotope response to altered CO_2 levels under controlled conditions are needed.

Another facet of the historical approach could be paleoecology, estimating growth conditions and plant behavior in geological periods of much higher CO_2 levels than at present. Early Tertiary conditions, 50 million years ago, may have included atmospheric CO_2 concentrations much higher than those of today. The application of stable carbon isotope techniques to fossil carbon, in conjunction with studies of stratigraphy and paleoclimatology, could be a promising approach.

The Ecotype Approach

All manipulation experiments have two great drawbacks: first, they are very expensive and need highly developed techniques, and second—and perhaps more importantly—they do not yield insights on long-term effects. In order to estimate long-term consequences of plant exposure to altered CO_2 we need experiments that last many years, if possible even centuries or millennia. Nature has undertaken such experiments in small areas on the globe. Investigations of plants that have a history of enhanced or depleted CO_2 supply over geological periods can tell us whether carbon relations of the plants, in relation to other metabolic

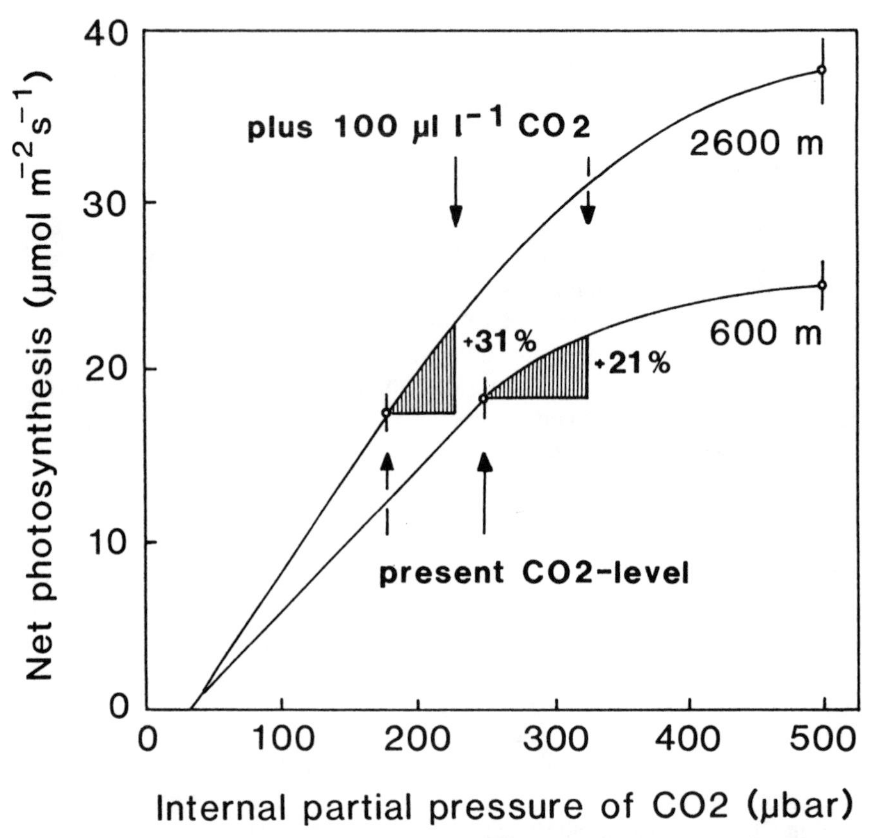

Fig. 3.4. The mean photosynthetic response to CO_2 in plant species from high altitude (low partial pressure of CO_2) and low altitude (high partial pressure of CO_2). Shaded triangles indicate the different potential responses to increased CO_2 levels (from Körner and Diemer 1987, reproduced with permission of Blackwell Scientific Publications, Edinburgh.)

processes, do exhibit homeostatic features—at least at the leaf level. These living documents of past and ongoing natural CO_2 experiments will be the most profitable tools in teaching us long-term trends of adaptation.

Meteorological surveys of consistently deviating CO_2 concentration patterns in terrestrial and aquatic systems are required. Altitudinal differences (pressure gradients) may be particularly useful because of their stability over very long periods (Körner 1992). Systematic physiological tests and growth experiments need to be undertaken in plant species typically found in such special habitats. For example, plants which evolved under low CO_2 regimes at high altitude in the Alps have a greater efficiency of CO_2 utilization. Thus, despite a reduction in partial pressure of CO_2 by more than 20%, they exhibit similar photosynthetic rates as plants from "high" CO_2 environments at low altitude (Fig. 3.4; Körner and Diemer 1987). Körner *et al.* (1988) showed that this behavior is found on a global scale. Ongoing experiments under controlled conditions have revealed that these response differences are lost in some species when grown under high CO_2, while other taxa preserve them (Körner and Diemer, in preparation).

Conclusions

Vegetation responses to elevated CO_2 will be smaller than can be anticipated from plant responses in the laboratory or in greenhouses. In some natural plant communities they may be negligible, since carbon supply does not appear to provide a major limit to growth at the ecosystem level. Agricultural crops are likely to respond with yield increases. Future research on the CO_2 response of plants needs to be intensified through methods that permit large-scale screening and that produce conclusions about the behavior of plant communities and ecosystems. Forests require particular attention, since they make up the largest portion of the global carbon budget. We need to work toward a reversal of the trend shown in Figure 3.1 if our primary objective is a realistic prognosis of future changes of vegetation in response to elevated CO_2. Funding priorities should be set by the location of a research proposal along the curve depicted in Figure 3.1 with greatest priority put on the weakest information (lower end of scale).

Acknowledgments

The work that led to this paper was funded by the Austrian Fonds zur Förderung der wissenschaftlichen Forschung, Vienna, project P-7262-B. I am indebted to Jay Arnone for many helpful comments.

References

Acock, B. and Allen, Jr., L.H. (1985). Crop responses to elevated carbon dioxide concentrations. In *Direct Effects of Increasing Carbon Dioxide on Vegetation*, ed. B.R. Strain and J.D. Cure, pp. 53–97. DOE/ER-0238, U.S. Department of Energy, Washington, D.C.

Bach, W. (1985). Der anthropogen gestörte Kohlenstoffkreislauf: Methoden zur Abschätzung der CO_2-Entwicklung in der Vergangenheit und in der Zukunft. *Düsseldorfer Geobotanisches Kolloquium* (ISSN 0176-0769), **2**, 3–23.

Bazzaz, F.A. (1979). The physiological ecology of plant succession. *Annual Review of Ecology and Systematics*, **10**, 351–71.

Bazzaz, F.A. (1990). The response of natural ecosystems to rising global CO_2 levels. *Annual Review of Ecology and Systematics*, **21**, 167–96.

Bazzaz, F.A. and Carlson, R.W. (1984). The response of plants to elevated CO_2. 1. Competition among an assemblage of annuals at different levels of soil moisture. *Oecologia* (Berlin), **62**, 196–8.

Billings, W.D., Peterson, K.M., Luken, J.O. and Mortensen, D.A. (1984). Interaction of increasing atmospheric carbon dioxide and soil nitrogen on the carbon balance of tundra microcosms. *Oecologia* (Berlin), **65**, 26–9.

Bolin, B., Degens, E.T., Kempe, E. and Kentner, P. ed. (1979). *The Global Carbon Cycle*. SCOPE Report 13. New York: John Wiley & Sons.

Carter, D.R. and Peterson, K.M. (1983). Effects of a CO_2-enriched atmosphere on the growth and competitive interaction of a C3 and a C4 grass. *Oecologia* (Berlin), **58**, 188–93.

Cure, J.D. (1985). Carbon dioxide doubling responses: a crop survey. In *Direct Effects of Increasing Carbon Dioxide on Vegetation*, ed. B.R. Strain and J.D. Cure, pp. 99–116. DOE/ER-0238, U.S. Department of Energy, Washington, D.C.

Curtis, P.S., Drake, B.G., Leadley, P.W., Arp, W.J. and Whingham, D.F. (1989). Growth and senescence in plant communities exposed to elevated CO_2 concentrations on an estuarine marsh. *Oecologia* (Berlin), **78**, 20–6.

DeLucia, E.H., Sasek, T.W. and Strain, B.R. (1985). Photosynthetic inhibition after long-term exposure to elevated levels of atmospheric carbon dioxide. *Photosynthesis Research*, **7**, 175–84.

Downton, W.J.S., Björkman, O. and Pike, C.S. (1980). Consequences of increased atmospheric concentrations of carbon dioxide for growth and photosynthesis of higher plants. In *Carbon Dioxide and Climate: Australian Research*, ed. G.I. Pearman, pp. 143–51. Australian Academy of Science, Canberra.

Downton, W.J.S., Grant, W.J.R. and Loveys, B.R. (1987). Carbon dioxide enrichment increases yield of Valencia orange. *Australian Journal of Plant Physiology*, **14**, 493–501.

Drake, B.G., Rogers, H.H. and Allen, Jr., L.H. (1985). Methods of exposing plants to elevated carbon dioxide. In *Direct Effects of Increasing Carbon Dioxide on Vegetation*, ed. B.R. Strain and J.D. Cure, pp. 11–31. DOE/ER-0238, U.S. Department of Energy, Washington, D.C.

Eamus, D. and Jarvis, P.G. (1989). The direct effects of increase in the global atmospheric CO_2 concentration on natural and commercial trees and forests. *Advances in Ecological Research*, **19**, 1–55.

Farquhar, G.D. (1980). Carbon isotope discrimination by plants: Effects of carbon dioxide concentration and temperature via the ratio of intercellular and atmospheric CO_2 concentrations. In *Carbon Dioxide and Climate: Australian Research*, ed. G.I. Pearman, pp. 105–10. Australian Academy of Science, Canberra.

Fick, G.W., Loomis, R.S. and Williams, W.A. (1975). Sugar beet. In *Crop Physiology—Some Case Histories*, ed. L.T. Evans, pp. 287–95. Cambridge: Cambridge University Press.

Freyer, H.D. and Francey, R.J. (1981). Recent C13/C12 trends in atmospheric CO_2 and tree rings. *Nature*, **293**, 679–80.

Goudriaan, J. and Ajtay, G.L. (1979). The possible effects of increased CO_2 on photosynthesis. In *The Global Carbon Cycle*, ed. B. Bolin, E. Degens, J. Kempe and P. Ketner, pp. 237–49. New York: John Wiley & Sons.

Idso, S.B. and Quinn, J.A. (1983). Vegetational redistribution in Arizona and New Mexico in response to a doubling of the atmospheric CO_2 concentration. *Climatological Publication* (Arizona State University), **17**, 2–52.

Kimball, B.A. and Idso, S.B. (1983). Increasing atmospheric CO_2: Effects on crop yield, water use and climate. *Agricultural Water Management*, **7**, 55–72.

Körner, Ch. (1988). Does global increase of carbon dioxide alter stomatal density? *Flora*, **181**, 253–57.

Körner, Ch. (1991). Some often overlooked plant characteristics as determinants of plant growth: A reconsideration. *Functional Ecology* 5:162–173.

Körner, Ch. (1992). Response of alpine vegetation to global climate change. *Catena*, (in press).

Körner, Ch. and Diemer, M. (1987). In situ photosynthetic responses to light, temperature and carbon dioxide in herbaceous plants from low and high altitude. *Functional Ecology*, **1**, 179–94.

Körner, Ch., Farquhar, G.D. and Roksandic, Z. (1988). A global survey of carbon isotope discrimination in plants from high altitude. *Oecologia* (Berlin), **74**, 623–32.

Küppers, M. (1985). Carbon relations and competition between woody species in a Central European hedgerow. III. Carbon and water balance on the leaf level. *Oecologia* (Berlin), **65**, 94–100.

La Marche, V.C., Graybill, D.A., Fritts, H.C. and Rose, M.R. (1984). Increasing atmospheric carbon dioxide: Tree ring evidence for growth enhancement in natural vegetation. *Science*, **225**, 1019–21.

Lemon, E.R. ed. (1983). *CO_2 and Plants: The Response of Plants to Rising Levels of Atmospheric Carbon Dioxide*. Boulder: Westview Press, Inc.

Morison, J.I.L. (1985). Sensitivity of stomata and water use efficiency to high CO_2. *Plant, Cell and Environment*, **8**, 467–74.

Nijs, I., Impens, I. and Behaeghe, T. (1989). Leaf and canopy responses of *Lolium perenne* to long-term-elevated atmospheric carbon dioxide concentration. *Planta*, **177**, 312–20.

Norby, R.J., O'Neill, E.G. and Luxmoore, R.J. (1986). Effects of atmospheric CO_2 enrichment on the growth and mineral nutrition of *Quercus alba* seedlings in nutrient-poor soil. *Plant Physiology*, **82**, 83–9.

Reekie, E.G. and Bazzaz, F.A. (1989). Competition and patterns of resource use among seedlings of five tropical trees grown at ambient and elevated CO_2. *Oecologia* (Berlin), **79**, 212–22.

Solomon, A.M. (1988) Ecosystem theory required to identify future forest responses to changing CO_2 and climate. In *Ecodynamics*, ed. W. Wolff, C.-J. Soeder and F.R. Drepper, pp. 258–74. Berlin: Springer-Verlag.

Solomon, A.M. and West, D.C. (1985). Potential responses of forests to CO_2-induced climate change. In *Characterization of Information Requirements for Studies of CO_2 Effects: Water Resources, Agriculture, Fisheries, Forests and Human Health*, ed. M.R. White, pp. 147–69. DOE/ER-0236, U.S. Department of Energy, Washington, D.C.

Strain, B.R. (1985). Background on the response of vegetation to atmospheric carbon dioxide enrichment. In *Direct Effects of Increasing Carbon Dioxide on Vegetation*, ed. B.R. Strain and J.D. Cure, pp. 1–10. DOE/ER-0238, U.S. Department of Energy, Washington, D.C.

Strain, B.R. (1987). Direct effects of increasing atmospheric CO_2 on plants and ecosystems. *Trends in Ecology and Evolution*, **2**(1), 18–21.

Strain, B.R. and Bazzaz, F.A. (1983). Terrestrial plant communities. In *CO_2 and Plants: The Response of Plants to Rising Levels of Atmospheric Carbon Dioxide*, ed. E.R. Lemon, pp. 177–222. Boulder: Westview Press, Inc.

Strain, B.R. and Cure, J.D. (1985). *Direct Effects of Increasing Carbon Dioxide on Vegetation*. DOE/ER-0238, U.S. Department of Energy, Washington, D.C.

Strain, B.R., and Cure, D.J. (1986). *Direct Effects of Atmospheric CO_2 Enrichment on Plants and Ecosystems: a Bibliography with Abstracts*. CDIC-13, Oak Ridge National Laboratory, Oak Ridge, Tennessee.

Tissue, D.T. and Oechel, W.C. (1987). Response of *Eriophorum vaginatum* to elevated CO_2 and temperature in the Alaskan tussock tundra. *Ecology*, **68**, 401–10.

von Caemmerer, S. and Farquhar, G.D. (1984). Effects of partial defoliation, changes of irradiance during growth, short-term water stress and growth at enhanced p(CO_2) on the photosynthetic capacity of leaves of *Phaseolus vulgaris* L. *Planta*, **160**, 320–9.

Williams, J. (1978). *Carbon Dioxide, Climate and Society*. Oxford: Pergamon Press.

Wittwer, S.H. (1984). Carbon dioxide levels in the biosphere: effects on plant productivity. *CRC-Critical Reviews in Plant Science*, **2**(3), 171–98.

Wong, S.C. and Dunin, F.X. (1987). Photosynthesis and transpiration of trees in a eucalypt forest stand: CO_2, light and humidity responses. *Australian Journal of Plant Physiology*, **14**, 619–32.

Woodward, F.I. (1987). Stomatal numbers are sensitive to increases in CO_2 from pre-industrial levels. *Nature*, **327**, 617–8.

Wray, S.M. and Strain, B.R. (1986). Response of two old field perennials to interactions of CO_2 enrichment and drought stress. *American Journal of Botany*, **73**, 1486–91.

Yokota, A. and Canvin, D.T. (1986). Changes of ribulose bisphosphate carboxylase/oxygenase content, ribulose bisphosphate concentration, and photosynthetic activity during adaptation of high-CO_2 grown cells to low-CO_2 conditions in *Chlorella pyrenoidosa*. *Plant Physiology*, **80**, 341–5.

4

Leaf Responses to the Environment and Extrapolation to Larger Scales

F.I. Woodward

Introduction

The existence of the major biomes of the world depends on the conversion of CO_2 and photons of solar radiation to carbohydrates by the process of photosynthesis. The efficiency of this process of solar energy conversion is low, reaching about 3% at a maximum and more typically falling to 1% or less (Woodward and Sheehy 1983; Lawlor 1987). The majority of photosynthesis takes place in the leaves, which are most productive when intercepting the direct solar beam. In such a position, leaves are also subjected to the full range and force of the aerial environment with rapid changes in wind speed, solar radiation, CO_2 concentration, and humidity. These variations will lead to changes in the net balance between the gains and losses of energy by the leaf, causing fluctuations in leaf temperature. Changes in leaf temperature will simultaneously influence the gradient for transpiration from the leaf and the rates of temperature-dependent processes such as photosynthesis, respiration, and carbohydrate translocation (Jones 1983). The first part of this chapter describes the typical short-term responses (seconds to minutes) of leaves to the environment.

The relevance of short-term physiological dynamics to global ecology lies in the need to scale these well-characterized responses to ecosystems where the responses are poorly characterized. This scaling can only be achieved through an understanding of basic processes. These processes are best understood at small scales of time and space. For example the net carbon balance (photosynthesis minus respiration) of leaves and the annual total of leaf transpiration are properties which scale up to ecosystem processes. The mean annual transpiration rate of a canopy of leaves is positively correlated with the density and stature of vegetation (Woodward 1987). The second part of this chapter presents a model for predicting maximum leaf area index (LAI) from climatic variables, using estimates of transpiration, as an example of the scaling-up process.

Table 4.1. Typical values of energy fluxes incident on both surfaces of a horizontal leaf (from Woodward and Sheehy 1983; and Monteith and Unsworth 1990).

	Midday	Midnight
S (W m^{-2})	1000	0
L_d (W m^{-2})	330	300
L_u (W m^{-2})	440	400

Leaf Responses to the Environment

Leaf Energy Balance

The starting point for considering rapid leaf dynamics is a consideration of the gains and losses of energy by leaves. Leaves are the primary source of carbohydrates for the majority of the food chains of the world. The manufacture of these carbohydrates depends on the interception and use of solar radiation over a restricted range of equable temperatures (about 0–40°C). The temperature of a leaf depends on the net gains of energy and on the temperature of the surrounding air.

Leaves gain energy by absorbing solar (0.3 to 3.0 μm waveband) and terrestrial (>3.0 μm: Woodward and Sheehy 1983; Monteith and Unsworth 1990) radiation. Gains by conduction are not significant. The net balance of radiation gain by a leaf, R_n, is defined as

$$R_n = (S - S \cdot r - S \cdot t) + (L_d + L_u) - 2\varepsilon\sigma T_k^4 \ , \qquad (4.1)$$

where S is solar radiation (W m^{-2}), r is reflectance to solar radiation (typically 0.2), and t is leaf transmittance (typically 0.15, so that leaf absorptivity to solar radiation is on average 0.65, in the 0.3–3-μm waveband). For a leaf which is fully exposed to the sun, L_d is the downward flux of terrestrial radiation from the atmosphere and L_u is the upward flux from other leaves or the soil. Typical values of energy fluxes incident on a leaf are given in Table 4.1. The last term in Equation 4.1 defines the loss of long-wave radiation by both surfaces of the leaf. This is defined by σ, the Stefan-Boltzmann constant (5.67×10^{-8} W m^{-2} K^{-4}), the leaf temperature T_k (in Kelvins) and ε, the emissivity of the leaf. The emissivity of the leaf is a measure of its capacity to emit radiation and is typically very high at about 0.95 (Jones 1983).

Dissipating Absorbed Energy—General

The net balance of radiation absorbed by the leaf (R_n) is exactly balanced by processes which dissipate the energy, as required by the laws of thermodynamics. When R_n and leaf temperature are unchanging the major processes of energy dissipation are described as

$$R_n = C + \lambda E \ , \tag{4.2}$$

where C is convective heat exchange ($J\ m^{-2}\ s^{-1}$ or $W\ m^{-2}$) and λE is evaporative heat exchange ($J\ m^{-2}\ s^{-1}$ or $W\ m^{-2}$). When the leaf is warmer than the surrounding air it loses heat to the air. This occurs when the warmer molecules of air in close proximity to the leaf move away and intermingle with the colder surrounding molecules. These warmer molecules near the surface of the leaf may also be rapidly carried away from the leaf surface by the stream of air moving across the leaf. When convection removes heat from the leaf it is positive in sign. At night convection may be negative in sign, indicating that the leaf gains heat from the surrounding air.

The fresh weight of a leaf may be 80–90% water (Jones 1983). Within the leaf, crucially important air spaces interconnect all the cells. The gases CO_2 and O_2, which are the substrates for photosynthesis and respiration, respectively, move 10 000 times more rapidly from cell to cell in the air spaces than could be achieved if the interconnections were through liquid water. Without the air spaces, rates of photosynthesis would be much less, even within a thin leaf. As the leaf contains so much liquid water there is continuous evaporation into the intercellular spaces. A thin leaf (say 1 mm or less) can be considered isothermal and the whole of the intercellular air space complex will be saturated with water vapor. The air outside the leaf is rarely saturated with water and so there will always be a tendency for water to evaporate from the leaf.

The leaf of a flowering plant controls this water loss by adjusting the opening of the stomatal pores in the epidermis. There is a trade-off involved in this response because closing the stomata limits the inward diffusion of CO_2 and therefore the rate of photosynthesis. Photosynthesis is at a maximum when the stomata are fully open, because the stomatal restriction to the inward diffusion of CO_2 is at a minimum. In contrast leaf turgor, and probably expansion growth, are at a maximum when the stomata are virtually closed, effectively preventing the loss of water vapor by transpiration. The control mechanism for stomatal opening often appears to adjust toward a compromise (optimization) between maximum CO_2 uptake and minimum water loss (Cowan and Farquhar 1977).

Transpiration of water from the leaf extracts heat (the latent heat of vaporisation, which at 20°C is 2454 $J\ g^{-1}$ of transpired water) and efficiently cools the leaf. In addition the process pulls water from the soil into the continuous xylem pathway from the root to the leaf. As for convection, when transpiration occurs away from the leaf, the sign is positive. On rather infrequent occasions transpiration may be negative in sign, i.e., a gain of latent heat. In such a case dew tends to form on the leaf surface, releasing the latent heat of transpiration.

Dissipating Absorbed Energy—Detail

The rate of convection may be defined (Jones 1983; Woodward and Sheehy 1983) as

$$C = g_a \cdot p_a \cdot c_p \cdot (T_k - T_a) \ . \tag{4.3}$$

The equation calculates the heat content of the air touching the leaf surface at leaf temperature T_k, and the heat content of the air above, at air temperature T_a. The direction of the heat loss is determined by the sign of the difference between these two quantities. The actual flux rate is finally controlled by the thickness of the boundary layer of the leaf.

In more detail, p_a is the density of dry air (1.2 kg m^{-3} at 20°C) and c_p is the specific heat capacity of air (1010 J kg^{-1} K^{-1}). The boundary layer of the leaf exists because the movement of the surrounding air is retarded by friction with the leaf surface. The thicker the boundary layer, the greater the resistance to convection. This thickness may be quantified as the boundary layer resistance (r_a, s m^{-1}), but the inverse of resistance, the boundary layer conductance (g_a, m s^{-1}), is also commonly used.

The advantage of using conductance is more obvious when it is applied to quantifying stomatal opening, when the two measures are positively correlated. Transpiration occurs by the loss of water vapor from the intercellular air spaces of the leaf to the substomatal cavity and out of the leaf through the stomata. The rate of loss increases with increased stomatal opening. The molecules of water vapor must also traverse the leaf boundary layer before reaching the surrounding, bulk air. The formal description of this process of transpiration (Jones 1983; Woodward and Sheehy 1983) is

$$E = (p_a \cdot c_p / 66) \cdot (e_{s,Tk} - e_a) \cdot g_l \ , \tag{4.4}$$

where g_l is the leaf conductance to water vapor (m s^{-1}), combining the stomatal conductance, g_s, and the boundary layer conductance, g_a, as

$$g_l = g_a \cdot g_s / (g_a + g_s) \ . \tag{4.5}$$

In Equation 4.4, $e_{s,Tk}$ is the saturation water vapor pressure (Pa) at leaf temperature T_k, and e_a is the water vapor pressure of the surrounding air, at air temperature T_a.

The influence of variations in the boundary layer and stomatal conductances on leaf temperature and the net radiation balance (R_n, Eqn. 4.1) are shown in Table 4.2 for midday and in Table 4.3 for midnight (all with the incident energy fluxes shown in Table 4.1, a leaf absorptivity of 0.65 for solar radiation, and an emissivity, ε, of 0.95).

A stomatal conductance, g_s, of 10 mm s^{-1} is typical of an amphistomatous leaf with well open stomata. The value of 2 mm s^{-1} is more typical of a hypostomatous or amphistomatous leaf with partially closed stomata (Jones 1983; Monteith and Unsworth 1990). A leaf (30-mm width) in a wind speed of 4 m s^{-1} would have, typically, a boundary-layer conductance of 100 mm s^{-1} (Grace 1977). The

Table 4.2. Predicted midday leaf temperature (°C) and net radiant balance[a] (W m^{-2}) at typical values of stomatal and boundary-layer conductances and the radiant fluxes shown in Table 4.1.

		Stomatal Conductance (g_s)	
		2 mm s^{-1}	10 mm s^{-1}
Boundary-Layer Conductance (g_a)	17 mm s^{-1}	35.8°C 440 W m^{-2}	30.5°C 506 W m^{-2}
	100 mm s^{-1}	24.5°C 576 W m^{-2}	23.4°C 589 W m^{-2}

[a]Air temperature 20°C and water vapor pressure 1800 Pa.

Table 4.3. Predicted midnight leaf temperature (°C) and net radiant balance[a] (W m^{-2}) at typical values of stomatal and boundary-layer conductances and the radiant fluxes shown in Table 4.1.

		Stomatal Conductance (g_s)
		<1 mm s^{-1}
Boundary-Layer Conductance (g_a)	17 mm s^{-1}	17.0°C -62 W m^{-2}
	100 mm s^{-1}	19.3°C -86 W m^{-2}

[a]Air temperature 20°C.

boundary-layer thickness increases as wind speed decreases, so that for the same leaf g_a decreases to 17 mm s^{-1} at a wind speed of 0.1 mm s^{-1}.

Both the boundary-layer and stomatal conductances influence leaf temperature (Tab. 4.2). However, when g_a is high (100 mm s^{-1}), leaf temperature is rather insensitive to large changes in g_s. When g_a is low, leaf temperature increases and is more sensitive to stomatal conductance. As a consequence of this higher leaf temperature, the long-wave emissivity of the leaf ($2\varepsilon\sigma T_K^4$ in Eqn. 4.1) increases and the net radiant balance decreases.

At night the stomata of nonsucculent plants are effectively closed, a feature which prevents latent heat loss. The net radiant gain of the leaf is then balanced solely by convective exchange (Tab. 4.3).

At night the net radiant balance of the leaf is negative in sign and the leaf tends to cool to below air temperature, a process known as radiative cooling. A consequence of this cooling is that the leaf gains heat by convection from the air. Leaf temperature is lowest at the lower wind speed and boundary-layer conductance. The boundary-layer conductance also decreases as leaf size increases (Grace 1977). In combination, and during cold winter periods, these features may cause radiative cooling of large evergreen leaves to lethal low temperatures (Woodward 1987).

Rapid Changes in Leaf Temperature

The energy balance of a leaf has so far been considered to be at equilibrium. In nature this is rarely the case because, for example, of rapid changes in irradiance and wind speed. The response of a leaf to these changes is slow because of its mass which, albeit often small, is nevertheless significant. During periods when the flux of energy to a leaf is changing, the net radiant balance is dissipated by an additional process, net heat storage, J, so that

$$R_n = C + \lambda E + J \tag{4.6}$$

and the rate of change in leaf temperature may be defined as

$$\frac{dT}{dt}K = J/(p_l \cdot c_{p,l} . l) \ , \tag{4.7}$$

where p_l is the leaf density (typically 700 kg m^{-3}), $c_{p,l}$ is the leaf specific heat capacity (typically 3800 J kg^{-1} K^{-1}; Jones 1983), and l is leaf thickness.

The responses of leaf temperature to a rapid (less than 1 s) increase in solar radiation from 670 to 1000 W m^{-2} are shown for leaves 0.2 mm thick (Fig. 4.1) and 2 mm thick (Fig. 4.2), with the same combinations of stomatal and boundary-layer conductances as shown in Table 4.2. The slow response of the 2-mm-thick leaf (Fig. 4.2) is obvious. The speed of response may be quantified as the response time, τ, which is the time taken for the leaf to change by 63% from an initial to a final value. The response time varies between 3 and 13 s for the thin leaf (Fig. 4.1) and between 34 and 131 s for the thick leaf (Fig. 4.2). The range of response times for a particular leaf is controlled by changes in the stomatal and boundary-layer conductances, a decrease in either conductance increasing the response time. The influence of the stomatal and boundary-layer conductances on the equilibrium leaf temperatures is shown by the different values of leaf temperature before the increase in irradiance.

Effects of Changes in Leaf Temperature on Transpiration and Leaf Water Status

An immediate consequence of the increase in leaf temperature with irradiance (Figs. 4.1 and 4.2) is that the rates of convection (Eqn. 4.3) and transpiration (Eqn. 4.4) may increase because the driving forces for both processes increase. This effect may be considered for the leaf which is 0.2 mm thick and with conductances of case A (Fig. 4.1). When the air temperature is 20°C and the air vapor pressure is 1800 Pa (relative humidity 77%), an increase in leaf temperature of only 1.2°C effectively increases the vapor pressure deficit (VPD) between the leaf and the air by about 50%. With no change in the conductances this would also cause a 50% increase in the rate of transpiration. A similar proportional increase in VPD and transpiration rate is observed for case B conductances (Fig. 4.1), although the absolute rate of transpiration, with smaller conductances, is less.

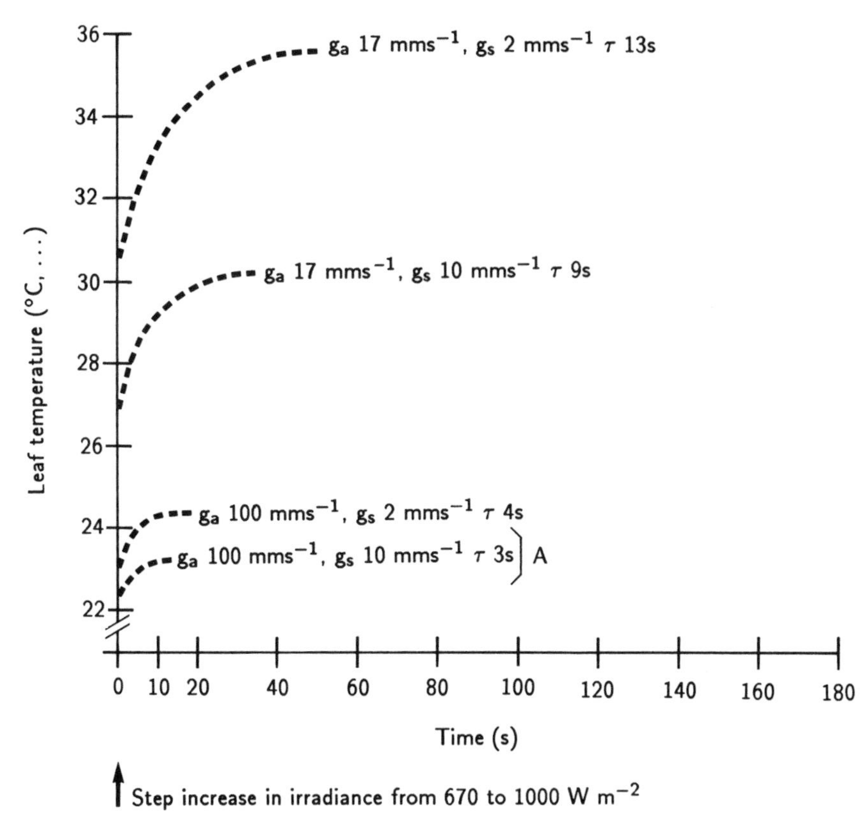

Fig. 4.1. Responses of leaf (thickness 0.2 mm) temperature to a step increase in irradiance.

These rapid increases in transpiration rate may not be immediately supported by the rate of water supply through the long xylem pathway from the water source in the soil. Such a supply problem is regularly seen when the water potential of the leaf declines during the period of maximum daily transpiration (Jones 1983). Species also differ in the rates at which the xylem can supply water. For example, species with xylem vessels constructed from tracheids (conifers) may be 40 or 50 times slower at conducting water than ring-porous hardwoods (e.g., ash, elm). Therefore there is a wide range of species-specific xylem conductivity and the potential for shortfalls of water supply to the leaf for replacing water lost by transpiration. When water loss from the leaf exceeds that supplied from the stem xylem, water is extracted from the leaf water store. For a leaf with a surface area of 1500 mm^2, this total water store is about 255 mm^3 for a leaf 0.2 mm thick and 2550 mm^3 for a leaf 2 mm thick.

In the example of case A conductances (Fig. 4.1) and a leaf 0.2 mm thick, if

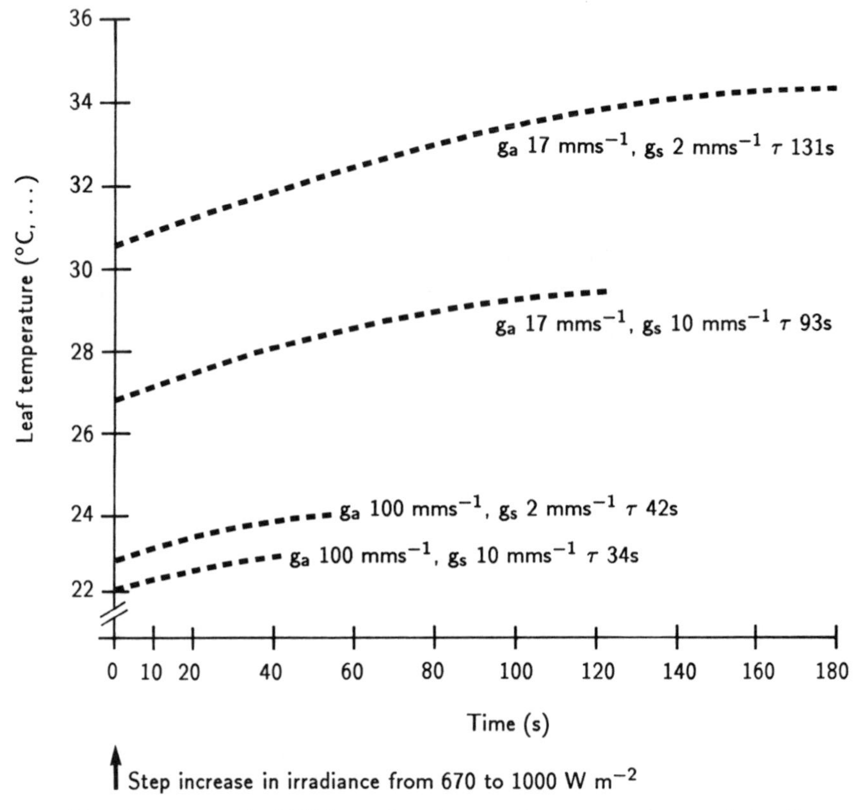

Fig. 4.2. Responses of leaf (thickness 2 mm) temperature to a step increase in irradiance.

it is assumed that the rate of transpiration at time zero (Fig. 4.1) is the maximum rate sustained by the xylem, then the increased rate of transpiration at the higher irradiance could be supported entirely from the leaf water store for a period of about 2 hours. This would increase to about 20 hours for the leaf which is 2 mm thick.

This simple case is more complex in reality. A leaf often loses turgor completely when its relative water content has fallen to only 65–75%. Therefore for the thin leaf, the 255 mm^3 of water available for transpiration is reduced to about 75 mm^3, which could only supply about half an hour of transpiration before the leaf was completely wilted. At this point the water potential of the leaf may have fallen to the range of -1 to -4 MPa. Further water loss may rapidly lead to death because the water potential decreases more rapidly as further water is lost from the wilted leaf (Jones 1983).

These excessive rates of transpiration are restricted by modulation of stomatal opening in response to VPD. As VPD increases the stomata close to some degree, and the overall stomatal conductance decreases (Schulze *et al.* 1972). For some species the control is so effective that the transpiration rate is constant across a wide range of VPDs (Schulze *et al.* 1972; Jones 1983). The VPD response of stomata therefore minimizes the use of the leaf water store during the day, tending to maximize leaf water potential and turgor.

Like leaf temperature, the stomata also have a finite response time on the order of 60–100 s, and during periods of rapid environmental change there could be significant use of the water store before the stomata exert maximum control. If these periods occur frequently during the day, e.g., when patches of cloud frequently obscure the sun, leaf turgor may fall to zero. The stomata of many species may close as a consequence of zero turgor, independently of any VPD response (Milburn 1979).

Effects of Changes in Leaf Temperature on Photosynthesis

The maximum rate of photosynthesis, A_{max}.

The rate of leaf photosynthesis will also track the changes in leaf temperature (Figs. 4.1 and 4.2) because it has a rapid response time (perhaps as little as 1 s for a well-illuminated leaf; Pearcy 1988). The temperature responses of photosynthesis depend on irradiance and the biochemical mechanism of photosynthesis. At irradiances of about 500 W m^{-2} and higher, the photosynthetic rate of species with C3 photosynthesis (species in which CO_2 is fixed by the enzyme ribulose bisphosphate carboxylase/oxygenase, rubisco, to form a three-carbon acid) is irradiance saturated. Photosynthesis is not usually light saturated until 1000 W m^{-2} in C4 species (where CO_2 is fixed in two stages, first by the enzyme phosphoenolpyruvate carboxylase, PEP carboxylase, which produces a four-carbon acid; this is then transported to the bundle sheath cells of the leaf, decarboxylated, and finally fixed by rubisco).

The maximum photosynthetic rate in C3 species, and the maximum observed in C4 species, show a marked sensitivity to temperature (Fig. 4.3a). All species possess an optimum temperature range (Baker *et al.* 1988) which is wider for the species from a cool climate (*Verbena*). In general the optimum temperature and often A_{max} tend to increase with the temperature of the local climate (Larcher 1980). The temperature responses of A_{max} will not be significantly influenced by man-induced increases in atmospheric CO_2 concentration. However, the absolute value will increase in C3 species (e.g., *Larrea* in Fig. 4.3a; Osmond *et al.* 1980), particularly within the short time scales being considered here. A_{max} in C4 species (e.g., *Tidestromia* in Fig. 4.3a) is insensitive to CO_2 concentration (Osmond *et al.* 1980).

The CO_2 concentration within plant canopies shows marked diurnal and short-term fluctuations both above and below the mean CO_2 partial pressure (Desjardins and Lemon 1974). It is therefore important to consider the responses of A_{max} to

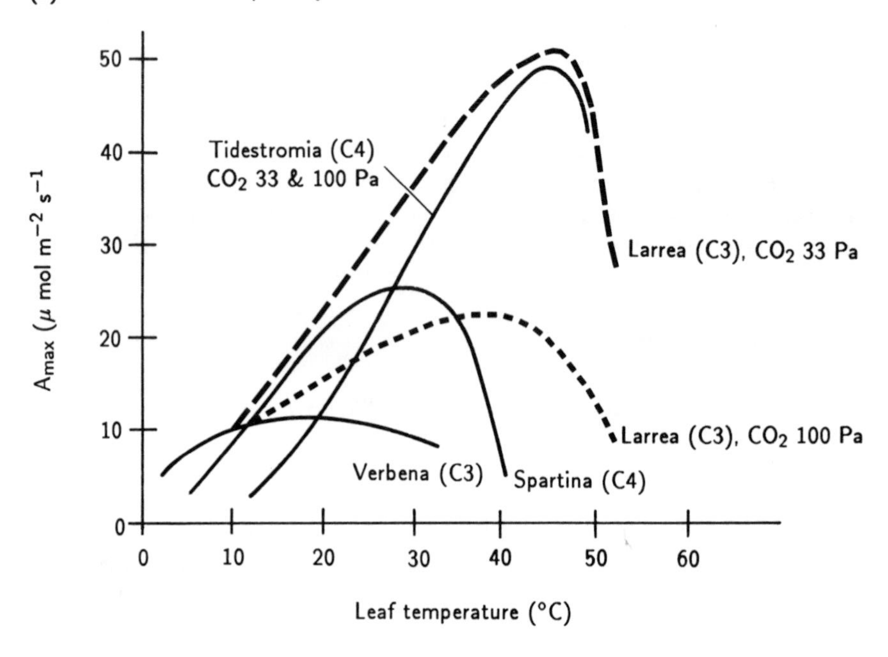

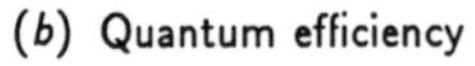

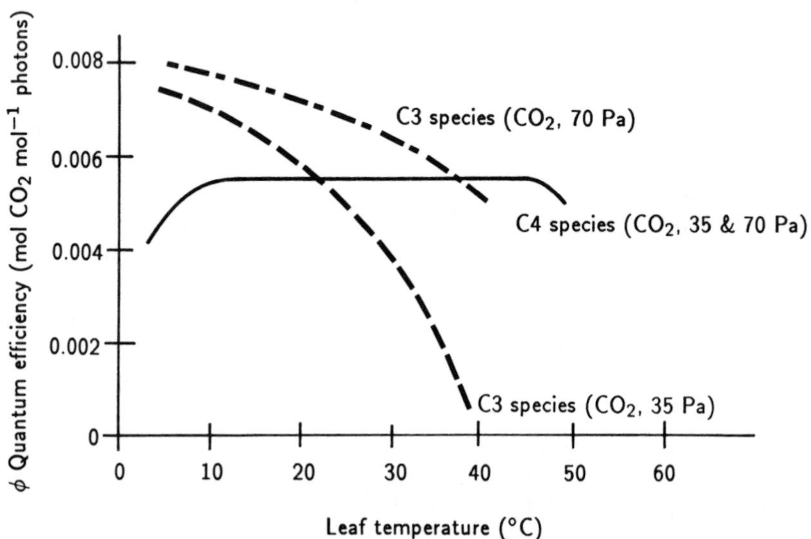

Fig. 4.3. Responses of photosynthesis to temperature. *a*. Maximum rate of photosynthesis, A_{max}. b. Quantum efficiency, ϕ.

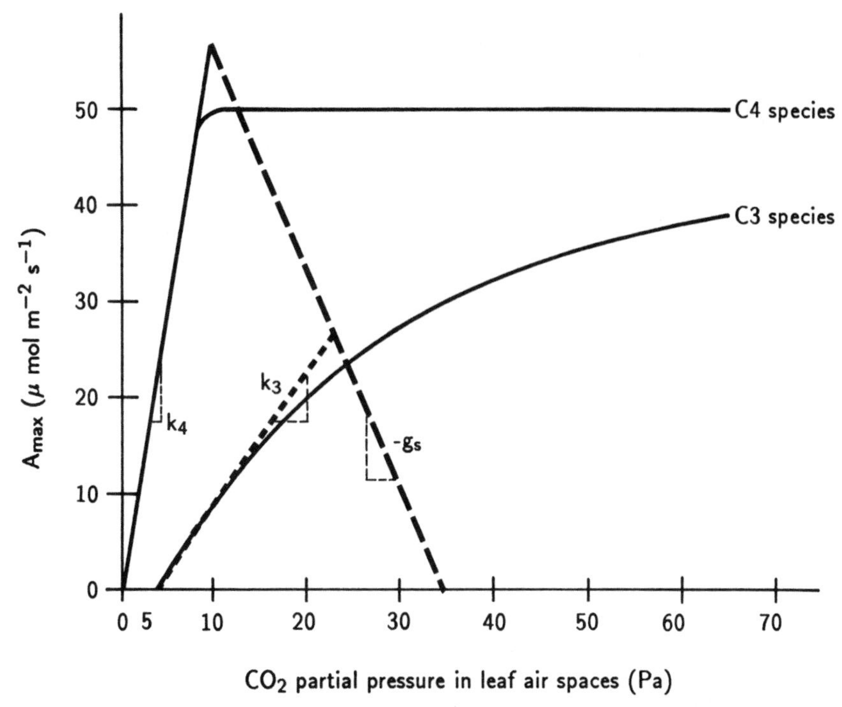

Fig. 4.4. Responses of A_{max} to CO_2 partial pressure (- - - -), carboxylation efficiency, k (— — —), stomatal conductance, g_s x − 1, with intercept on the X-axis at CO_2 partial pressure of the air.

these variations. The typical C3 and C4 responses are quite different (Fig. 4.4). In C4 species, PEP carboxylase captures CO_2 very efficiently and once the CO_2 partial pressure within the air spaces of the leaf reaches about 10 Pa (about one-third of ambient air), the mechanism is CO_2 saturated and no further CO_2 response is observed as the CO_2 partial pressure is increased.

In contrast, C3 species show a continued increase in A_{max} up to the limits shown in Figure 4.4. This CO_2 response reflects the fact that rubisco is not CO_2 saturated at current CO_2 levels. This occurs in part because the CO_2 partial pressure is too low to saturate rubisco, which has a rather low affinity for CO_2. In addition, the process of photorespiration (the fixing of oxygen by rubisco, ultimately releasing CO_2) competitively inhibits CO_2 fixation. The rate of photo-respiration increases with temperature (Farquhar 1988), decreasing the potential for CO_2 fixation. An increase in the CO_2 partial pressure favors an increase in CO_2 fixation over photorespiration. In addition to the effect of photorespiration, the amounts of active rubisco and the initial acceptor of CO_2, ribulose bisphosphate (RuBP), influence the response of A_{max} to CO_2 (Farquhar and von Caemmerer 1982). At CO_2 partial pressures below about 20 Pa (Fig. 4.4), A_{max} is controlled primarily by the amount of active rubisco and the CO_2 partial pressure. This rate of carboxylation is shown as k_3 for C3 species (and k_4 for C4 species) in Figure 4.4. The coefficient k is also known as the carboxylation efficiency. As the CO_2 partial pressure within the leaf, usually termed the intercellular partial pressure, p_i, increases, the response of A_{max} in C3 species becomes increasingly curved. This reflects a slow transition from the primarily rubisco-activation-limited region to the region where RuBP supply, by the Calvin cycle, is increasingly limiting (Farquhar and von Caemmerer 1982). At low temperatures, e.g., below 5°C in C3 species and 10°C in C4 species, RuBP supply may become limiting because of low production rates.

The activity of rubisco appears to be controlled by the enzyme rubisco activase (Salvucci 1989) and it is thought that rubisco activity is adjusted so that the steady-state levels of RuBP are in excess of those required to saturate rubisco. The biochemical processes are not yet clarified, but it appears that at current partial pressures of CO_2, C3 photosynthesis is controlled at the boundary between the limits due to rubisco activity and RuBP supply.

Any changes in stomatal conductance, due to changes in VPD, for example, will also influence A_{max}. In Figure 4.4 the supply function of CO_2 through the stomata is shown as the dotted line, which is the stomatal conductance, g_s, times -1. If the intercept of the CO_2 response of A_{max} and p_i, the CO_2 compensation point (Γ), is known, then A_{max} is determined by the combination of a demand function (Eqn. 4.8) by rubisco or PEP carboxylase and a supply function (Eqn. 4.9) through the stomata (Farquhar and Sharkey 1982) as

$$A_{max} = k\,(p_i - \Gamma)\,/\,P \ , \tag{4.8}$$
$$A_{max} = g_s(p_a - p_i)/\,P \ , \tag{4.9}$$

where P is atmospheric pressure and p_a is the CO_2 partial pressure of the air surrounding the leaf. The CO_2 compensation point, Γ, is positively correlated with the rate of photorespiration and in C3 species is typically on the order of 4 Pa, and markedly temperature sensitive (Farquhar 1988). In contrast, Γ in C4 species is close to zero, indicating the lack of a detectable rate of photorespiration in species with this photosynthetic pathway.

Combining Equations 4.8 and 4.9 it is possible to calculate p_i, at the point of intersection of the supply and demand functions in Figure 4.4, as

$$p_i = \frac{(\Gamma * k) + (p_a * g_s)}{(k + g_s)} \ . \tag{4.10}$$

A_{max} may then be calculated by substitution in either Equation 4.8 or 4.9. This estimate of A_{max} is likely to be an overestimate of the actual A_{max} due to nonlinearities of both the C3 and C4 responses.

A further consideration is the CO_2 response of stomatal conductance, which occurs in a large number of species (Meidner and Mansfield 1968; Jarvis and Morison 1981). An increase in CO_2 partial pressure (either p_a or p_i) causes a reduction in g_s. However, this decrease in g_s leads to a reduction in the rate of CO_2 diffusion into the leaf. As a consequence p_i decreases because of photosynthetic uptake. As p_i falls g_s will be expected to increase. This feedback mechanism is thought, by some, to be important both as a controller of stomatal conductance and as a mechanism for optimizing rates of transpiration and photosynthesis (Cowan and Farquhar 1977; Farquhar *et al.* 1978). However, in the short term of seconds, the stomata are slow to respond to changes in p_i (response times greater than 60 s), so p_i will show marked variation in a fluctuating environment (Jarvis and Morison 1981).

Models have been constructed for calculating the rates of RuBP supply which, in C3 species, are probably the dominant factor in reducing actual A_{max} from that predicted from Equations 4.8–4.10. Such models incorporate irradiance, which controls RuBP production through the electron transport pathway (Farquhar and von Caemmerer 1982). In these models, increasing irradiance will increase the value of p_i at which the transition from rubisco to RuBP limitation occurs. However, a truly mechanistic model needs to predict the behavior of rubisco activase, a feature which has not yet been realized.

Photosynthesis at Low Irradiance

The rate of photosynthesis at low irradiance may be simply calculated by the product of the quantum efficiency, ϕ (moles of CO_2 fixed per mole (i.e., Avogadro's number) of photons) and irradiance. In C3 and C4 species this predicts the rate of photosynthesis up to an irradiance of about 150 W m^{-2} (Fig. 4.5). At greater irradiances the actual photosynthetic rate is rather less because of the action of the biochemical controls on A_{max} as mentioned above.

A knowledge of ϕ is important for predicting the contribution to plant or canopy photosynthesis by shaded leaves. In C4 species ϕ is rather insensitive to both temperature and the CO_2 partial pressure (Fig. 4.3b). In contrast C3 species show a marked sensitivity of ϕ to both temperature and CO_2 (Fig. 4.3b). The primary cause of this sensitivity is the response of photorespiration, which diminishes ϕ by the competitive inhibition of CO_2 fixation by rubisco. Therefore as photorespiration increases with temperature (Lawlor 1987), ϕ decreases. The amelioration of this temperature effect by an increased CO_2 partial pressure (Fig. 4.3b) also reflects the increase in CO_2 fixation by rubisco, relative to O_2 fixation in photorespiration.

Shade plants growing on a forest floor, or even shaded leaves on a sun plant, may benefit photosynthetically at low irradiance from the typically elevated CO_2 partial pressures in these environments (Woodward *et al.* 1991).

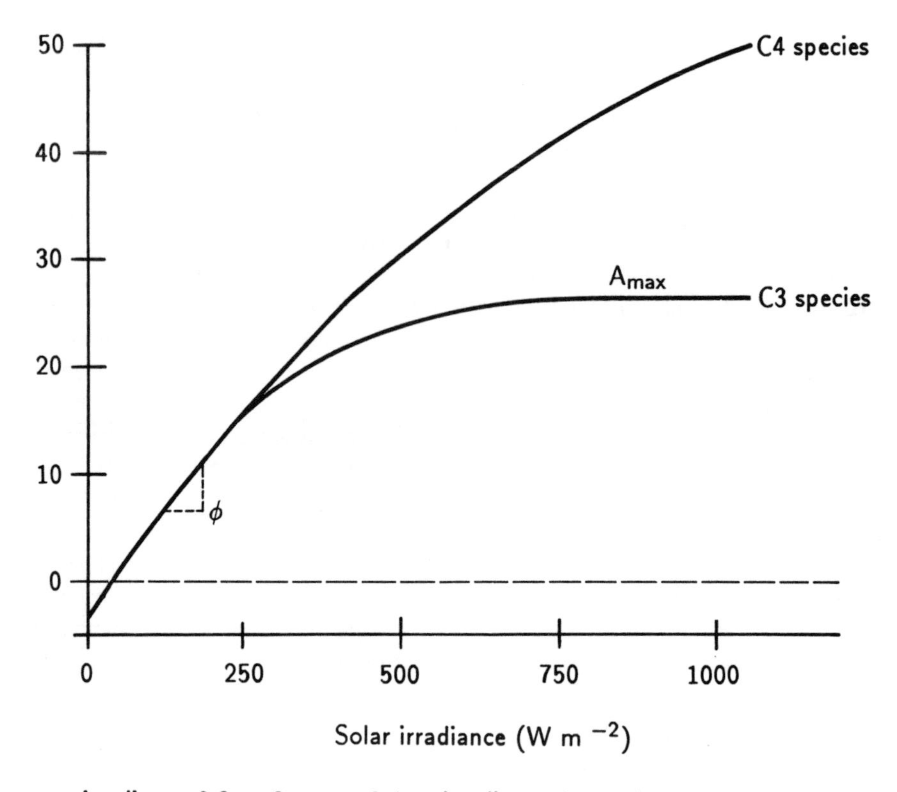

Fig. 4.5. Irradiance responses of photosynthesis, A_{max} (the maximum rate of photosynthesis) and ϕ (the quantum efficiency) (from Woodward and Sheehy 1983).

Predicting Photosynthesis

The high and low irradiance responses of photosynthesis, A, for both C3 and C4 species may be incorporated (Fig. 4.5) in the following simple equation (from Landsberg 1986):

$$A = \frac{I * \phi * A_{max}}{(I * \phi) + A_{max}} - R. \qquad (4.11)$$

The equation is a reasonable predictor of A at any irradiance, I, in combination with predictions of A_{max} from Equations 4.8–4.10. Typical irradiance responses of C3 and C4 photosynthesis curves, with the same value of ϕ (ca. 25°C, Fig. 4.3b), are shown in Figure 4.5.

The daily total of dark respiration, averaged to the same time basis as A (e.g., seconds), is shown by R and in this simple formulation is seen as the debit side of photosynthesis. This is not the case in reality, where respiration is the source of high-energy compounds and carbon substrates for growth and for maintaining cellular integrity. It is not known whether these respiratory processes of synthetic and maintenance respiration occur in the leaf in the light (Lawlor 1987). They are different processes from photorespiration, which only occurs in the light.

Leaf Responses to Increased Irradiance

Sun Leaves

When a leaf which has developed in moderate to high irradiance is suddenly subjected to a large increase in irradiance (e.g., Figs. 4.1 and 4.2), its photochemical rate of photosynthesis may become saturated (Fig. 4.5). As a consequence more quanta are being absorbed by the chlorophyll antennal system than can be used photochemically. In this situation the chlorophyll system may become damaged or functionally changed (Cleland *et al.* 1986; Woodward 1990a), a process known as photoinhibition. In some cases photoinhibition is so strong that the photosynthetic response is severely reduced, requiring protein synthesis for repair (Greer *et al.* 1986). However, an array of mechanisms are available within the chloroplast for minimizing photoinhibitory damage (Demmig and Winter 1988).

In spite of the protective mechanisms it has been found (Drake 1990) that at current levels of CO_2, leaves may be regularly photoinhibited during high irradiance in the natural environment for significant periods of the day. When the same plants are grown at double the partial pressure of CO_2, photoinhibition is strongly reduced and the photosynthetic rate is, consequently, significantly increased. This rather unexpected observation suggests that increases in atmospheric CO_2 partial pressures may stimulate photosynthetic rate by processes in addition to those described in Figures 4.3 and 4.4.

Shade Leaves

Leaves which develop in a predominantly shade environment contain less rubisco than sun-developed leaves but generally have a greater development of the chlorophyll system (Evans 1989), in order to maximize photon capture at low irradiance. For this reason shade leaves, when subjected to high irradiance sunflecks, may suffer greater photoinhibition than sun leaves (Cleland *et al.* 1986). During long periods of low irradiance (prior to sunflecks), rubisco is only partially activated. This prevents rubisco from sequestering metabolites that are required for other processes (Salvucci 1989). However, the photosynthetic rate may be slow to increase during a sunfleck, until rubisco is activated. Stomatal opening is also sensitive to irradiance, decreasing as irradiance diminishes (Jones 1983).

Shade leaves are therefore regulated downward in terms of photosynthesis in low irradiance. When these leaves are subjected to a sudden increase in irradiance during a sunfleck they take a finite time to respond (Fig. 4.6; from Pearcy 1988). During the first few seconds of the sunfleck (Fig. 4.6), in which the irradiance increases from 6 to 520 μmol m^{-2} s^{-1}, there is a rapid increase in the photosynthetic rate (Fig. 4.6a), mirrored by a fall in the intercellular CO_2 partial pressure (Fig. 4.6c). Photosynthesis is quick to respond, rapidly depleting the CO_2 partial pressure within the leaf. This initial, rapid photosynthetic response is followed by a slower response, which probably indicates the response time of complete rubisco activation and the limitation of CO_2 supply through the slowly opening stomata. Stomatal opening is slow (Fig. 4.6b) and is incomplete even after 1000 s.

If the irradiance is again reduced to 6 μmol m^{-2} s^{-1} (Fig. 4.6) then the photosynthetic rate falls rapidly, more so than g_s (Pearcy 1988). In a rapidly ensuing sunfleck the photosynthetic rate saturates within about 100 s. This rapid response shows the carryover of rubisco activation from the previous sunfleck. This process of the light-priming of rubisco maximizes the capacity of a shade leaf to make photosynthetic gains out of a rapid series of sunflecks, a response often seen in the field when sunflecks are clumped in occurrence (Chazdon and Fetcher 1984). Leaves can also photosynthesize after the end of a sunfleck, using RuBP synthesized in the previous sunfleck (Sharkey *et al.* 1986).

Extrapolating Leaf Responses to Larger Scales

So far this chapter has introduced a series of rapid physiological responses by leaves to a changing environment. Scaling up to longer time-scales would require the inclusion of a range of whole-plant responses because these will influence the leaf response. For example the short-term response of photosynthesis to increased CO_2 concentration is an increase in photosynthesis (Figs. 4.3 and 4.4). However, this short-term stimulation may diminish or even be negated in the long term because of the lack of an active sink for photosynthetic products in other parts of the plant (Sage *et al.* 1989; Woodward *et al.* 1991).

Over the longer term the supply of soil nutrients, such as nitrogen, may be limiting. This will exert a strong effect on the rate of photosynthesis (Woodward 1990b) and on the pattern of allocation of photosynthetic products within the plant (Osmond *et al.* 1980; Jones 1983). Changes in allocation patterns will influence rates of organ growth and the subsequent capacity of the plant to capture resources, perhaps in competition with neighboring plants.

It is clear that scaling up to the plant and population level, both temporally and spatially, for predicting dynamic responses to a changing environment is extremely difficult at present because of a lack of knowledge of the mechanisms

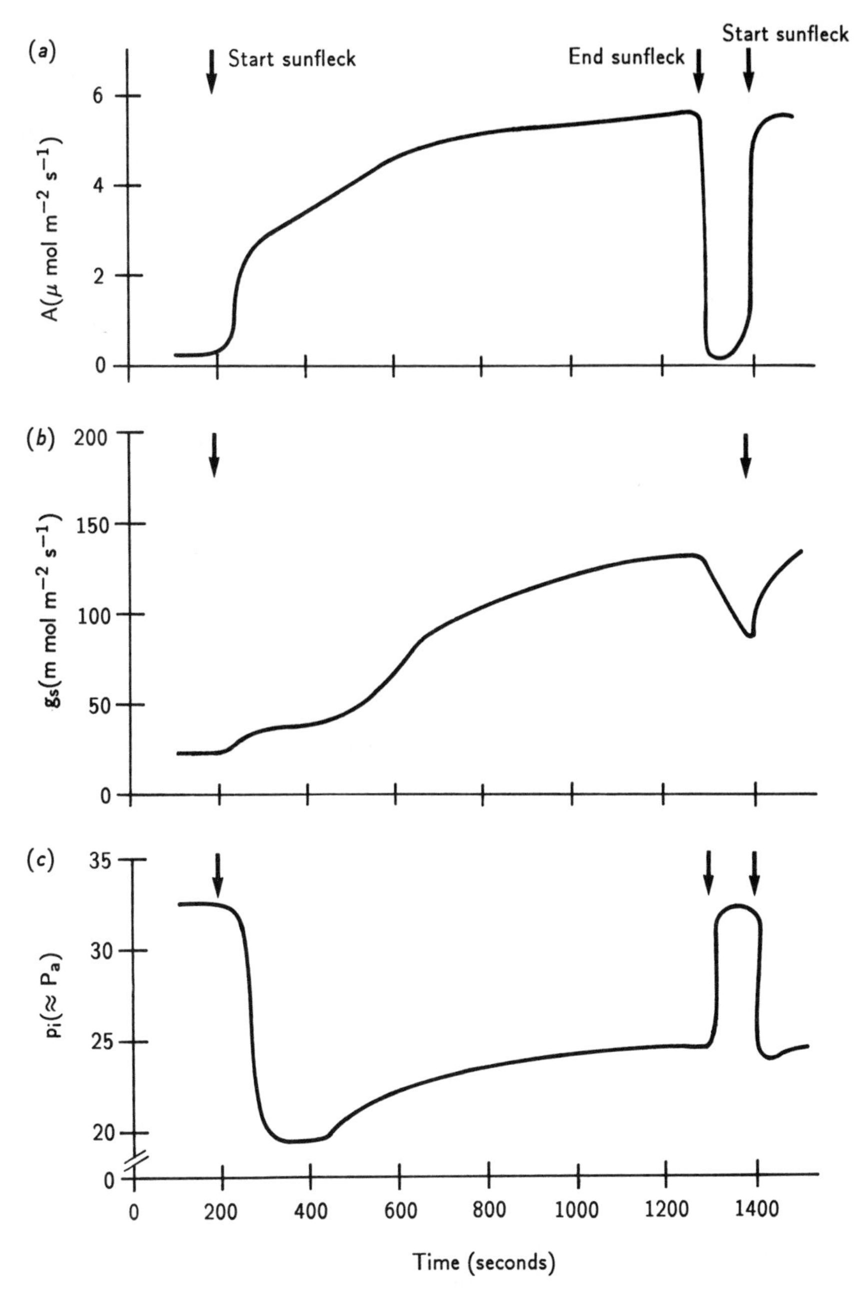

Fig. 4.6. Response of gas exchange in *Argyrodendron* (shade plant) to sunflecks (from Pearcy 1988). *a*. Photosynthesis. *b*. Stomatal conductance. *c*. Intercellular partial pressure of CO_2 (p_i).

behind the responses and the interactive processes at larger scales (Woodward and Diament 1991). However, it is the aim of the remainder of this chapter to describe one example of successful scaling up from leaf-level responses to global ecology. A model which can be used to predict maximum canopy LAI from climatic indices, using estimated transpiration by vegetation, is presented. As the number of leaves in a vegetation canopy increases, the net photosynthetic capacity increases (Landsberg 1986), with consequences for the local, regional, and global carbon budget. Thus knowledge of LAI values is central to studies of global vegetation dynamics.

Estimating Leaf Transpiration

The influence of the environment on leaf transpiration has been discussed earlier in this chapter. A plant's sensors of the evaporative environment are the stomata; in addition, the presence of the leaf or plant impedes local air flow, creating a leaf or plant boundary layer which also influences transpiration. The environmental driving forces for transpiration and the plant responses are incorporated in the general energy balance equation given earlier (Eqn. 4.6):

$$R_n = C + \lambda E + J \ ,$$

where R_n is the net input of radiation to a leaf, C is convective heat exchange, λE is evaporative heat exchange (by transpiration through stomata or evaporation from water films), and J is heat storage. J is small in magnitude and can usually be ignored. Thus Equation 4.6 may be simplified and expanded to show the driving forces and controlling conductances :

$$R_n = (\chi_l - \chi_a)\cdot\frac{g_s \cdot g_a}{g_s + g_a} + p\cdot c_p\cdot(T_l - T_a)\cdot g_a . \quad (4.12)$$

In this equation, χ_l and χ_a are the absolute humidities (g m^{-3}) of the air within a leaf (or the leaves of a plant or canopy) and of the surrounding air, respectively, and g_a and g_s are the boundary layer and stomatal conductances to water vapor, respectively. (Earlier in this chapter, the gradient of absolute humidity is also expressed as the water vapor pressure deficit (VPD):

$$VPD = (e_{s(Tl)} - e) \ , \quad (4.13)$$

where $e_{s(Tl)}$ is the saturated water vapor pressure (Pa) in the leaves and e is the water vapor pressure of the air.)

The first term in Equation 4.12 calculates transpiration, and also the rate of evaporation of a film of water on a leaf surface, when g_s is equal to zero. The second term in the equation estimates the convective flux from a leaf or canopy of leaves. p is the density of the air (g m^{-3}) and c_p is the specific heat of air (J

$g^{-1} K^{-1}$); T_l and T_a are the temperature of the leaf and of the surrounding air, respectively.

Given measurements or estimates of climate—that is, of R_n, air temperature, vapor pressure, and wind speed, it is possible to predict leaf transpiration from Equation 4.12. Knowledge of g_a and g_s is also required.

Estimating Canopy Transpiration

In principle Equation 4.12 could be applied to whole-plant canopies or larger areas of vegetation, but in practice the process of scaling up leads to a simplification of this equation. The simplest case is for a large area of moderately uniform vegetation, which is well supplied with water. Using the Priestley-Taylor approximation (Priestley and Taylor 1972), transpiration may be estimated as

$$\lambda E = \frac{1{\cdot}26{\cdot}s{\cdot}R_n}{s+\delta} \ , \qquad (4.14)$$

where s is the slope of the curve relating the saturation vapor pressure of water to temperature (Pa K^{-1}) and δ is the psychrometric constant (Pa K^{-1}).

This equation has the merit of extreme simplicity and requires no information about the underlying vegetation. However, forests differ in their capacity to transpire water and in their responses when soil water is limiting; therefore the approximation requires more biological information. Such an approach is satisfied by the Penman-Monteith approach to predicting transpiration (Monteith 1965). In this case canopy estimates of g_a and g_s are required for predicting transpiration as

$$\lambda E = \frac{s{\cdot}R_n + p{\cdot}c_p{\cdot}(e_{s(Ta)} - e){\cdot}g_a}{(s + (g_a/g_s)} \ , \qquad (4.15)$$

where $e_{s(Ta)}$ is the saturated water vapor pressure (Pa) of the air. Jones (1983) and Paw U and Gao (1988) discuss the various simplifications involved in this derivation. A major feature is the exclusion of leaf temperature in the VPD term (compare Eqns. 4.13 and 4.15) and its replacement by air temperature.

Equation 4.15 may be applied to canopies differing in leaf area index (LAI), a feature which influences g_a and g_s. In a plant canopy consisting of layers of leaves, transpiration from each layer occurs simultaneously, and in direct relationship to the LAI in each layer. Therefore in a canopy of i layers of leaves, the canopy stomatal conductance, g_s, is defined as

$$g_s = \sum (g_{s,i}{\cdot}L_i) \ , \qquad (4.16)$$

where L_i is the leaf area index and $g_{s,i}$ the stomatal conductance of layer i. Transpiration occurs from each leaf layer through the boundary layer of each leaf layer and through a boundary layer between the leaf layers. Unfortunately it is not easily possible to calculate the boundary-layer conductance between the leaf layers. However, it is likely that the conductance is high and can be ignored for coarse-scale calculations of canopy transpiration. Therefore canopy g_a may be calculated in the same way as canopy g_s, by using Equation 4.16 and a knowledge of the wind-speed profile within the canopy. This too is poorly known (Landsberg 1986), so the estimate of canopy g_a will also be poor on this account.

Predicting Canopy Leaf Area Index

The Penman-Monteith equation (Eqn. 4.15) may be used to predict transpiration by canopies differing in LAI. The reverse process, of predicting the maximum or mean LAI for which transpiration at a specified site and in a specified climate is balanced by precipitation, is the aim of this section. Woodward (1987) outlined the philosophy of this approach in detail. In essence, it is assumed that over 1 year the precipitation which percolates through the leaf canopy and to the roots of the plants is balanced by transpiration through the leaves of the canopy and by evaporation of surface films of water collected on the leaves. Too high an LAI will lead to periods in which drought and leaf abscission occur. Too low an LAI leads to a surplus of water and to an increase in leaf area expansion and therefore LAI.

The first requirement for the prediction of LAI is a generalized profile of the leaf layers (strata) in the canopy. These profiles are shown in Figure 4.7 (from Woodward 1987). The individual strata are used to calculate canopy g_a and g_s as described above. Woodward (1987) only considered the effect of irradiance on g_s, but if sufficient microclimatic detail is known all of the responses described earlier in this chapter can be used.

The decline in irradiance through the canopy is also shown in Figure 4.7 and this can be used to predict $g_{s,i}$. If wind-speed profiles through the canopy are available, then $g_{a,i}$ (the boundary-layer conductance of layer i) can be calculated to provide an estimate of canopy g_a. In general wind-speed profiles are not available and so Woodward (1987) suggests using a mean speed of 0.4 m s^{-1} within the canopy and a measured meteorological station wind speed for the top layer of the canopy in order to derive canopy g_a.

The Penman-Monteith equation, in combination with the LAI profiles (Fig. 4.7), has been used to estimate site LAI and the results are shown in Figure 4.8. Two climatic stations are shown: a dry site (Brisbane, Australia) and a wet site (Sydney, Australia). For each month of the year, starting with the highest LAI (LAI 9), the balance between incoming precipitation and outgoing evapotranspiration (evaporation plus transpiration) is shown as the resulting soil water deficit. At Brisbane, the soil water deficit increases throughout the year, because evapotranspiration exceeds precipitation. From December onward the soil water content is recharged by increased precipitation. However, for LAI 9 this recharge is

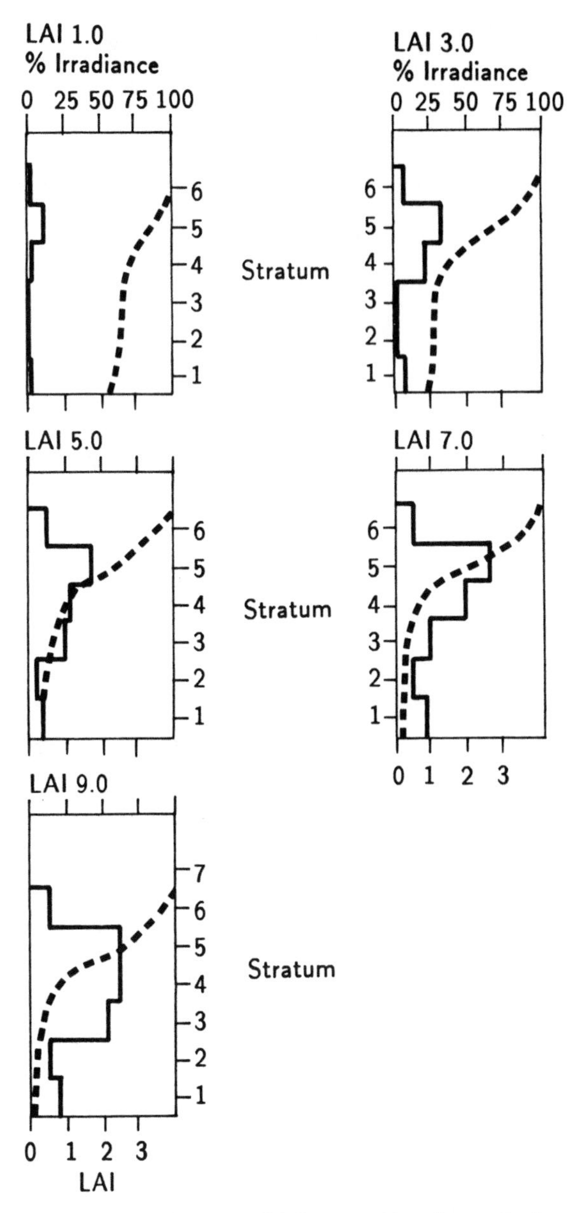

Fig. 4.7. Leaf area (solid lines) and irradiance (broken lines) profiles through canopies differing in leaf area index (LAI).

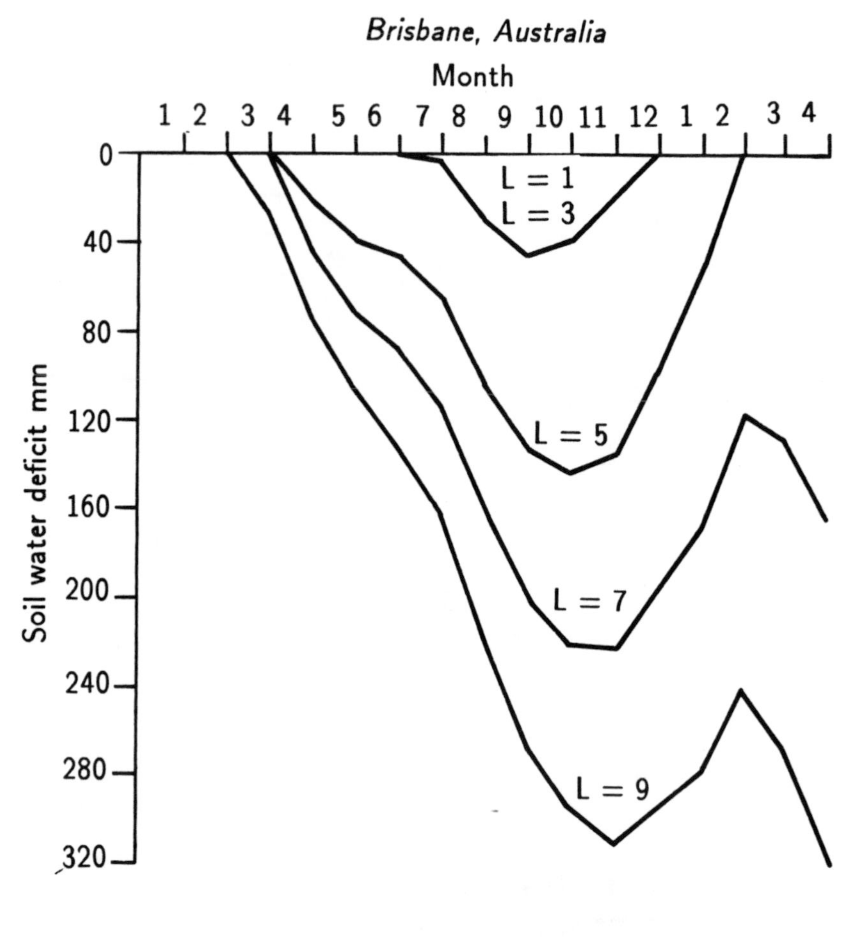

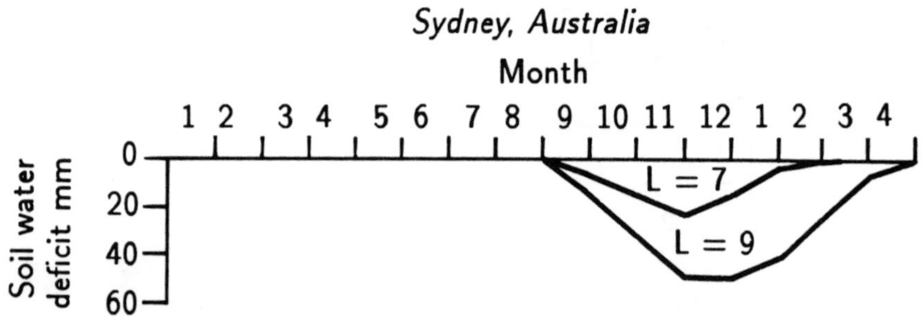

Fig. 4.8. Predicted monthly soil water deficit for soils at two sites in Australia: a. Brisbane. b. Sydney.

incomplete by the onset of the next dry season. The model then predicts an increased drying of the soil. Clearly, for Brisbane, an LAI of 9 exceeds the carrying capacity of the local precipitation.

At the other extreme of LAI (1–3), the canopy fails to evaporate a large fraction of the incoming precipitation and so it is assumed that the canopy could develop a greater LAI; for Brisbane this is satisfied at an LAI of about 5. For Sydney the local precipitation can sustain an LAI of greater than 9. In this case it is possible that the upper limit of LAI is determined by irradiance.

An additional issue is raised by the model predictions shown in Figure 4.8. This is the degree of soil water deficit which can be endured by a plant canopy without it losing leaves by drought-caused abscission. This information is in poor supply and so it is difficult to add any precision. However, Woodward (1987) suggested that once the soil water deficit exceeded 80 mm of rainfall equivalent, leaf abscission was probable. Therefore the LAI of 5 for Brisbane may show some leaf abscission during the period from August to December. It is also likely, however, that species of plants occurring in areas of regular droughts may be more capable of enduring lower soil water deficits than plants from more mesophytic climates (Larcher 1980). In addition the nature of the soil texture (e.g., sand or clay) will also influence the plant response. It is clear that much more research is required on this topic.

Mapping Predictions of LAI

Using the model described above, and bearing in mind the lack of information on some aspects of it, a global map of LAI has been produced (Fig. 4.9). A maximum value of g_s for an individual leaf has been assumed throughout to be 100 mmol m^{-2} s^{-1}. The possibility of leaf abscission during a dry season has not been allowed in this run of the model.

Through comparison with global vegetation maps (see Cramer and Leemans, this volume) it may be seen that the absolute value of LAI is a good predictor of vegetation type. The forests of the equatorial and tropical regions, and of the warm and cool temperate and boreal zones, are all clearly defined. Similarly the areas of desert, grassland, and shrub are all accurately predicted.

A consequence of preventing leaf abscission during the dry season is that the annual average of LAI in strongly seasonal climates, such as the monsoon climate of India, is lower than expected. An unrealistically high LAI is predicted for the tundra zone, where there is a small evaporative demand which is easily satisfied by incoming precipitation.

The two problems outlined above may be incorporated into the model predictions by allowing leaf abscission to occur when the soil water deficit exceeds 80 mm. The LAI for such a climate would then be the maximum that could be reached in the wet season.

The reason that the tundra region fails to develop a high LAI is because the

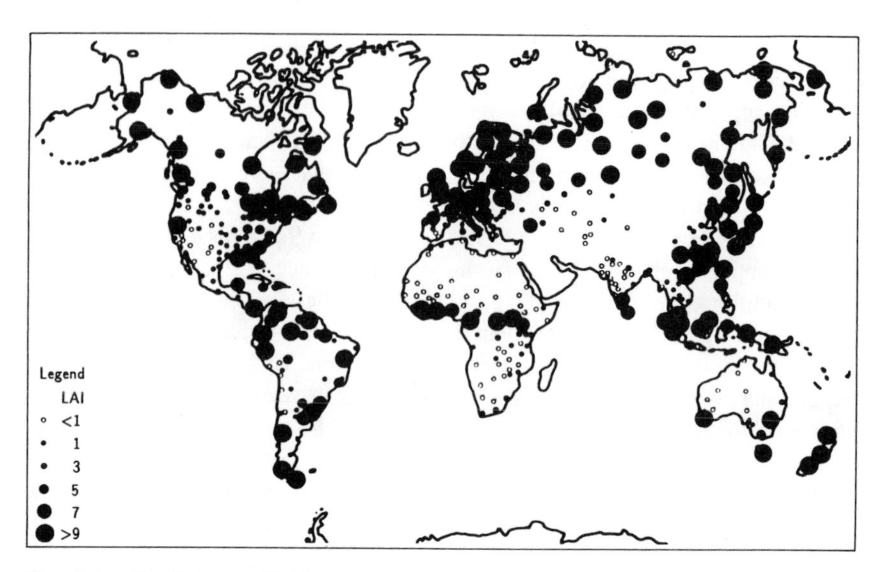

Fig. 4.9. Predictions of LAI.

growing season is short. The warmth and duration of the growing season may be defined in terms of degree days (the number of days during which the temperature exceeds some threshold temperature, taken to be 0°C for tundra). Woodward (1987) has chosen an upper degree-day threshold of 600 to define the warm-temperature limits of tundra, and therefore the cold-temperature limits of the adjacent boreal forest.

A final refinement to the model is to allow leaf abscission to occur when the absolute winter temperatures fall to between −15 and −40°C. Woodward (1987) has argued that in such a climate the dominant vegetation should be a broadleaf deciduous forest (assuming sufficient precipitation). Such a vegetation type would drop its leaves when the mean minimum temperature falls below 0°C and develop new leaves when the temperature exceeds the same threshold.

The effects of these modifications on the model's predictions are shown in Figure 4.10. There are distinct improvements in the predictions (compare with the vegetation map in Cramer and Leemans, this volume), particularly for the seasonal environments.

The Importance of Vegetation Height

The model for predicting LAI works well for most of the terrestrial vegetation types of the world (Woodward 1987). The value of its mechanistic basis is that novel changes in the environment may also be included. One important requirement is the prediction of the direct effect of the expected increases in the concentration of atmospheric CO_2. Woodward (1990b) has described a method

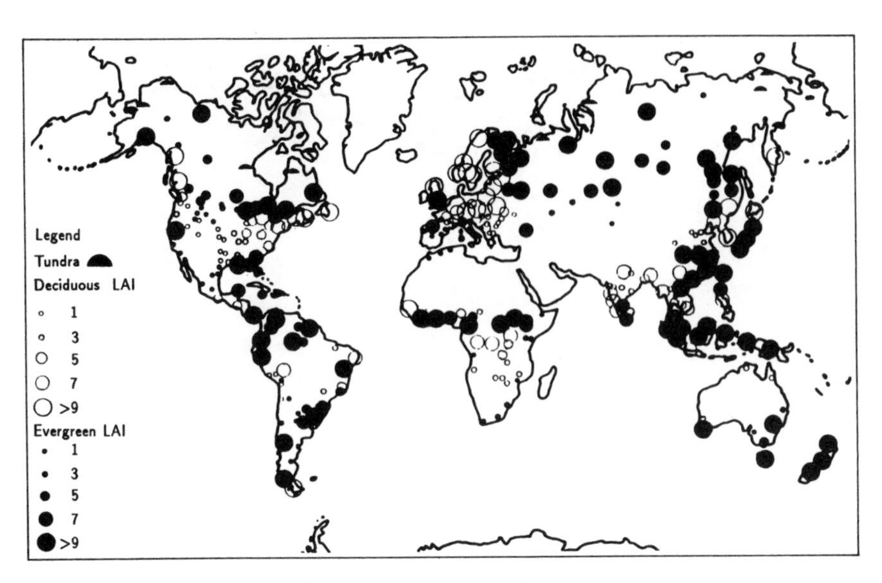

Fig. 4.10. Predictions of LAI and vegetation physiognomy.

for predicting the influence of CO_2 concentration on LAI, through its effect in reducing stomatal conductance. However, the responses of a short grassland and a tall forest are rather different (Fig. 4.11). For grassland, over a wide range of LAI, a doubling in the CO_2 concentration causes only a small reduction in canopy transpiration of about 4%. For forest the reduction is greater, at about 10%. The differences between the two vegetation types occur because of marked differences in aerodynamic coupling between the canopies and the air above. The forest is well coupled, with a rough aerodynamic surface (Jarvis and McNaughton 1986). The transpiration rate of the canopy therefore responds strongly to the water-vapor-pressure deficit of the air. In contrast, the grassland has a smooth aerodynamic surface and is poorly coupled to the air above. The grass may significantly humidify the air within the canopy, and, as a consequence, transpiration is more strongly controlled by the net radiant balance than by the vapor-pressure deficit of the air (Jarvis and McNaughton 1986).

Given that LAI is predicted as the intercept of the relationships between throughfall, canopy transpiration, and LAI (Fig. 4.11), the direct effect of CO_2 is small for grassland but may be large (e.g., throughfall 1 in Fig. 4.11) for forest. Therefore general models for predicting the hydrological balance of global vegetation types will require a knowledge of vegetation height for an adequate prediction under different atmospheric concentrations of CO_2.

The mechanistic predictions of LAI can be used to infer vegetation type (Woodward 1987; Woodward and McKee 1991). This inference has only been achieved by correlation (Fig. 4.12) on the basis of present-day observations, in

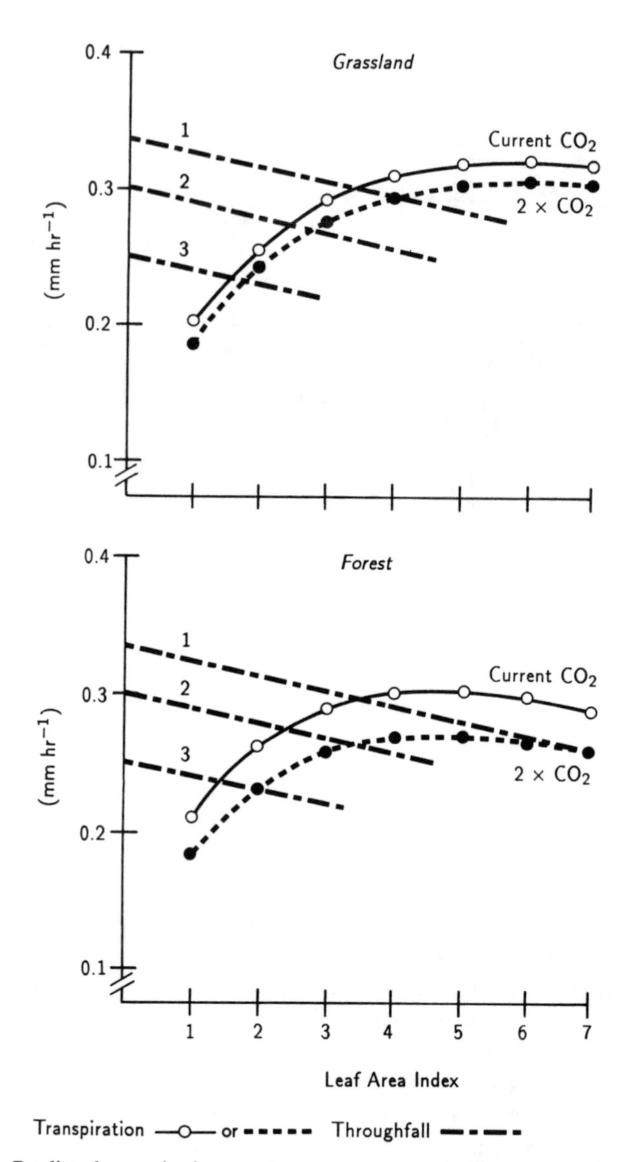

Fig. 4.11. Predicted transpiration rate (——o—— current CO_2 concentration, - - -●- - - doubled CO_2 concentration) for grassland and forest canopies over a range of LAI. Three estimates of precipitation throughfall (—·—·—·—) are shown for predicting LAI. Throughfall decreases with LAI because of an increase in leaf interception of precipitation and subsequent evaporation. At the current CO_2 concentration, the predicted LAI is the intercept of the throughfall line and the canopy transpiration rate (——o——). Under a doubled CO_2 concentration the predicted LAI is the intercept between the throughfall line and canopy transpiration (- - -●- - -) (from Woodward 1990b).

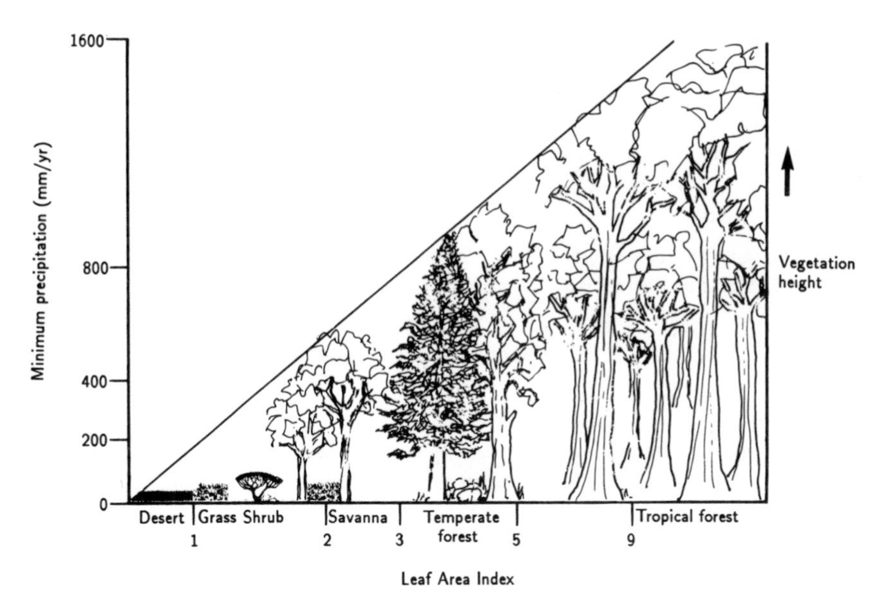

Fig. 4.12. The relationships between global averages of minimum precipitation, LAI and vegetation height (from Woodward and McKee 1991).

which the greater the LAI the taller and more forestlike the vegetation. Although the relationship between precipitation and vegetation height and type is approximately linear, this is clearly not the case for LAI (Fig. 4.12). Given the general applicability of the correlation between LAI and vegetation type (Woodward 1987), and the likelihood of a height-dependent response to CO_2 (Fig. 4.11), the first step in determining the mechanistic basis of the correlation between LAI and vegetation type must be to predict vegetation height. This will immediately allow predictions of forests, shrubland and dwarf grassland. The variable effect of CO_2 concentration on vegetation transpiration (Fig. 4.11) may then be incorporated into a general vegetation model. For further discussion on the significance and prediction of tree height, see Friend (this volume).

Concluding Comments

In this chapter I have examined the short-term responses of leaves to changes in environmental variables and given an example of the scaling up of such responses to a global level. Such scaling-up studies are likely to be of increasing importance in the prediction of the response of the world's vegetation to global environmental change. The responses of individual leaves will be very sensitive to environmental change, in a predictable manner. What is uncertain is precisely how these responses will influence larger-scale ecosystem processes. This will be one of the major challenges facing ecophysiologists in the future.

References

Baker, N.R., Long, S.P. and Ort, D.R. (1988). Photosynthesis and temperature, with particular reference to effects on quantum yield. In *Plants and Temperature*, ed. S.P. Long and F.I. Woodward, pp. 347–75. Society for Experimental Biology Symposium, Vol. XXXXII. The Company of Biologists, Ltd., Cambridge U.K.

Chazdon, R.L. and Fetcher, N. (1984). Photosynthetic light environments on a lowland tropical forest in Costa Rica. *Journal of Ecology*, **72**, 553–64.

Cleland, R.E., Melis, A. and Neale, P.J. (1986). Mechanism of photoinhibition: Photochemical reaction centre inactivation in system II of chloroplasts. *Photosynthesis Research*, **9**, 79–88.

Cowan, I.R. and Farquhar, G.D. (1977). Stomatal function in relation to leaf metabolism and environment. In *Integration of Activity in the Higher Plant*, pp. 471–505. Society for Experimental Biology Symposium, 31. Cambridge, UK: Cambridge University Press.

Demmig, B. and Winter, K. (1988). Light response of CO_2 assimilation, reduction state of Q, and radiationless energy dissipation in intact leaves. *Australian Journal of Plant Physiology*, **15**, 151–62.

Desjardins, R.L. and Lemon, E.R. (1974). Limitations of an eddy-correlation technique for the determination of the carbon dioxide and sensible heat fluxes. *Boundary-Layer Meteorology*, **5**, 475–88.

Drake, B.G. (1989). Effects of Elevated Carbon Dioxide on Chesapeake Bay Westlands. IV. Ecosystem and Whole Plant Responses, April-November, 1988. US DOE/CO_2 Greenbook 051. Atmospheric and Climate Research Division, ER-76, U.S. Department of Energy, Washington, D.C.

Evans, J.R. (1989). Photosynthetic and nitrogen relationships in leaves of C3 plants. *Oecologia*, **78**, 9–19.

Farquhar, G.D. (1988). Models relating subcellular effects of temperature to whole plant responses. In *Plants and Temperature*, ed. S.P. Long and F.I. Woodward, pp. 395–409. Society for Experimental Biology Symposium, Vol. XXXXII. The Company of Biologists, Cambridge, U.K.

Farquhar, G.D., Dubbe, D.R. and Raschke, K. (1978). Gain of the feedback loop involving carbon dioxide and stomata: Theory and measurement. *Plant Physiology*, **62**, 406–12.

Farquhar, G.D. and Sharkey, T.D. (1982). Stomatal conductance and photosynthesis. *Annual Review of Plant Physiology*, **33**, 317–45.

Farquhar, G.D. and von Caemmerer, S. (1982). Modelling of photosynthetic response to environmental conditions. In *Encyclopedia of Plant Physiology. Physiological Plant Ecology II*, ed. O. Lange, P.S. Nobel, C.B. Osmond and H. Ziegler. Vol. 12B, pp. 549–87. Berlin: Springer-Verlag.

Grace, J. (1977). *Plant Response to Wind*. London: Academic Press.

Greer, D.H., Berry, J.A. and Björkman, O. (1986). Photoinhibition of photosynthesis in intact bean leaves: Role of light and temperature, and requirement for chloroplast-protein synthesis during recovery. *Planta*, **168**, 253–60.

Jarvis, P.G. and McNaughton, K.G. (1986). Stomatal control of transpiration: Scaling up from leaf to region. *Advances in Ecological Research*, **15**, 1–49.

Jarvis, P.G. and Morison, J.I.L. (1981). The control of transpiration and photosynthesis by the stomata. In *Stomatal Physiology*, ed. P.G. Jarvis and T.A. Mansfield, pp. 247–79. Society for Experimental Biology, Seminar Series, 8. Cambridge, U.K.: Cambridge University Press.

Jones, H.G. (1983). *Plants and Microclimate*. Cambridge: Cambridge University Press.

Landsberg, J.J. (1986). *Physiological Ecology of Forest Production*. London: Academic Press.

Larcher, W. (1980). *Physiological Plant Ecology*. 2nd ed. New York: Springer-Verlag.

Lawlor, D.W. (1987). *Photosynthesis: Metabolism, Control and Physiology*. Harlow, Essex, U.K.: Longman.

Meidner, H. and Mansfield, T.A. (1968). *Physiology of Stomata*. London: McGraw-Hill.

Milburn, J.A. (1979). *Water Flow in Plants*. London: Longman.

Monteith, J.L. (1965). Evaporation and environment. In *The State and Movement of Water in Living Organisms*, ed. C.E. Fogg, pp. 205–34. Symposium of the Society for Experimental Biology, Vol. 19. Cambridge: Cambridge University Press.

Monteith, J.L. and Unsworth, M.H. (1990). *Principles of Environmental Physics*. 2nd ed. London: E. Arnold.

Osmond, C.B., Björkman, O. and Anderson, D.J. (1980). *Physiological Processes in Plant Ecology. Toward a Synthesis with Atriplex*. New York: Springer-Verlag.

Paw U, K.T. and Gao, W. (1988). Applications of solutions to non-linear energy budget equations. *Agricultural and Forest Meteorology*, **43**, 121–45.

Pearcy, R.W. (1988). Photosynthetic utilisation of lightflecks by understory plants. *Australian Journal of Plant Physiology*, **15**, 223–38.

Priestley, C.H.B. and Taylor, R.J. (1972). On the assessment of surface heat flux and evaporation using large-scale parameters. *Monthly Weather Review*, **100**, 81–92.

Sage, R.F., Sharkey, T.D. and Seemann, J.R. (1989). Acclimation of photosynthesis to elevated CO_2 in five C3 species. *Plant Physiology*, **89**, 590–6.

Salvucci, M.E. (1989). Regulation of rubisco activity in vivo. *Physiologia Plantarum*, **77**, 164–71.

Schulze, E.D., Lange, O.L., Buchbom, U., Kappen, L. and Evenari, M. (1972). Stomatal responses to changes in humidity in plants growing in the desert. *Planta*, **108**, 259–70.

Sharkey, T.D., Seemann, J.R. and Pearcy, R.W. (1986). Contribution of metabolites of photosynthesis to post illumination CO_2 assimilation in response to lightflecks. *Plant Physiology*, **82**, 1063–8.

Woodward, F.I. (1987). *Climate and Plant Distribution*. Cambridge: Cambridge University Press.

Woodward, F.I. (1990a). From ecosystems to genes: The importance of shade tolerance. *Trends in Ecology and Evolution*, **5**, 111–5.

Woodward, F.I. (1990b). Global change: Translating plant ecophysiological responses to ecosystems. *Trends in Ecology and Evolution*, **5**, 308–11.

Woodward, F.I. and Diament, A.D. (1991). Functional approaches to predicting the ecological effects of global change. *Functional Ecology*, 5:202–212.

Woodward, F.I. and McKee, I.F. (1991). Vegetation and climate. *Environment International*, 17:535–546.

Woodward, F.I. and Sheehy, J.E. (1983). *Principles and Measurements in Environmental Biology*. London: Butterworths.

Woodward, F.I., Thompson, G.B. and McKee, I.F. (1991). The effects of elevated concentrations of carbon dioxide on individual plants, populations, communities and ecosystems. *Annals of Botany*, 67 (supplement 1):23–38.

5

The Prediction and Physiological Significance of Tree Height

A.D. Friend

Introduction

In this chapter the potential limits on the maximum height of trees are examined with a view toward making maximum height predictable from biological and physical principles. The maximum height of trees is species-dependent. Within a species, increased availability of resources such as nutrients, water, and irradiance leads to increased height, but why is it that species do not grow higher than they do, even in the most favorable sites? Is there a genetic component to the maximum height of species? If not, then what environmental factor or factors limit their height?

With increased height come advantages to a plant such as a greater share of irradiance, if other species are also competing. This is probably the dominant force responsible for height growth and has been used to model optimal tree height growth (Iwasa *et al.* 1985; King 1990). King (1990) concluded that trees in single-species, even-aged stands should not gain any advantage by growing taller than 87% of their theoretical break-even height, that is, the height at which the tree ceases to have enough photosynthate available for wood production, and hence growth. It is this break-even height, and its prediction, which are the focus of this chapter.

As well as an increased share of irradiance, increased height may make a species more obvious to potential pollinators; if a species is anemophilous it will increase pollen dispersal. Also, increased height may increase seed and fruit dispersal distances. However, it also has costs. These include physiological costs such as the construction and maintenance of supporting tissues, increased susceptibility to windthrow and breakage, and the increased difficulty of getting water and nutrients to the leaves. Ecological costs might include greater visibility to potential herbivores.

Many theories of tree growth include the cost of maintenance respiration of the supporting tissues as an assumption restricting height growth, and tree size, to some maximum (Whittaker and Woodwell 1967; Kramer and Kozlowski 1979; Waring and Schlesinger 1985; Givnish 1988). As the live sapwood and phloem volume increases, so does total tree respiration, until no more excess carbon is available for growth: all is used for maintenance and replacement of dead organs. Givnish (1988) showed that the maximum height of *Liriodendron tulipifera* could be predicted from considerations of maintenance and construction costs relative to the amount of incident irradiance on the canopy. However, a potential cost ignored from such models has been the reduced water potential experienced by leaves as height increases. As the vertical distance between a leaf and the soil from which it must draw its water increases, so must the energy required to take water to the leaf. This is because of the effect of gravity and the resistance within the plant to water flow. The energy is provided by the leaf in the form of an increase in solute concentration and/or a reduction in turgor potential, reducing the chemical potential of water in the leaf to below that in the soil. Hence water will move from a point of low to one of high chemical (water) potential, causing an increase in the total entropy of the system. The reduction in leaf water potential necessary for this to happen has consequences for the photosynthesis of leaves.

Another possible limit to the height of trees is the purely physical aspect. A very tall tree experiencing horizontal forces may fall, either by uprooting or by snapping.

Hence there are three possible reasons for the maximum height of trees: water uptake, support tissue maintenance, and structural forces. These will now be discussed, and the evidence for and against each hypothesis will be assessed.

Water Uptake

Zimmermann (1971) wrote that "long-distance transport problems are . . . among the most important problems in tree physiology". For leaves to fix CO_2 from the atmosphere they must lose water through their stomata if the leaf–to–air-vapor-pressure deficit (δ_l) is greater than zero. If this water is not replaced the leaf water potential (ψ_l) will fall, which will have adverse effects on photosynthesis and leaf growth. For leaves at the top of a tree to be supplied with water they must possess a sufficiently negative water potential to overcome the effects of gravity and stem resistance on water flow through the conducting vessels. However, a low ψ_l can reduce a leaf's photosynthetic potential. Many of the components of photosynthesis have been shown to be affected by ψ_l, particularly the Calvin cycle (Graan and Boyer 1990).

Zimmermann (1971) investigated the supply of water to tree canopies in some detail. If all the stomata on a tree are closed then there will exist a hydrostatic pressure gradient of 0.01 MPa m^{-1} between the roots and the leaves. Thus the water potential at the top of a 100-m-tall tree would be 1 MPa below that at the bottom if the osmotic potential were the same. For water to move up the tree, the ψ_l in the highest leaves would thus have to be at least 1 MPa lower than the soil water potential (ψ_s) to overcome gravity. However, when the stomata open, and water starts to move up the tree, the resistance to flow must also be overcome. Zimmermann (1971) calculated that this requires an additional gradient of between 0.005 and 0.01 MPa m^{-1}. This gradient is independent of pore diameter. (See below.) From field measurements, Zimmermann (1971) concluded that a total gradient of about 0.015 MPa m^{-1} can be expected under conditions of high transpiration, and one of 0.01 MPa m^{-1} when transpiration is very low or zero. It should be remembered that there will be some distance to be added to allow for the height of the base of the tree above the average point at which water enters the root system, and for nonvertical water movement along branches and roots which will be subject to a resistance of at least 0.005 MPa m^{-1} when transpiration is occurring through these organs.

Thus a 100-m-tall tree, growing in soil with a ψ_s equal to -0.1 MPa, would have leaves operating with a maximum ψ_l lower than -1.6 MPa. Since very few trees grow taller than 100 m, this value of ψ_l can be taken as defining the limit of leaf viability. If this is done, a soil with a growing season ψ_s equal to -1.2 MPa could support trees no higher than 27 m. However, it should also be considered that as soil dries the difference in ψ_s between the root and the soil increases (Nobel 1983). This is due to the exponential increase in the resistance to flow of soil water with decreasing soil water content. The rate of this increase in soil resistance is dependent on soil type. In soils with small pore sizes and high water-retention capacities, such as clays, there is a greater increase in resistance to flow for a given decrease in water content. Consequently the water content at which wilting is thought to occur is higher than for more porous soils.

Leaf water potentials which might be thought to be of consequence in limiting height growth are difficult to predict accurately. As leaf water potential falls, stomata tend to close: in addition, low leaf water potentials have direct effects on photosynthesis (Hall 1982; Schulze and Hall 1982). The range and degree of response vary between species and may even vary within individuals between different times of the year (e.g., Lakso 1979). Generally, a reduction in stomatal conductance and photosynthesis is detected between ψ_l values of about -0.5 MPa and -1.0 MPa, and there follows a more-or-less linear decline in stomatal conductance and mesophyll capacity. Photosynthesis is negligible in many mesophytic trees below a leaf water potential of between -1.5 MPa and -2.0 MPa (e.g., Hinkley *et al.* 1975; Waring *et al.* 1981). As well as ψ_l, leaf turgor potential (ψ_p) must also be considered. This is because turgor is necessary for the growth of leaves; it provides the energy for cell expansion (Tyree and Jarvis 1982; Cleland 1986). If turgor potential cannot be maintained above some threshold value then leaf expansion cannot occur. However, this threshold value is difficult to measure directly (Cosgrove *et al.* 1984; Boyer *et al.* 1985). Under conditions of high day transpiration, leaf expansion might be expected to occur only at night (as long as temperatures are high enough), when the stomata will be more-or-less closed (e.g., Williams and Biddiscombe 1965; but see Taylor and Davies 1986). Under these conditions there will be the reduced hydrostatic pressure gradient of -0.01 MPa m^{-1} up the stem, allowing the maintenance of a higher leaf turgor. With closed stomata, growth will only be possible in the leaves at the top of a 100-m-tall tree if their ψ_l is more than 1.0 MPa, plus a threshold value, lower than the root vessel water potential. This threshold value has been measured as 0.18 MPa in birch (*Betula pendula*) and 0.37 MPa in maple (*Acer pseudoplatanus*) (Taylor and Davies 1986).

If one assumes that the tallest trees in the world are growing in a soil containing water at field capacity, and that this soil is of a loam type, and if the effect of height on ψ_l is limiting any further height growth, what might one expect maximum heights to be under other conditions?

It was stated above that a water potential gradient of -0.015 MPa m^{-1} exists up a tree under conditions of high transpiration. A loam soil at field capacity has a water potential of about -0.11 MPa. Thus if the resistance across the root membranes and leaf petioles amounts to a further drop of 0.2 MPa (Nobel 1983) then the leaves at the top of a 100-m-tall tree will be operating at a ψ_l below -1.81 MPa. This, then, can be taken as a possible limit to a leaf being able to achieve net positive photosynthesis and agrees well with observations (e.g., Hinkley *et al.* 1975). If soil water potential at the root surfaces is changed then the predicted maximum height of trees will be as given in Figure 5.1.

If it is assumed that nighttime leaf water potentials are the most important for growth, then since no water flow is occurring, soil type and resistance will be unimportant. If this is the case the critical ψ_l will be about -1.11 MPa. The predicted maximum tree heights for this situation are also given in Figure 5.1. It is clear that since the critical ψ_l is higher for leaf expansion, the effect of reducing the soil water potential is greater.

This approach to the problem of predicting the maximum height of trees allows an evaluation of the potential importance of soil water in determining tree height. The critical ψ_l values are the most problematical aspects of this approach. This is particularly the case when one considers the large absolute differences between species and the differences in their abilities to acclimate in response to low soil water availability. Rates of cell expansion have been found to be reduced by any fall in ψ_l below -0.2 MPa, with complete cessation of growth occurring between -0.4 MPa and -1.2 MPa, depending on species and acclimation (Jones 1986). These values are somewhat higher than those that cause critical reductions in photosynthesis. However, since the growth of each leaf need occur only once, whereas photosynthesis must continue over much longer periods (in most plants), short-term favorable periods for soil water availability might allow rapid leaf expansion, and hence subsequent photosynthesis under less favorable conditions.

Leaf water potential is linearly related to the rate of evaporation of water from

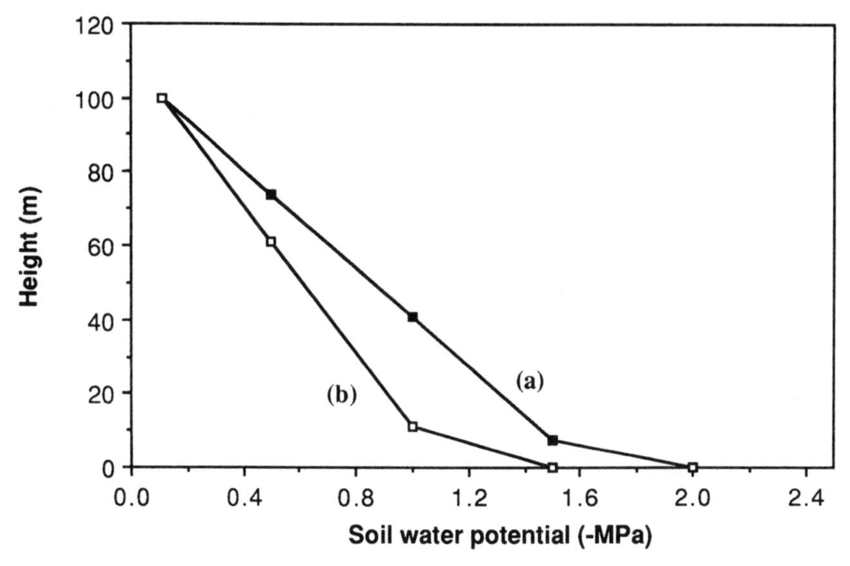

Fig. 5.1. Predicted maximum tree heights at different growing-season root-surface soil water potentials. a. Maximum height if photosynthesis is limiting. b. Maximum height if leaf growth is limiting.

a leaf (Givnish 1986) and at a given stomatal conductance is consequently linearly related to δ_l. Thus even though it would appear that nighttime consequences of ψ_l for leaf expansion may be critical in dry soils, high daytime δ_l values may reduce ψ_l values to those critical for photosynthesis.

The effect of transpiration on ψ_l is given by Jones (1986) as

$$\psi_l = \psi_s - E\,R_{sp} \tag{5.1}$$

where E is the rate of transpiration ($\text{mol}_{(H_2O)}\ \text{m}^{-2}\ \text{s}^{-1}$) and R_{sp} is the sum of all the soil and plant resistances to water flow ($\text{m}^2\ \text{s}^1\ \text{mol}_{(H_2O)}{}^{-1}$).

For a leaf at the top of a 100-m-tall tree transpiring from a soil with a water potential of -0.11 MPa, the critical ψ_l for photosynthesis was estimated as -1.81 MPa due to changes in water potential necessary to pull water up the trunk. This can be incorporated into Equation 5.1 to give

$$\psi_l = \psi_s - E\,R_{sp} - 0.015\,H \tag{5.2}$$

where H is the height (m) of the leaf from where the soil water enters the root. If we assume that the maximum stomatal conductance of the leaves at the top of the tree is 2.75 mm s^{-1} (Jones 1986), then at 20°C this is equivalent to a conductance to water of 0.113 $\text{mol}_{(H_2O)}\ \text{m}^{-2}\ \text{s}^{-1}$. Transpiration can be modelled as (Farquhar and Sharkey 1982)

$$E = g_w (e_i - e_a)/(P - (e_i + e_a)/2) \tag{5.3}$$

where P is atmospheric pressure (Pa), g_w is total conductance (leaf and boundary) to water ($mol_{(H_2O)}$ m^{-2} s^{-1}), e_i is leaf air-space vapor pressure (Pa), and e_a is ambient vapor pressure (Pa), so that $\delta_l = e_i - e_a$. If it is assumed that P is 0.1 MPa, and that the air in the leaf is at 100% relative humidity, then we need only to estimate R_{sp}. Using the value given by Givnish (1986), the responses given in Figure 5.2 are obtained.

If one compares Figures 5.1 and 5.2, it appears that leaf expansion during the night may be critical for the limitation of tree height. However, this will depend on the prevalent temperatures and soil water contents at the critical periods. Leaf–to–air-vapor-pressure deficits probably have little influence on tree height. Even if the root conductance to water were half the value assumed in Figure 5.2, the maximum height at a δ_l of 2.11 kPa would still be 82.4 m if maximum stomatal conductance were maintained. Maximum heights calculated thus may be minima since E would fall as the stomata close in response to an increase in δ_l. However, it is possible that this may be counteracted by the subsequent stomatal limitation of photosynthesis.

The accuracy of the predictions presented here depends on how realistic the critical values of ψ_l for photosynthesis and growth are. With respect to the former the figure of -1.81 MPa corresponds well with the start of rapid abscisic acid accumulation in low-water–conditioned *Gossypium hirsutum* (Bradford and Hsiao 1982). This figure also agrees well with other studies of the direct effects of low ψ_l on photosynthesis (Bradford and Hsiao 1982). Hence it is at least theoretically possible that the absolute upper limit of tree height is caused by a reduced supply of water to the upper leaves. It should be reiterated that this phenomenon occurs to the same extent independently of xylem or tracheid vessel diameter. Consequently both gymnosperms and angiosperms can reach heights of up to 100 m (Zimmermann 1971).

As vessel diameter increases, resistance falls and peak velocity of water flow increases, though the risk of embolism is increased (Zimmermann 1978). Thus species with wider vessels might be expected to reach greater heights. However, larger vessel diameters will not necessarily increase the potential height of a tree if the leaf area increases in proportion; the demand for water increases with the rate of its supply. This is seen to occur.

There is a close correlation between sapwood area and leaf area within species, but the ratio varies between species (e.g., Kaufmann and Troendle 1981; Waring *et al.* 1982). It seems likely that between-species variation in the ratio of sapwood area to leaf area might be explained, at least in part, by variation in vessel diameter. This idea is supported by the results of Whitehead *et al.* (1984), who found that the difference in this ratio of sapwood area to leaf area between *Picea sitchensis* and *Pinus contorta* could be explained by the difference in the respective sapwood permeabilities. Another cause of variation is the amount of

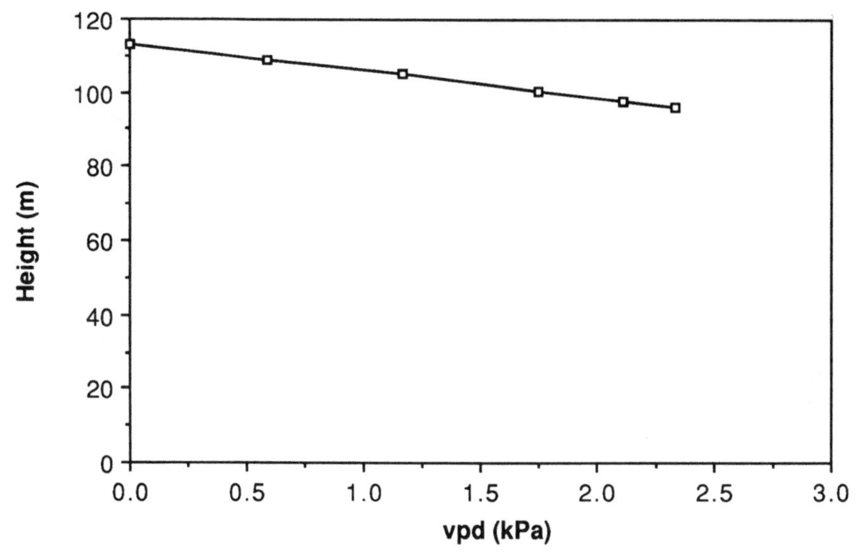

Fig. 5.2. Predicted maximum height of trees at different leaf–to–air-vapor-pressure deficits (vpd) if the critical ψ_l is -1.81 MPa. Both air and leaf temperature are assumed to be 20°C, soil water potential is -0.11 MPa, leaf conductance to water is 0.113 mol m^{-2} s^{-1}, boundary-layer conductance is 0.5 $mol_{(H_2O)}$ m^{-2} s^{-1} (total = 0.092 $mol_{(H_2O)}$ m^{-2} s^{-1}), and P is 0.1 MPa.

sapwood used for storage of carbohydrates and nutrients and that used for water transport. Sapwood area and pore diameter affect the amount of water supplied to the canopy per unit time and hence restrict the leaf area and weight that can be supported (Waring and Schlesinger 1985).

If the variation in vessel diameter, and thus in efficiency of water transport, results in the between-species variation in the sapwood area/leaf area ratio, it would appear that factors other than just water supply might have to be considered to explain the variation in maximum height between species. To achieve this an integrated approach is required which simultaneously considers all of the gains and losses associated with height growth.

The Use of a Model of Tree Growth Incorporating Support Tissue Maintenance to Predict Maximum Height

Although not related to height alone, the size of trees might be limited by the metabolic costs associated with support and storage tissues, and these costs may change in response to a changing climate (Ryan 1990a). Implicit in gap models of forests (Shugart 1984) is the assumption that as a tree grows under favorable conditions, its size eventually becomes limited by support costs. To ascertain the validity of this it is necessary to know the annual carbon dynamics of the tree as a whole and how they vary as the tree grows. Waring and Schlesinger (1985) argued that the energy costs for construction and maintenance of wood limit total growth, and hence by implication, limit maximum height. However, it is difficult to measure maintenance respiration in the stems, branches, and roots of large living trees.

It is logical to assume that the two most important factors determining the amount of maintenance respiration in a tree are the temperature and the dry weight of living tissue. To estimate maintenance respiration rates in a tree, *Pinus contorta* (lodgepole pine) was chosen as an example, although some of the parameters for regressions had to be taken from *P. rigida*. From the regression equations of Whittaker and Woodwell (1968) it can be calculated that a mature *P. rigida* tree with a diameter at breast height (dbh) of 120 cm will have a stem wood dry weight of 4642 kg to support, of which 2285 kg is sapwood (from *P. contorta*: Pearson *et al.* 1984). If it is assumed that 5% of the sapwood is living (Ryan 1990b), then living cells in the stem sapwood amount to some 114 kg (dry weight). Branch dry weight will be 2601 kg (Whittaker and Woodwell 1968), about 56% of that in the stem wood. Projected leaf area is calculated from sapwood area as 749 m^2 (Waring *et al.* 1982), which at a specific leaf area (area of leaf per unit weight) of 3.17 $m^2\ kg^{-1}$ gives a leaf weight of 236 kg (Kaufmann and Troendle 1981). If the root/shoot ratio is taken as 0.174 (*P. rigida*: Whittaker and Woodwell 1968), then the root dry weight will be 1205 kg. If we assume that the same proportions of root and branch are alive as in the stem, then total nonleaf live biomass will be 208 kg. This value can then be used to calculate maintenance respiration if the temperature is known (Ryan 1990b). For this example it is 0.177 kg C d^{-1} at 20°C.

A model was written which grew a tree such as this one by using physiology-based equations. Height was calculated by using the following relationship: height (cm) $= 137 + 39.452(\text{dbh}) - 0.039452(\text{dbh})^2$, with dbh in cm (after Shugart 1984). Thus a tree with a dbh of 120 cm would have a height of 43 m. Maintenance respiration was calculated as being proportional to temperature and the estimated amount of nonleaf live tissue; leaf maintenance respiration was included in the calculation of daily photosynthesis. The physiological equations and environmental parameter calculations were based on the Forest-BGC model of Running and Coughlan (1988). Stem dbh was incremented annually from carbon fixed daily, by allometry. A Foret-type shell (Shugart 1984) was built around the daily physiological model to follow the growth of a single tree in one 100-m^2 plot.

The physiological equations used in this model did not originally include an influence of height on leaf water potential. However, it was introduced by reducing the daily maximum canopy water potential by 0.015 MPa per meter of height. Since photosynthesis is calculated for the entire crown at once, the height used for this function was calculated as the point halfway between the top and the bottom of the crown. This would probably underestimate the effect of height

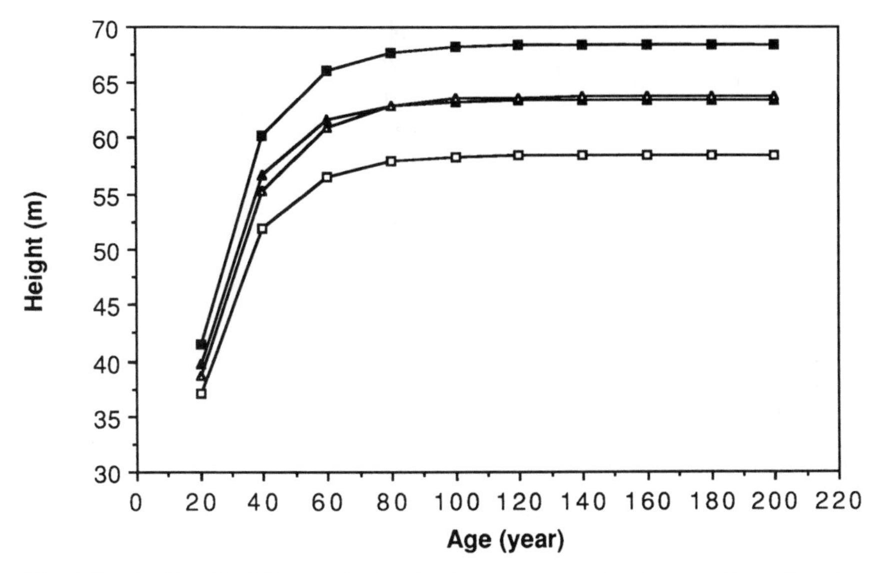

Fig. 5.3. Predicted height growth of a pine tree; age zero when tree was 5 cm diameter at breast height (dbh) and 3 m tall, for all simulations. Open squares: maintenance respiration and height-related leaf water potential (ψ_l) effects included; closed squares: neither of these effects included; open triangles: without maintenance respiration, but height-related ψ_l effects included; closed triangles: maintenance respiration, but no height-related ψ_l effects.

since the higher, better-lit, leaves are more important for carbon gain. The maintenance respiration coefficient used was 0.000 204 kg C (kg live wood)$^{-1}$ d^{-1} at 0°C, with a Q_{10} of 2.04 (Ryan 1990b). This coefficient may well vary between species, depending on factors such as the amount of nitrogen in the live cells (Ryan 1990a).

The model was used to estimate the relative importance of maintenance respiration and leaf water potential in determining the maximum size and height of the simulated pine tree. The tree was grown for 200 years, from an initial diameter of 5 cm and height of 3 m, under an idealized climate in which the conditions every day were such that soil water caused the daily maximum leaf water potential to be −0.5 MPa before the influence of height. The mean temperature was set at 16.5°C. Short-wave irradiance was 30 000 kJ m^{-2} d^{-1}, and relative humidity was 99%. The combined effects of reduced leaf water potential with height, and maintenance respiration, reduce the predicted maximum height of the tree from 68.38 m to 58.50 m (Fig. 5.3). The predicted height of the tree without these two limitations is constrained because the size of the plot is finite and hence limits the effectiveness of increases in leaf area. This situation is mirrored in real forests in which the leaf area of an individual tree remains almost constant once canopy closure has occurred, unless surrounding trees die (Waring and Schlesinger

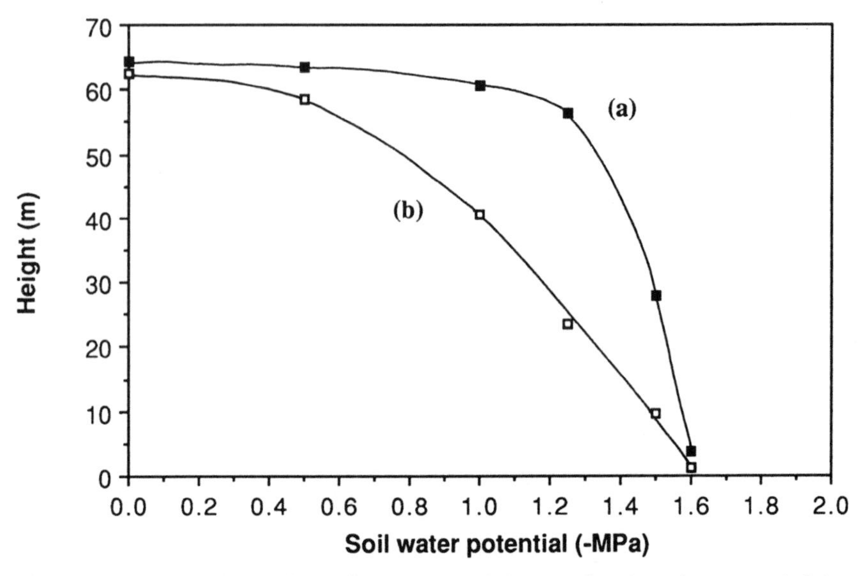

Fig. 5.4. Effect of root-surface soil water potential on predicted maximum tree height. Without (a) and with (b) an effect of height on leaf water potential included in the model.

1985). Doubling the size of the plot enables the tree to reach 68.50 m, with the maintenance respiration and height effects included. However, despite this somewhat artificial restriction on the absolute size of the tree, comparison of the relative importance of maintenance respiration and leaf water potential is still valid. This shows that 48% of the relative reduction in predicted maximum height under this 'climate' is due to the effect of height on leaf water potential and 52% is because of maintenance respiration. The effect of leaf water potential might have been even greater if its influence on leaf expansion could have been included.

It was suspected that the soil water status would have an effect on the relative importance of leaf water potential in limiting height. To test this the soil water potential was varied, and the results are presented in Figure 5.4. They show that the effect of predicted height-related leaf water potential costs is greatest at intermediate levels of soil water. At a soil water potential of −1.25 MPa, the effect of height on leaf water potential alone is to reduce the maximum height of the tree by 58%. The effect of soil water potential on predicted maximum tree height as calculated by this method is very similar to that in Figure 5.1. In the model used for the predictions in Figure 5.4, the leaf water potential at stomatal closure was set at −1.6 MPa (Running and Coughlan 1988). The effect of soil water potential would have been greater if soil water conductivity had also been included.

Under warmer climates maintenance respiration may be more important. This was tested by varying the mean temperature (Fig. 5.5). Increased temperature

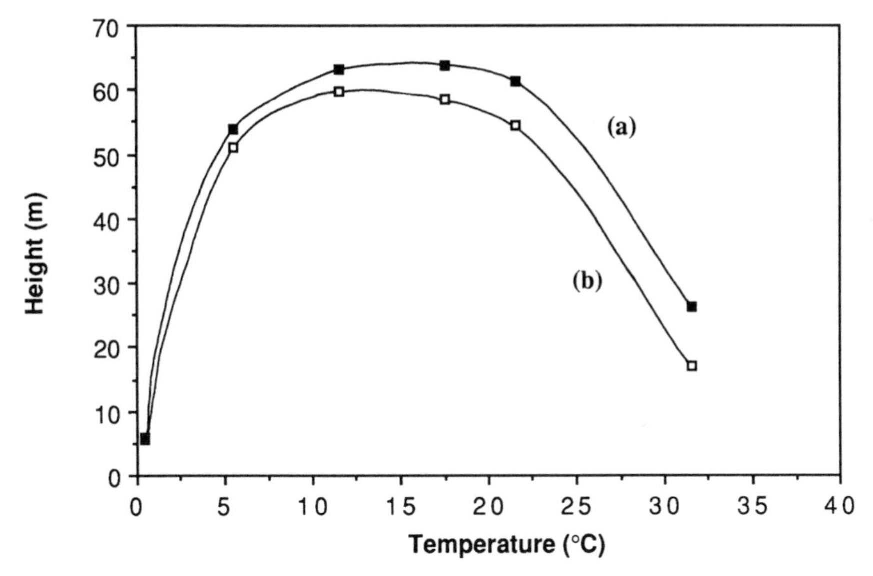

Fig. 5.5. Effect of temperature on predicted maximum tree height without (a) and with (b) nonleaf maintenance respiration included in the model.

did increase the relative effect of maintenance respiration, though above 20°C the change was slight. These predictions suggest that the effect of height on leaf water potential might be of substantial importance to the limitation of tree size and height. It should be stated that the temperature which drives maintenance respiration is that of the sapwood, which is somewhat less variable than the air temperature, and may well be higher than air temperature during the night. It is, however, predictable (Ryan 1990b).

Ryan and Waring (1990) compared the predictions of the Forest-BGC model with field measurements of productivity in three subalpine lodgepole pine forests of different ages. Their version of this model included the effects of maintenance respiration, but not height, and gave predictions close to the measured gross primary productivity (GPP) values in the two younger forests. In the 233-year-old forest, however, the predictions were 36% higher than the values actually found, suggesting that factors not included in Forest-BGC must be operating in the oldest stand. This, together with estimates of actual maintenance respiration rates, indicated that the effect of reduced leaf water potentials with height must be responsible for the reduced GPP of the oldest forest. Increased maintenance respiration rates were initially assumed to be responsible for the reduction in productivity in the oldest stand, but this was discounted since the sapwood volume per unit ground area increased by only 22% between the 40- and the 233-year-old forests, yet the GPP fell by 71%.

Pearson *et al.* (1987) also found that the productivity of lodgepole pine forests fell as they got older, and taller. Direct evidence for the importance of reduced water availability in leaves at the top of trees was found in Scots pine (*Pinus sylvestris*) by Mattson-Djos (1981) and in Douglas fir (*Pseudotsuga menziesii*) by Klein *et al.* (1976). In addition, Waring (1970) found a correlation between the maximum height of Douglas fir along a summer moisture gradient and predawn leaf water potentials in Oregon.

Mechanical Limitations

Niklas (1989) considered that the vertical growth of trees is limited because of mechanical instability or failure in very tall trees. McMahon (1973, 1975) discussed the evidence for the height of trees being limited by elastic buckling. He found a positive correlation between maximum stem diameter and maximum height of trees in the United States. No trees were found to be taller than the height at which buckling would be expected to take place, and he took this to imply that the proportions of trees are limited by the mechanical properties of wood. The trees in fact clustered around a height one-quarter of that expected to produce buckling for a given diameter. The tallest tree for its diameter was still two-thirds of its buckling height. The work of McMahon (1973) suggests that height could be predicted from diameter by using a two-third power of diameter. However, his compilation of data shows a variation in height for trees of, say, a diameter of 100 cm, of between 4.2 m and 46 m. This does not appear to be a good basis for prediction and does not support the contention that the largest, and tallest, trees are limited by buckling considerations.

Conclusions

There is strong evidence, from both models and field data, that tree height is limited both by maintenance respiration and by the largely ignored effect of height on leaf water potential. It seems that at least half of the limitation may be due to the latter. This conclusion has many implications for the modeling of forest productivity.

It seems possible that the variation in the maximum height of different species might be explained by such factors as differences in the proportion of live cells in the stems, branches, and roots and in the live cell constituents causing different respiration coefficients. The role of differences in the efficiency of water supply to the leaves is unknown. Both of these features of trees need further study, the results of which will be required for the mechanistic modelling of forests in response to climate change. The model results presented here suggest that changes in soil water content may be critical.

Acknowledgments

I am grateful to Allen Solomon and the International Institute for Applied Systems Analysis in Laxenburg, Austria, for providing the impetus for this work, and to Hank Shugart for providing facilities enabling it to be completed. I am also indebted to Peter Mitchell for comments and criticisms of an earlier draft.

References

Boyer, J.S., Cavalieri, A.J. and Schulze, E.-D. (1985). Control of the rate of cell enlargement: excision, wall relaxation, and growth-induced water potentials. *Planta,* **163**, 527–43.

Bradford, K.J. and Hsiao, T.C. (1982). Physiological responses to moderate water stress. In *Encyclopedia of Plant Physiology (NS) Volume 12B. Physiological Plant Ecology II: Water Relations and Carbon Assimilation*, ed. O.L. Lange, P.S. Nobel, C.B. Osmond and H. Ziegler, pp. 263–324. Berlin: Springer-Verlag.

Cleland, R.E. (1986). The role of hormones in wall loosening and plant growth. *Australian Journal of Plant Physiology,* **13**, 93–103.

Cosgrove, D.J., van Volkenburgh, E. and Cleland, R.E. (1984). Stress relaxation of cell walls and the yield threshold for growth. Demonstration and measurement by micro-pressure probe and psychrometer techniques. *Planta,* **162**, 46–54.

Farquhar, G.D. and Sharkey, T.D. (1982). Stomatal conductance and photosynthesis. *Annual Review of Plant Physiology,* **33**, 317–45.

Givnish, T.J. (1986). Optimal stomatal conductance, allocation of energy between leaves and roots, and the marginal cost of transpiration. In *On the Economy of Plant Form and Function*, ed. T.J. Givnish, pp. 171–213. Cambridge: Cambridge University Press.

Givnish, T.J. (1988). Adaptation to sun and shade: a whole plant perspective. *Australian Journal of Plant Physiology,* **15**, 63–92.

Graan, T. and Boyer, J.S. (1990). Very high CO_2 partially restores photosynthesis in sunflower at low water potentials. *Planta,* **181**, 378–84.

Hall, A.E. (1982). Mathematical models of plant water loss and plant water relations. In *Encyclopedia of Plant Physiology (NS) Volume 12B. Physiological Plant Ecology II: Water Relations and Carbon Assimilation*, ed. O.L. Lange, P.S. Nobel, C.B. Osmond and H. Ziegler, pp. 231–61. Berlin: Springer-Verlag.

Hinkley, T.M., Schroeder, M.D., Roberts, J.E. and Bruckerhoff, D.N. (1975). Effect of several environmental variables and xylem pressure potential on leaf surface resistance in white oak. *Forest Science,* **21**, 201–11.

Iwasa, Y., Cohen, D. and León, J.A. (1985). Tree height and crown shape, as results of competitive games. *Journal of Theoretical Biology,* **112**, 279–97.

Jones, H.G. (1986). *Plants and Microclimate*. 2nd ed. Cambridge: Cambridge University Press.

Kaufmann, M.R. and Troendle, C.A. (1981). The relationship of leaf area and foliage biomass to sapwood conducting area in four subalpine forest tree species. *Forest Science,* **27**, 477–82.

King, D.A. (1990). The adaptive significance of tree height. *American Naturalist,* **6**, 809–28.

Klein, J.R., Reed, K.L., Waring, R.H. and Stewart, M.L. (1976). Field measurements of transpiration in Douglas-fir. *Journal of Applied Ecology,* **13**, 273–83.

Kramer, P.J. and Kozlowski, T.T. (1979). *Physiology of Woody Plants*. New York: Academic Press.

Lakso, A.N. (1979). Seasonal changes in stomatal response to leaf water potential in apple. *Journal of the American Society for Horticultural Science,* **104**, 58–60.

Mattson-Djos, D. (1981). The use of pressure-bomb and porometer for describing plant water stress in tree seedlings. In *Proceedings of a Nordic Symposium on Vitality and Quality of Nursery Stock*, ed. P. Puttonen, p. 57. Department of Silviculture, University of Helsinki, Finland.

McMahon, T. (1973). Size and shape in biology. *Science,* **179**, 1201–4.

McMahon, T.A. (1975). The mechanical design of trees. *Scientific American,* **233**, 92–102.

Niklas, K.J. (1989). The cellular mechanics of plants. *American Scientist,* **77**, 344–9.

Nobel, P.S. (1983). *Biophysical Plant Physiology and Ecology*. San Francisco: W. H. Freeman.

Pearson, J.A., Fahey, T.J. and Knight, D.H. (1984). Biomass and leaf area in contrasting lodgepole pine forests. *Canadian Journal of Forest Research,* **14**, 259–65.

Pearson, J.A., Knight, D.H. and Fahey, T.J. (1987). Biomass and nutrient accumulation during stand development in Wyoming lodgepole pine forests. *Ecology,* **68**, 1966–73.

Running, S.W. and Coughlan, J.C. (1988). A general model of forest ecosystem processes for regional applications. I. Hydrological balance, canopy gas exchange and primary production processes. *Ecological Modelling,* **42**, 125–54.

Ryan, M.G. (1990a). Effects of climate change on plant respiration. *Ecological Applications,* 1:157–167.

Ryan, M.G. (1990b). Growth and maintenance respiration in stems of *Pinus contorta* and *Picea engelmannii*. *Canadian Journal of Forest Research,* **20**, 48–57.

Ryan, M.G. and Waring, R.H. (1992). Maintenance respiration and stand development in a subalpine lodgepole pine forest. *Ecology,* (in press).

Schulze, E.-D. and Hall, A.E. (1982). Stomatal responses, water loss and CO_2 assimilation rates of plants in contrasting environments. In *Encyclopedia of Plant Physiology (NS) Volume 12B. Physiological Plant Ecology II: Water Relations and Carbon Assimilation*, ed. O.L. Lange, P.S. Nobel, C.B. Osmond and H. Ziegler, pp. 181–230. Berlin: Springer-Verlag.

Shugart, H.H. (1984). *A Theory of Forest Dynamics*. New York: Springer-Verlag.

Taylor, G. and Davies, W.J. (1986). Leaf growth of *Betula* and *Acer* in simulated shadelight. *Oecologia,* **69**, 589–93.

Tyree, M.T. and Jarvis, P.G. (1982). Water in tissues and cells. In *Encyclopedia of Plant Physiology (NS) Volume 12B. Physiological Plant Ecology II: Water Relations and Carbon Assimilation*, ed. O.L. Lange, P.S. Nobel, C.B. Osmond and H. Ziegler, pp. 35–77. Berlin: Springer-Verlag.

Waring, R.H. (1970). Matching species to site. In *Regeneration of Ponderosa Pine*, ed. R.K. Hermann, pp. 54–61. Forest Research Laboratory, Oregon State University, Corvallis.

Waring, R.H., Rogers, J.J. and Swank, W.T. (1981). Water relations and hydrologic cycles. In *Dynamic Properties of Forest Ecosystems*, ed. D.E. Reichle, pp. 205–64. London: Cambridge University Press.

Waring, R.H. and Schlesinger, W.H. (1985). *Forest Ecosystems*. Orlando: Academic Press.

Waring, R.H., Schroeder, P.E. and Oren, R. (1982). Application of the pipe model theory to predict canopy leaf area. *Canadian Journal of Forest Research,* **12**, 556–60.

Whitehead, D., Edwards, W. R. N. and Jarvis, P. G. (1984). Conducting sapwood area, foliage area, and permeability in mature trees of *Picea sitchensis* and *Pinus contorta*. *Canadian Journal of Forest Research*, **14**, 940–7.

Whittaker, R.H. and Woodwell, G.M. (1967). Surface area relations of woody plants and forest communities. *American Journal of Botany,* **54**, 931–9.

Whittaker, R.H. and Woodwell, G.M. (1968). Dimension and production relations of trees and shrubs in the Brookhaven Forest, New York. *Journal of Ecology,* **56**, 1–25.

Williams, C.N. and Biddiscombe, E.F. (1965). Extension growth of grass tillers in the field. *Australian Journal of Agricultural Science,* **16**, 14–22.

Zimmermann, M.H. (1971). Transport in the xylem. In *Trees: Structure and Function*, ed. M.H. Zimmermann and C. L. Brown, pp. 169–220. New York: Springer-Verlag.

Zimmermann, M.H. (1978). Structural requirements for optimal water conduction in tree stems. In *Tropical Trees as Living Systems*, ed. P.B. Tomlinson and M.H. Zimmermann, pp. 517–32. Cambridge: Cambridge University Press.

6

Uncertainties in the Terrestrial Carbon Cycle

W.M. Post

Introduction

Over the past several decades, significant progress has been made in measuring and understanding the global carbon cycle and in developing methods for projecting future changes in atmospheric CO_2 concentration. During this time, a natural starting point was to check the balance sheet that accounts for all carbon as it exchanged between the major global carbon reservoirs. While it is possible to achieve a balance for a single instant in time (for example, the year 1980: see Detweiler and Hall 1988; Houghton *et al.* 1987), it is not possible with current information to balance carbon fluxes for decade-or-longer time periods. The inability to account for all carbon exchanges indicates an insufficient knowledge of global carbon cycle processes. In this chapter, I outline the scale of the discrepancies involved and offer hypotheses concerning previously underappreciated carbon fluxes that suggest new research directions. These hypotheses postulate global vegetation change at several time scales as a plausible reason for our inability to "balance" the global carbon cycle over long time periods.

Overview of Carbon Cycle Fluxes

Atmosphere

Fossil fuel burning is clearly one major cause of the recent atmospheric CO_2 increase. This is suggested by the fact that observed increases in atmospheric CO_2 parallel the cumulative input of fossil fuel CO_2 to the atmosphere (Fig. 6.1). The amount of fossil fuel burned has increased at a rate of 4.3% per year since the beginning of the Industrial Revolution, except for brief periods during the Great Depression and the world wars. The amount of carbon contributed to the

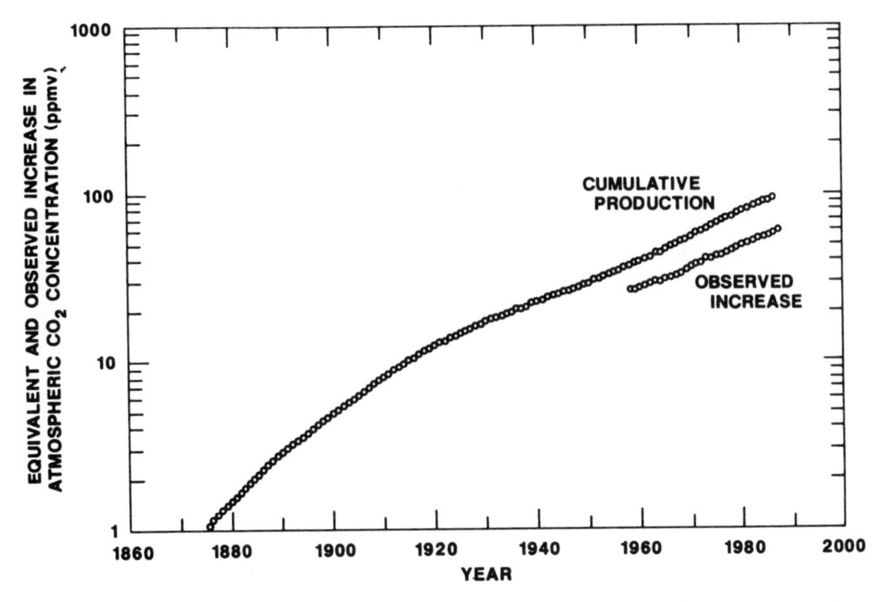

Fig. 6.1. The cumulative production of CO_2 since 1860 expressed as the equivalent atmospheric concentration (1 Pg C = 0.471 ppmv CO_2) compared with the observed increase in the mean annual concentration. The 1860 atmospheric concentration was assumed to be 290 ppmv.

atmosphere from this source is now about 5.5 Pg C · yr^{-1} (1 Pg = 10^{15} g). Fossil fuel burning represents a substantial flow of carbon in the form of CO_2 to the atmosphere, annually accounting for approximately 0.75% of the current atmospheric CO_2 amount of 737 Pg C.

Since accurate and regular measurements of atmospheric CO_2 began in 1958 at Mauna Loa, Hawaii, the atmospheric concentration has grown from 315 ppmv to 349 ppmv in 1987, an increase of 34 ppmv. (See Figure 2.1; Keeling 1988.) Over the entire period from 1860 to 1987, enough fossil fuel carbon was released into the atmosphere to raise the atmospheric concentration of CO_2 by 92 ppmv. The observed increase, assuming an 1860 atmospheric concentration of 290 ppmv (Oeschger et al. 1985), is 59 ppmv. The difference between the two is largely due to net uptake of fossil fuel CO_2 by other reservoirs in the global carbon cycle. The net rise in atmospheric CO_2 concentration is about 60% of the fossil fuel flux. But there were surely additional inputs from the biosphere to the atmosphere, so there had to be correspondingly larger removals elsewhere.

The annual fossil fuel flux of CO_2 to the atmosphere is quite small compared with annual fluxes of CO_2 between the atmosphere and terrestrial ecosystems and the world's oceans. The net primary production of the terrestrial biosphere is approximately 60 Pg · yr^{-1}, almost balanced, in all probability, by a yearly equivalent amount of decomposition from undisturbed ecosystems (Olson *et al.*

1983) plus an additional net flux of CO_2 from the conversion of natural vegetation to managed use on the order of 0.9–2.9 Pg · yr^{-1} (Houghton *et al.* 1987). Uptake by oceans is approximately 93 Pg C · yr^{-1}, primarily in high-latitude regions, and release by oceans is 90 Pg · yr^{-1}, primarily in tropical and subtropical latitudes (Peng *et al.* 1983; Siegenthaler and Oeschger 1987). Thus, while the cumulative effect of fossil fuel burning is a significant perturbation of the atmosphere, we cannot determine the effect of fossil fuel burning on atmospheric CO_2 concentrations without understanding the effect of the other large fluxes on atmospheric CO_2 levels.

The fluxes between the major carbon reservoirs are continually changing because of human activities and the increasing atmospheric CO_2 levels. Estimates presented here are valid for the early 1980s and cannot be viewed simply as components of a budget of the global carbon cycle at equilibrium (Peng 1986; Detweiler and Hall 1988). The global carbon cycle is responding to strong transient forcings that make such a depiction invalid except for perhaps a single moment in time. Equilibrium depictions will be increasingly invalid if global climatic change affects the processes that regulate the amount of CO_2 in the atmosphere. Let us examine major components of the global carbon cycle that may be affected by global change.

Ocean

The net flux of CO_2 between the atmosphere and the ocean is determined by the difference between the partial pressure of CO_2 in the atmosphere and the partial pressure of CO_2 in surface seawater. The net flux of CO_2, F, can be expressed by the formula

$$F = E \cdot [\mathrm{pCO_2(air)} - \mathrm{pCO_2(seawater)}] \ , \qquad (6.1)$$

where E is the gas transfer coefficient and pCO_2 is the partial pressure of CO_2. The gas transfer coefficient is a nonlinear function most sensitive to surface wind speed. E is especially nonlinear at high wind speeds where turbulent mixing and air bubble formation occur. The pCO_2 of seawater is a complex function sensitive to salinity, temperature, total CO_2 (i.e., including all carbonate forms), and alkalinity. Total CO_2 is in turn governed by ocean carbonate chemistry and biological activity. As atmospheric CO_2 increases and all other factors remain constant, there is a tendency for the oceans to take up CO_2 at an increasing rate. This rate increase has been confirmed for the time period of 1860 to present by direct measurements of ocean chemistry and modeling studies utilizing these data (Broecker and Peng 1982; Siegenthaler and Oeschger 1987).

Predicted global climate change makes it unlikely that factors other than atmospheric pCO_2 will remain constant in the future. In particular, dramatic changes in CO_2 uptake by the oceans are likely to result from changes in wind patterns over high-latitude oceans, where most CO_2 is taken up, and from changes in general ocean circulation. Ocean circulation changes may have been responsible for a sudden increase of 80 ppmv in the atmospheric CO_2 concentration at about 10 000 years ago, while climate was warming from glacial conditions to interglacial conditions (Broecker and Peng 1986; Broecker 1987). The effect of altered wind patterns on ocean CO_2 exchange is a direct effect that can be computed by using the concepts in Equation 6.1 provided a reliable relationship between wind stress and gas exchange can be obtained. The effect of changes in ocean circulation on atmospheric CO_2 involves many processes including salinity gradients, nutrient availability, particulate organic matter fluxes, biological production, carbonate compensation by calcite dissolution in sediments, and others. These relationships are not well known at present (Peng and Broecker 1984; Bolin 1986).

Terrestrial Vegetation

Changes in terrestrial vegetation will be important in determining future concentrations of CO_2. I will first summarize current knowledge of the role of the terrestrial biosphere in the global carbon cycle and then discuss how future global change, both directly by increased atmospheric CO_2 and indirectly through CO_2-induced climate changes, will alter the feedbacks between the terrestrial biosphere and the global carbon cycle.

An early simplification in exploring the effect of the terrestrial biosphere on the global carbon cycle was to assume that in undisturbed or "natural" ecosystems, net primary production (NPP) is matched, on average, by decomposition, resulting in approximation to steady state, or a net ecosystem production (NEP) of 0.0 (Houghton *et al.* 1983, 1985; Detweiler *et al.* 1986) if averaged over large-enough areas. If the simplification was valid, natural ecosystems could be ignored and attention could be focused on ecosystems in transition due to man's activities. The most dramatic activities that result in a net release of CO_2 to the atmosphere include conversion of unmanaged vegetation to agricultural use and to urban areas. Activities that result in a net uptake of CO_2 include abandonment from agricultural use and subsequent plant succession to an undisturbed vegetation, or deliberate afforestation.

Activities that perturb natural ecosystems release CO_2 to the atmosphere at several time-scales. An important example is shifting cultivation in tropical regions, which is responsible for most of the current land-clearing flux of CO_2. Shifting cultivation involves clearing native vegetation and planting agricultural crops for several years. The agricultural plot is then abandoned, resulting in growth of undisturbed secondary vegetation for one to several decades before burning and replanting again. During shifting cultivation much of the ecosystem carbon is released immediately by burning. Remaining aboveground organic material decomposes within 1–5 years. The active fraction of soil organic matter

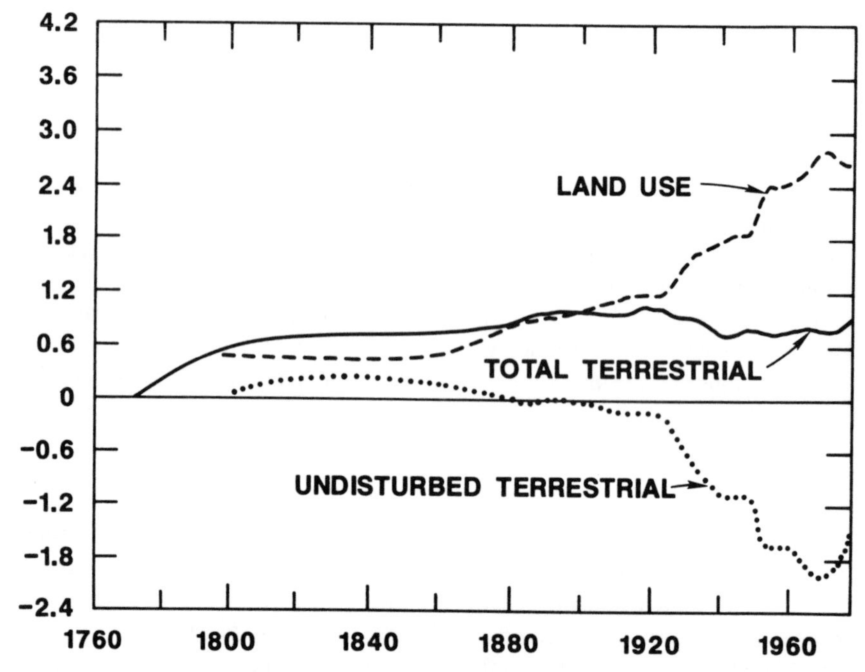

Fig. 6.2. The total terrestrial net carbon exchange with the atmosphere needs to be resolved into at least two parts—net flux due to certain land-use changes and net flux due to all other changes to terrestrial ecosystems, labeled here as "undisturbed terrestrial" response. The total terrestrial and land-use net fluxes may be estimated indirectly by using ocean model–atmosphere record deconvolution and historical reconstruction, respectively. The undisturbed terrestrial net flux is obtained by subtracting land-use flux from total flux. Different deconvolutions and reconstructions will yield different estimates of undisturbed terrestrial net fluxes. The total terrestrial curve (solid line) is from the Siegenthaler and Oeschger (1987) outcrop diffusion model, and the land-use curve (dashed line) is from Houghton *et al.* (1983). The resulting undisturbed terrestrial curve (dotted line) indicates that, to be consistent with these two studies, the undisturbed terrestrial biosphere must have been a small net source of CO_2 before 1900 and an increasingly stronger sink of CO_2 since 1900.

is released more slowly over a 5–50-year time span. Recovery of ecosystem carbon after abandonment is a long process occurring on the order of decades to centuries. Thus, the current net land-use flux rate depends on the previous land use (Chan *et al.* 1980; Houghton *et al.* 1983; Oeschger and Heimann 1983; Richards *et al.* 1983).

Historical reconstruction of the land-use flux of carbon to the atmosphere has been completed using primary data sources by Richards *et al.* (1983) and Houghton *et al.* (1983). Figure 6.2 presents the results of the later analysis of the net terrestrial flux due to land clearing, taking into account the different sources and

Table 6.1. Balance sheet for 1860–1980, showing exchanges between the three major global carbon sources and sinks to the atmosphere.

	Exchange amount (Pg C)
Sources	
Fossil fuel	150–190
Land use	135–228
Sink	
Ocean uptake	78–40
Total predicted increase	207–378
Observed increase	129–172

sinks and their associated time-scales of response. Houghton *et al.* (1983) estimate that over the period of 1860 to 1980, the terrestrial biosphere has been a net source of 135–228 Pg C to the atmosphere.

The amount of carbon from fossil fuel contributed to the atmosphere from 1860 to 1980 is reasonably well known. The ocean uptake of carbon can be estimated from ocean models. Table 6.1 provides a balance sheet of carbon exchanges between the terrestrial biosphere, oceans, and atmosphere. Adding the lower bound estimates of the releases from fossil fuel and land clearing to the atmosphere and subtracting the highest estimate of ocean uptake should approximate the lowest prediction of the total increase of carbon in the atmosphere since 1860. Similarly, adding the highest release estimates and subtracting the lowest uptake estimate should give an upper bound on the expected atmospheric increase since 1860. Calculated in this way, the atmosphere should have gained 207–378 Pg C between 1860 and 1980.

However, the observed atmospheric increase for this period (assuming an 1860 concentration of 290 ppmv) is 129–172 Pg C. It is usually assumed that since the ranges of predicted and observed increases in atmospheric carbon do not even overlap, there must be an error in one or more of the estimates. This assumption may or may not be correct. Since the historic (Neftel *et al.* 1985) and recent (Keeling 1988) atmospheric CO_2 concentrations and the thermodynamics of surface ocean exchange of CO_2 are known with a high degree of certainty, it is widely assumed that the land-use flux estimate is in error and should be reduced (Peng *et al.* 1983; Harrington 1987; Siegenthaler and Oeschger 1987). Recent work also indicates that initial ecosystem carbon estimates used in calculating land-use flux may have been too high (Olson *et al.* 1983; Richards *et al.* 1985). Lowering these estimates may result in lowering the land-use flux estimate. But the latter analyses also suggest that an even larger problem may lie in the simplifying assumption that human influence was negligible on the many ecosystems which were not simply converted to crops or cycles of shifting cultivation (Olson 1981, 1982).

The uncertainties in available land-use estimates are generally regarded as being too great to allow use of these estimates in balancing the global carbon cycle. Initially, the lack of any information concerning the land-use flux of carbon, and later a mistrust of these estimates, led to development of an alternative to historical reconstruction for estimating the flux. This approach infers the possible land-use flux as the difference between fossil fuel emissions, changes in atmospheric CO_2, and the uptake of atmospheric CO_2 specified by an ocean model (Peng *et al.* 1983; Emanuel *et al.* 1984; Peng and Freyer 1986; Peng 1986; Siegenthaler and Oeschger 1987). The inferred land-use carbon flux from such a deconvolution depends on the ocean model assumptions.

In general, land-use estimates from deconvolution do not agree with those derived from historical reconstruction. Figure 6.2 shows the results of one deconvolution study (using the outcrop diffusion model of Siegenthaler and Oeschger 1987) plotted against one reconstructed history from land-use flux estimated from primary data. The historic patterns of the two methods are qualitatively different. The reconstruction estimate shows a more-or-less exponential increase in net carbon released by land use since 1900. This trend is largely due to the increased rate of tropical forest clearing over the last 50 years. The deconvolution estimate indicates that the land-use flux has been relatively constant over the last 160 years, except for slightly higher rates at the beginning of the 20th century. The outcrop diffusion ocean model can allow as much as 153 Pg from terrestrial sources; however, over half of this must be released before 1900 to be consistent with the atmospheric CO_2 record from ice-core air bubbles and recent atmospheric CO_2 measurements.

Enting and Mansbridge (1987) used a linear programming technique to determine if any conventional ocean model (i.e., one that is linear time-invariant in terms of CO_2 uptake) could account for the rate of land-use release estimated by Houghton *et al.* (1983) and be consistent with atmospheric CO_2 concentrations observed at Mauna Loa and measured in ice-core air bubbles. They conclude that the discrepancy between CO_2 concentrations obtained from ice-cores and the best fit that is possible using land-use releases from Houghton *et al.* (1983), allowing complete freedom in the choice of a linear steady-state model of ocean CO_2 uptake, is too large to claim consistency. In the conclusion of their paper, Enting and Mansbridge (1987) mention four possibilities that could explain the discrepancy: (1) systematic bias in the ice-core data with the bias changing sign twice over the last 180 years; (2) incorrect source strength estimates in either the terrestrial biotic component or, less probably, the fossil carbon component; (3) assumption of preindustrial steady state of all components of the global carbon cycle; or (4) occurrence of significant nonlinear effects in ocean uptake of CO_2. There is insufficient evidence at this time to consider the first possibility. The other three possibilities are actively being pursued. Several lines of evidence suggest that complexities of ocean dynamics associated with the global circulation of ocean waters are important in considering CO_2 uptake. Such considerations are leading to development of three-dimensional general ocean circulation models that will be capable of nonlinearities suggested by the fourth possibility. The second and third possibilities are being considered, although much work needs to be done. Resolution of these issues will be important in understanding the operation of the global carbon cycle and in predicting future atmospheric CO_2 concentrations. They are also related to issues concerning global vegetation change and to the core issue of whether the biosphere and its major parts can be "sustained" under rapidly changing cultural and environmental conditions.

Global Vegetation Change and Carbon-Cycle Balance

There is an alternative to the idea that the so-called land-use flux estimates are incorrect. We can accept the Houghton *et al.* (1983) estimates, and others, of the cropping land-use flux and any future refinements or corrections as only a partial component of the entire terrestrial biosphere net flux with the atmosphere. To do this we need to abandon the concept that natural terrestrial ecosystems currently approximate steady-state conditions and perhaps also abandon the concept that these ecosystems were in or near equilibrium before 1860. To visualize the implications of eliminating this equilibrium assumption, let us assume that the Siegenthaler and Oeschger (1987) deconvolution accurately depicts not the land-use flux but the total terrestrial biosphere's net exchange of CO_2 with the atmosphere, of which cropping land use is only a component. The difference between the Houghton *et al.* (1983) land-use estimate and the Siegenthaler and Oeschger (1987) terrestrial biota deconvolution, then, is the net flux of CO_2 between the atmosphere and natural terrestrial ecosystems. This is shown as the dotted line in Figure 6.2.

Given the above assumptions, the undisturbed terrestrial biosphere was a net source of carbon to the atmosphere before 1900. Since then it has been an increasingly stronger sink of atmospheric carbon. Details of the possible natural terrestrial ecosystem net flux will be different depending on the ocean model used for the deconvolution and future refinements or modifications in the magnitude of the land-use flux. The general pattern, however, will remain the same since almost all deconvolutions indicate that the total net terrestrial flux has remained relatively level or has shown a decreasing trend since 1900; crop land-clearing reconstructions will probably always show an increasing flux since 1900 because of the recent accelerated rate of land use in the tropical regions (Richards *et al.* 1983; Emanuel *et al.* 1985a; Houghton *et al.* 1987), even if further analyses reduce the total land-use flux (e.g., Detweiler and Hall 1988).

There are a number of mechanisms that might account for such an undisturbed terrestrial biosphere response. These can be grouped into three general classes: (1) a direct effect of increasing atmospheric CO_2 levels; (2) transient responses due to nonequilibrium conditions; and (3) an indirect effect resulting from climate change associated with changes in CO_2 levels. Each of these classes of mechanisms is considered below.

Direct CO_2 Fertilization

An increase in biomass due to the fertilizing effect of increased CO_2 concentrations is plausible. Rapid changes in photosynthesis and water-use efficiency due to changes in stomatal control under enhanced CO_2 have been observed in laboratory experiments (Rogers *et al.* 1983; Strain and Cure 1985). In addition to these laboratory experiments, some researchers believe that CO_2-enhanced growth can be observed in natural vegetation (La Marche *et al.* 1984). Several global carbon cycle models have incorporated such a fertilizer effect (Bacastow and Keeling 1973; Esser 1987).

There is, however, considerable uncertainty in extrapolating the laboratory experiments to natural ecosystems where many interacting factors such as nutrient and water dynamics, competition, and long-term carbon allocation must be considered. Unequivocal empirical information on growth responses in natural ecosystems is lacking except for results from microcosm experiments in a tundra ecosystem (Billings *et al.* 1983, 1984; Oechel and Strain 1985). In these tundra microcosm experiments, initial positive growth response to CO_2 enrichment disappeared after a relatively short acclimatization period. Thus, while growth enhancements because of CO_2 fertilization at a global scale are possible, they represent only a hypothesis that has not yet been confirmed or rejected. Extrapolation of chamber and microcosm results to a global response is not warranted at this time.

Long-Term Vegetation Transient

The assumption that undisturbed ecosystems were at steady state 200 years ago and have remained at steady state since then is made primarily for convenience in modeling and because of the difficulty in detecting small changes in net ecosystem storage at a global scale (Houghton *et al.* 1985). There is evidence of carbon accumulation in 'mature' systems, not only in wetlands (Armentano *et al.* 1984) and in boreal and tundra ecosystems (Miller 1981; Billings *et al.* 1982), but also in some tropical forests (Sanford *et al.* 1984; Lugo and Brown 1986; Saldarriaga and West 1986).

It can be argued that such fluxes cannot be large enough to be important every year (at least 0.1 Pg C yr^{-1}). The possible changes in net terrestrial carbon storage over time periods of a few hundred years inferred from analyses of ice-core carbon isotope measurement (Gammon *et al.* 1985) are, however, sufficient to invalidate the assumption of steady state in models of the last 200 years (Houghton *et al.* 1985). Even if environmental conditions regulating undisturbed ecosystem dynamics have remained constant over the last 200 years, long-term transient responses resulting from successional processes and species migration could still be causing changes in carbon storage.

Current Climatic Forcing

Global natural vegetation may already be responding to ongoing climate change. There has been a detectable change in climatic data, although it may not be large enough to be attributed to increased levels of CO_2 in the atmosphere. Of particular interest to global vegetation change and to a program to study these changes is the hypothesis that vegetation is already responding to these changes whatever the cause. Over the last century, global average temperatures have risen by 0.5C (Hansen *et al.* 1981). It is not known whether this increase is large enough to account for a substantial portion of the possible natural vegetation response indicated in Figure 6.2. More fundamentally, it is not known whether climate warming is even sufficient to cause a net increase in carbon uptake and storage by various terrestrial ecosystems. Perhaps increased turnover rate of recent detritus, peat, or lake sediments might offset any increase in net primary production.

An ecosystem responds to a shift in climate with an alteration of many processes and feedbacks that operate on many time scales. The potential ecosystem responses to climate changes are presented diagrammatically in Figure 6.3. The balance between decomposition-respiration and primary production determines the net storage and release of carbon to the atmosphere. Climate change affects these two basic processes in different ways and thus they must be considered separately. Decomposition has a secondary effect of releasing nutrients, a process which enhances primary production and strengthens the feedbacks involving primary production. Decomposition-respiration is the fastest-responding process (subannual time response). A warmer climate will generally increase the rate of decomposition-respiration, releasing additional CO_2 to the atmosphere. This positive feedback process may have a large influence on future atmospheric CO_2 levels if ecosystems with large pools of dead organic matter (i.e., peatlands, boreal, and tundra ecosystems) are affected by climatic warming.

Primary production can respond to climate change on a time-scale corresponding to the length of the growing season. Increases in primary production reduce atmospheric CO_2, so enhancement of primary production results in a negative feedback between vegetation and atmospheric CO_2. If drought results from climate change, this negative feedback will be weakened, resulting in less CO_2 being taken up. Periodic climatic events, such as the El Niño–Southern Oscillation events, produce detectable changes in both vegetation uptake of CO_2 and atmospheric CO_2 concentration (Cleveland *et al.* 1983; Bacastow, *et al.* 1985; Enting 1987; Fung *et al.* 1987). D'Arrigo *et al.* (1987) found that for a period of nearly a decade, tree production (measured as tree-ring growth increment) in the boreal zone could be related to the CO_2 drawdown (year maximum − year minimum CO_2 concentration) measured in that region. Remote sensing of production and

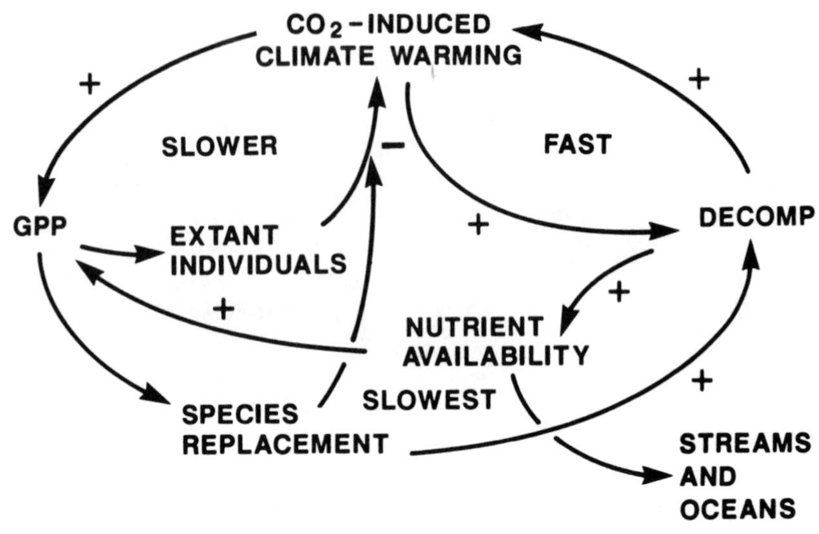

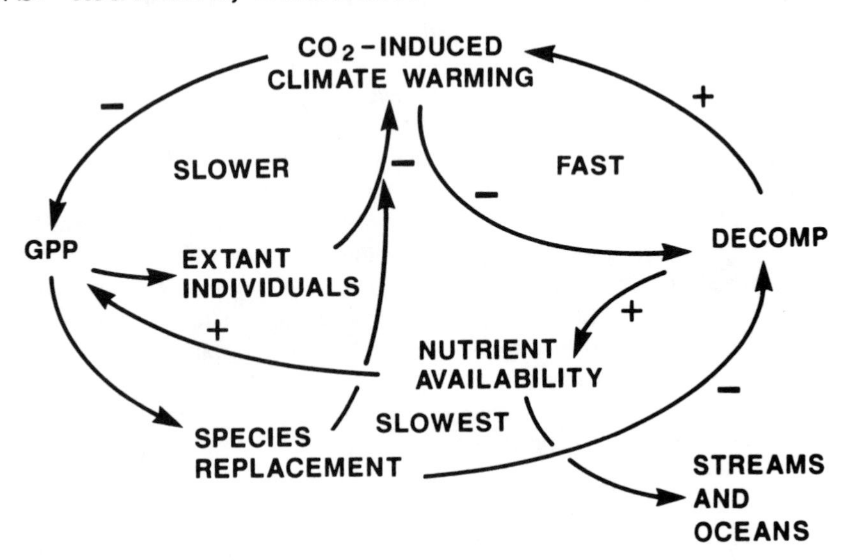

Fig. 6.3. Feedbacks in terrestrial ecosystem responses to CO_2-induced climate change. Arrows with plus signs (+) indicate processes that have positive effects or increase the rates of other processes. Arrows with minus signs (−) indicate processes that have the opposite effects. GPP represents gross primary production.

modeling of interannual changes in atmosphere-biosphere exchanges will provide a basis for evaluating the contribution of land biota to atmospheric CO_2 variations.

Longer-term changes in climate can affect successional processes that result in species replacement. Such species replacements will also alter ecosystem primary productivity directly by replacement of species with higher or lower rates of productivity (Solomon 1986) or indirectly by replacement of species that alter the nutrient cycling of the ecosystem (Pastor and Post 1986, 1988). Since species composition has a strong effect on nutrient cycling, this introduces a feedback at decade-long time-scales that again depends on the presence or absence of drought. If the replacing taxa enhance soil nutrient availability by shedding more easily decomposable litter, productivity is enhanced. This will be the case when soil moisture availability is high, allowing succession of mesic species. When soil moisture is low, succession will result in dominance by drought-tolerant species producing slowly decomposing litter, resulting in lower soil nutrient availability and lower productivity.

Conclusions and Recommendations

From the perspective of global carbon cycling, the challenge of a research program which investigates global vegetation change is to develop research tools to directly determine the response of undisturbed or "natural" ecosystems to atmospheric CO_2 concentrations, climatic patterns, and other global processes. Such processes may cause a transient response of these ecosystems which, in turn, will produce measurable effects on atmospheric CO_2 over subannual-to-decade time-scales.

Our present inability to balance all the global carbon cycle fluxes over the period 1860 to the present may result from overlooking the potential dynamic response of undisturbed terrestrial vegetation to transient environmental conditions. The range of the dynamic response needed over this time period varies from a carbon flux of $+0.5$ to -2 Pg C · yr^{-1} depending on various assumptions. This range represents less than 3% of the terrestrial annual primary production or decomposition-respiration rates. The cumulative changes in carbon storage in undisturbed terrestrial vegetation necessary to accommodate such fluxes fall below our current abilities to detect such storage changes. For the purposes of understanding the global carbon cycle and predicting future atmospheric CO_2 levels, knowledge of global vegetation change in response to CO_2 and climate changes is essential.

The direct response of terrestrial vegetation to increased atmospheric CO_2 may already be taking place. However, attributing the current imbalance in our conceptualization of global carbon cycle fluxes over the period 1860–1980 entirely to CO_2 fertilization requires a 13% increase in NPP resulting from a 17% increase in CO_2 (Esser 1987). Such an enthusiastic conclusion is not warranted by current direct observations or laboratory and field microcosm experiments.

The actual magnitude of the past, current, and future CO_2 fertilization effect in undisturbed ecosystems remains an important unmeasured quantity.

Shifts in global terrestrial carbon storage, because of changes in the balance between annual production and decomposition-respiration induced by small climate shifts, are likely and may help to explain the magnitude of the inferred undisturbed terrestrial vegetation response shown in Figure 6.2. This speculation cannot be confirmed or rejected with present information. Remote sensing measurements of interannual global or regional productivity changes, and carbon isotope and CO_2 measurements from locations influenced by terrestrial vegetation, show promise as useful techniques and need further development. Interpretation and extrapolation of these techniques will require complex atmospheric and vegetation process models.

In addition to the complex interactions and feedbacks in terrestrial ecosystems involved in determining their individual responses to climate change, the spatial distribution of terrestrial ecosystems in relationship to the spatial heterogeneity of climate changes and geological constraints must also be considered (Emanuel *et al.* 1985b; Pastor and Post 1988). Applications of the necessary geographically oriented tools for collecting, overlaying, interpreting, and modeling spatially heterogeneous vegetation, soils, and geological data and processes need to be greatly expanded.

Species composition shifts in the past century are less likely an explanation of the imbalance in our calculations of the global carbon cycle. Species shifts occur at the decade-to-century time-scales of ecosystem succession and species migration. However, when projected CO_2-induced climate changes become significant in the future (MacCracken and Luther 1985), such feedbacks will become significant. It is highly unlikely that positive and negative feedbacks will exactly cancel each other. One or the other will prevail and cause deviations from current trends in atmospheric CO_2.

Acknowledgments

Research supported by the National Science Foundation Ecosystems Studies Program under Interagency Agreement BSR–8315185 with the U.S. Department of Energy, and the U.S. Department of Energy Carbon Dioxide Research Division under contract DE-AC05–84OR21400 with Martin Marietta Energy Systems, Oak Ridge National Laboratory. This paper is Publication No. 3638, Environmental Sciences Division, Oak Ridge National Laboratory.

References

Armentano, T.V., Menges, E.S., Molofsky, J. and Lawler, D.J. (1984). *Carbon exchange of organic soil ecosystems of the world.* Holcomb Research Institute Paper No. 27. Indianapolis, Indiana: Butler University.

Bacastow, R. and Keeling, C. (1973). Atmospheric carbon dioxide and radiocarbon in the natural carbon cycle: II. Changes from A.D. 1700 to 2070 as deduced from a geochemical model. In *Carbon and the Biosphere*, ed. G.M. Woodwell and E.V. Pecan, pp. 86–135. Washington, D.C.: U.S. Atomic Energy Commission.

Bacastow, R.B., Keeling, C.D. and Whorf, T.D. (1985). Seasonal amplitude increase in atmospheric CO_2 concentration at Mauna Loa, Hawaii, 1959–1982. *Journal of Geophysical Research*, **90**, 529–40.

Billings, W.C., Luken, J.O., Mortensen, D.A. and Peterson, K.M. (1982). Arctic tundra: A source or sink for atmospheric carbon dioxide in a changing environment. *Oecologia*, **53**, 7–11.

Billings, W.C., Luken, J.O., Mortensen, D.A. and Peterson, K.M. (1983). Increasing atmospheric carbon dioxide: Possible effects on arctic tundra. *Oecologia*, **58**, 286–9.

Billings, W.C., Peterson, K.M., Luken, J.O. and Mortensen, D.A. (1984). Interaction of increasing atmospheric carbon dioxide and soil nitrogen on the carbon balance of tundra microcosms. *Oecologia*, **65**, 26–9.

Bolin, B. (1986). How much CO_2 will remain in the atmosphere? In *The Greenhouse Effect, Climatic Change, and Ecosystems*, ed. B. Bolin, B.R. Döös, R.A. Warrick and J. Jaeger, pp. 93–155. SCOPE 29. Chichester: John Wiley.

Broecker, W.S. (1987). Unpleasant surprises in the greenhouse? *Nature*, **238**, 123–6.

Broecker, W.S. and Peng, T.-H. (1982). *Tracers in the Sea*. Palisades, New York: Eldigio Press.

Broecker, W.S. and Peng, T.-H. (1986). Carbon cycle: Glacial to interglacial changes in the operation of the global carbon cycle. *Radiocarbon*, **28**, 309–27.

Chan, Y.H., Olson, J.S. and Emanuel, W.R. (1980). Land use and energy scenarios affecting the global carbon cycle. *Environment International*, **4**, 189–206.

Cleveland, W.A., Freeny, A.E. and Graedel, T.E. (1983). The seasonal component of atmospheric CO_2: Information from new approaches to the decomposition of seasonal time series. *Journal of Geophysical Research*, **88**, 10934–46.

D'Arrigo, R.D., Jacoby, G.C. and Fung, I.Y. (1987). Boreal forests and atmosphere-biosphere exchange of carbon dioxide. *Nature*, **329**, 321–3.

Detweiler, R.P. and Hall, C.A.S. (1988). Tropical forests and the global carbon cycle. *Science*, **239**, 42–7.

Detweiler, R.P., Hall, C.A.S. and Bogdanoff, P. (1986). Land use change and carbon exchange in the tropics: II. Estimates for the entire region. *Environmental Management*, **9**, 335–44.

Emanuel, W.R., Fung, I., Killough, G.G., Moore, B. and Peng, T.-H. (1985a). Modeling the global carbon cycle and changes in the atmospheric carbon dioxide levels. In *Atmospheric Carbon Dioxide and the Global Carbon Cycle*, ed. J.R. Trabalka, pp. 141–73. DOE/ER–0239. Washington, D.C.: Carbon Dioxide Research Division, U.S. Department of Energy.

Emanuel, W.R., Killough, G.G., Post, W.M. and Shugart, H.H. (1984). Modeling terrestrial ecosystems in the global carbon cycle with shifts in carbon storage capacity by land-use change. *Ecology*, **65**, 970–83.

Emanuel, W.R., Shugart, H.H. and Stevenson, M.P. (1985b). Climatic change and the broad-scale distribution of terrestrial ecosystem complexes. *Climate Change*, **7**, 29–43.

Enting, I.G. (1987). The interannual variation in the seasonal cycle of carbon dioxide concentration at Mauna Loa. *Journal of Geophysical Research*, **92**, 5497–504.

Enting, I.G. and Mansbridge, J.V. (1987). The incompatibility of ice-core CO_2 data with reconstructions of biotic CO_2 sources. *Tellus*, **39B**, 318–25.

Esser, G. (1987). Sensitivity of global carbon pools and fluxes to human and potential climatic impacts. *Tellus*, **39B**, 245–60.

Fung, I., Prentice, K., Matthews, E., Lerner, J. and Russell, G. (1987). Three-dimensional tracer study of atmospheric CO_2: Response to seasonal exchanges with the terrestrial biosphere. *Journal of Geophysical Research*, **88**, 1281–94.

Gammon, R.H., Sundquist, E.T. and Fraser, P.J. (1985). History of carbon dioxide in the atmosphere. In *Atmospheric Carbon Dioxide and the Global Carbon Cycle*, ed. J.R. Trabalka, pp. 25–62. DOE/ER–0239. Washington, D.C.: Carbon Dioxide Research Division, U.S. Department of Energy.

Hansen, J., Johnson, D., Lacis, A., Lebedeff, S., Lee, P., Rind, D. and Russell, G.J. (1981). Climate impact of increasing atmospheric carbon dioxide. *Science*, **213**, 957–66.

Harrington, J.B. (1987). Climatic change: A review of causes. *Canadian Journal of Forest Research*, **17**, 1313–39.

Houghton, R.A., Boone, R.D., Fruci, J.R., Hobbie, J.E., Melillo, J.M., Palm, C.A., Peterson, B.J., Shaver, G.R., Woodwell, G.M., Moore, B., Skole, D.L. and Myers, N. (1987). The flux of carbon from terrestrial ecosystems to the atmosphere in 1980 due to changes in land use: Geographic distribution of global flux. *Tellus*, **39B**, 122–39.

Houghton, R.A., Hobbie, J.E., Melillo, J.M., Moore, B., Peterson, B.J., Shaver, G.R. and Woodwell, G.M. (1983). Changes in the carbon content of terrestrial biota and soils between 1860 and 1980: Net release of CO_2 to the atmosphere. *Ecological Monographs*, **53**, 235–62.

Houghton, R.A., Schlesinger, W.H., Brown, S. and Richards, J.F. (1985). Carbon dioxide exchange between the atmosphere and terrestrial ecosystems. In *Atmospheric Carbon Dioxide and the Global Carbon Cycle*, ed. J.R. Trabalka, pp. 113–40. DOE/ER–0239. Washington, D.C.: Carbon Dioxide Research Division, U.S. Department of Energy.

Keeling, C.D. (1988). *Atmospheric CO_2 concentrations—Mauna Loa, Hawaii 1958–1987*. NDP–001/R1. Oak Ridge, Tennessee: Carbon Dioxide Information and Analysis Center, Oak Ridge National Laboratory.

La Marche, V.C. Jr., Graybill, D.A., Fritts, H.C. and Rose, M.R. (1984). Increasing atmospheric carbon dioxide: Tree-ring evidence for growth enhancement in natural vegetation. *Science*, **225**, 1019–21.

Lugo, A.E. and Brown, S. (1986). Steady state terrestrial ecosystems and the global carbon cycle. *Vegetatio*, **68**, 83–90.

MacCracken, M.C. and Luther, F.M. ed. (1985). *The Potential Climatic Effects of Increasing Carbon Dioxide*. DOE/ER–0237. Washington, D.C.: U.S. Department of Energy.

Miller, P.C. (1981). *Carbon Balance in Northern Ecosystems and the Potential Effect of Carbon Dioxide Induced Climate Change*. Washington, D.C.: U.S. Department of Energy.

Neftel, A., Moore, E., Oeschger, H. and Stauffer, B. (1985). Evidence from polar ice cores for the increase in atmospheric CO_2 in the past two centuries. *Nature*, **315**, 45–7.

Oechel, W. and Strain, B.R. (1985). Native species responses to increased carbon dioxide concentration. In *Direct Effects of Increasing Carbon Dioxide on Vegetation*, ed. B.R. Strain and J.D. Cure, pp. 117–54. DOE/ER–0238. Washington, D.C.: U.S. Department of Energy.

Oeschger, H. and Heimann, M. (1983). Uncertainties of predictions of future atmospheric CO_2 concentrations. *Journal of Geophysical Research*, **88**, 1258–62.

Oeschger, H., Stauffer, B., Finkel, R. and Langway, C.C. Jr. (1985). Variations of the CO_2 concentration of occluded air and of anions and dust in polar ice cores. In *The Carbon Cycle and Atmospheric CO_2: Natural Variations from Archean to Present*, ed. E.T. Sundquist and W.S. Broecker, pp. 132–42. Monographs of the American Geophysical Union **32**.

Olson, J.S. (1981). Carbon balance in relation to fire regimes. In *Fire Regimes and Ecosystem Properties*, ed. H.A. Mooney, T.M. Bonnicksen, N.L. Christensen, J.E. Lotan and W.A. Reiners, pp. 327–78. GTR WO–26. Washington, D.C.: Forest Service, U.S. Department of Agriculture.

Olson, J.S. (1982). Earth's vegetation and carbon dioxide. In *Carbon Dioxide Review: 1982*, ed. W. Clark, pp. 388–98. New York: Oxford University Press.

Olson, J.S., Watts, J.A. and Allison, L.J. (1983). *Carbon in Live Vegetation of Major World Ecosystems*. ORNL–5862. Oak Ridge, Tennessee: Oak Ridge National Laboratory.

Pastor, J. and Post, W.M. (1986). Influence of climate, soil moisture, and succession on forest carbon and nitrogen cycles. *Biogeochemistry*, **2**, 3–27.

Pastor, J. and Post, W.M. (1988). Response of northern forest to CO_2-induced climate change. *Nature*, **334**, 55–9.

Peng, T.-H. (1986). Land use change and carbon exchange in the tropics: II. Estimates for the entire region: Comment. *Environmental Management*, **10**, 573–5.

Peng, T.-H. and Broecker, W.S. (1984). Ocean life cycles and the atmospheric CO_2 content. *Journal of Geophysical Research*, **89**, 8170–80.

Peng, T.-H., Broecker, W.S., Freyer, H.D. and Trumbore, S. (1983). A deconvolution of the tree ring-based $\delta^{13}C$ record. *Journal of Geophysical Research*, **88**, 3609–20.

Peng, T.-H. and Freyer, H.D. (1986). Revised estimates of atmospheric CO_2 variations based on the tree-ring $\delta^{13}C$ record. In *The Changing Carbon Cycle: A Global Analysis*, ed. J.R. Trabalka and D.E. Reichle, pp. 151–9. New York: Springer-Verlag.

Richards, J.F., Haynes, E.S. and Hagen, J.R. (1985). Changes in the land history and human productivity in Northern India, 1870–1970. *Agricultural History*, **59**, 523–48.

Richards, J.P., Olson, J.S. and Rotty, R.M. (1983). *Development of a Data Base for Carbon Dioxide Releases Resulting from Conversion of Land to Agricultural Uses*. ORAU/IEA–82–10(M), ORNL-TM–8801. Oak Ridge, Tennessee: Oak Ridge National Laboratory.

Rogers, H.H., Bingham, G.E., Cure, J.D., Smith, J.M. and Surano, K.A. (1983). Responses of selected plant species to elevated carbon dioxide in the field. *Journal of Environmental Quality*, **12**, 42–4.

Saldarriaga, J.G. and West, D.C. (1986). Holocene fires in the northern Amazon basin. *Quaternary Research*, **26**, 358–66.

Sanford, R.L. Jr., Saldarriaga, J.G., Clark, K.E., Uhl, C. and Herrera, R. (1984). Amazon rain forest fires. *Science*, **227**, 53–5.

Siegenthaler, U. and Oeschger, H. (1987). Biospheric CO_2 emissions during the past 200 years reconstructed by deconvolution of ice core data. *Tellus*, **39B**, 140–54.

Solomon, A.M. (1986). Transient response of forest to CO_2-induced climate change: Simulation modeling experiments in eastern North America. *Oecologia*, **68**, 567–79.

Strain, B.R. and Cure, J.D. ed. (1985). *Direct Effects of Increasing Carbon Dioxide on Vegetation*. DOE/ER–0238. Washington, D.C.: U.S. Department of Energy.

7

Coupling of the Atmosphere with Vegetation

Philippe Martin

Introduction

More than two-thirds of the Earth is covered with oceans. Nonetheless, as pointed out by Verstraete and Dickinson (1986), continental surfaces provide much of the spatial and temporal variability that makes the weather and climate. To a large extent, vegetation determines the physical characteristics of ice-free continental surfaces, and hence key climatic parameters such as albedo, surface energy fluxes, and so on (Shukla and Mintz 1982; Dickinson 1984; Dickinson and Hanson 1984; Wilson *et al.* 1987). Conversely, there exists a strong connection between vegetation and climate (e.g., Walter 1973; Woodward 1987).

Locally, the most perceivable effect of vegetation is on surface hydrology, since the vegetation controls transpiration, determines interception losses, affects infiltration in the soil, and thus greatly influences runoff (Hall 1971; Rutter 1975). With regard to transpiration, which is vaporization at the leaf surface of water extracted from the soil by the plant, the physiological responses of the vegetation and its physical characteristics determine the partitioning between the sensible and latent heat at the surface of the Earth. This, in turn, affects atmospheric motion and the water balance. Hence, the determination of the control by vegetation of the evaporative flux is of interest to atmospheric scientists, hydrologists, plant physiologists, and ecologists alike.

To explore atmosphere/vegetation interactions, a formal (if not arbitrary) formulation of the coupling between the atmosphere and vegetation would be useful. Such a pedagogical tool was devised by McNaughton and Jarvis (1983). Indeed, MacNaughton and Jarvis derived an expression for the elasticity of evaporation with respect to canopy conductance, i.e., the fractional change in the evaporative flux from a canopy, dE/E,[1] caused by a fractional change in canopy conductance,

[1]A list of the variables, with their definitions and units, may found in Table 7.1.

Table 7.1. Variable list

Variable	Definition	Units
c_p	Molar specific heat of dry air at constant pressure	$J\ mol^{-1}\ K^{-1}$
D	Saturation vapor pressure deficit of the ambient air	Pa
E	Flux of evaporated water	$mol\ m^{-2}\ s^{-1}$
E_0	Flux of evaporated water calculated without radiative coupling	$mol\ m^{-2}\ s^{-1}$
E_{gr}	Flux of evaporated water calculated with radiative coupling	$mol\ m^{-2}\ s^{-1}$
E_{eq}	"Equilibrium evaporation rate"	$mol\ m^{-2}\ s^{-1}$
$\tilde{E}_{eq}$	Recalculated "equilibrium evaporation rate"	$mol\ m^{-2}\ s^{-1}$
E_{imp}	"Imposed transpiration rate"	$mol\ m^{-2}\ s^{-1}$
g_b	Aerodynamic boundary-layer conductance	$mol\ m^{-2}\ s^{-1}$
g_c	Canopy conductance	$mol\ m^{-2}\ s^{-1}$
g_r	Radiative heat transfer "conductance"	$mol\ m^{-2}\ s^{-1}$
G	Ground heat flux	$W\ m^{-2}$
H	Sensible heat flux	$W\ m^{-2}$
R_n	Net radiation	$W\ m^{-2}$
R_{ni}	Isothermal net radiation	$W\ m^{-2}$
s	Derivative of the saturation vapor pressure with respect to temperature	$Pa\ K^{-1}$
T_a	Air temperature	K
T_c	Canopy temperature	K
γ	Psychrometric constant	$Pa\ K^{-1}$
ε	Ratio of s over γ	—
λ	Molar latent heat of vaporization	$J\ mol^{-1}$
Λ	Leaf area index	$m^2\ m^{-2}$
Ω_c	Decoupling factor without radiative coupling	—
$\tilde{\Omega}_c$	Decoupling factor with radiative coupling	—
σ	Stefan-Boltzmann constant	$W\ m^{-2}\ K^{-4}$

dg_c/g_c, to analyze the coupling between vegetation and the atmosphere. Under the assumption that leaf net radiation remains unchanged, it was concluded that one cannot expect a fractional change in stomatal resistance to cause a proportional change in leaf or canopy transpiration, especially for vegetation with low aerodynamic roughness. However, as stomata close, transpiration decreases, while the temperature of sunlit leaves and the associated outgoing long-wave radiation from the leaf increase, and thus transpiration and leaf net radiation do change.

The goals of this chapter are to present the pedagogical tool developed by Jarvis and McNaughton; to examine the assumptions upon which it rests, considering in particular the importance of the stomatal feedback which was (consciously) left out (1986:40); to suggest an alternative derivation for the elasticity of evaporation to conductance; and to discuss the theoretical and practical implications in terms of the role of vegetation and the estimation of latent heat flux.

Evaporation from a Canopy

As pointed out by Jones (1983:187), there are two approaches to taking the temperature gradient between the air and the leaf into account when calculating the evaporation from a canopy. One can either set up an iterative computing procedure (e.g., Gates and Papain 1971) or utilize the concept of "isothermal net radiation" (Monteith 1973:163). Since the objective is to derive an analytical solution, the isothermal net radiation approach is adopted here to consider the effects of small temperature differences on the control of transpiration by vegetation.

In latent heat flux calculations, a canopy can be modeled as a "big leaf". The big-leaf approximation is standard in micrometeorological models (e.g., Rosenberg *et al.* 1983), boundary-layer schemes (e.g., Stewart and de Bruin 1985), and even land-surface parameterizations for atmospheric general circulation models (e.g., Deardorff 1978).

The big-leaf approximation rests on the assumption that the canopy air is well-mixed and it ignores radiative transfers within the canopy. Despite temperature differences within the canopy of up to 12°C (Miller 1971) and great heterogeneity among leaf resistances (Halldin and Lindroth 1986), this approach still yields reliable estimates (e.g., Calder 1977; Slabbers 1977).

The Energy Balance

Assuming that leaves have a negligible heat capacity, the energy balance of a big leaf can be written as

$$R_n = S_n + L_a - \varepsilon_c \sigma T_c^4 = H + \lambda E + G + M \tag{7.1}$$

where R_n is the net radiation per unit area of big leaf, in Wm^{-2}; S_n is the net flux of solar radiation received per unit area, in Wm^{-2}; L_a is the downward flux of long-wave radiation per unit area, in Wm^{-2}; ε_c is the emissivity of the canopy (i.e., the efficiency with which the canopy is able to radiate); σ is the Stefan-Boltzmann constant, in $Wm^{-2}\ K^{-4}$; T_c is a leaf temperature corresponding to the mean temperature of the big leaf, in K; λ is the molar latent heat of vaporization at big-leaf temperature, in $J\ mol^{-1}$; E is the flux of evaporated water, in mol $m^{-2}s^{-1}$; G is the ground heat flux, in Wm^{-2}; M is the net heat flux stored in biochemical reactions, in Wm^{-2} (here, it is assumed that M is negligible); and H is the sensible heat flux, in Wm^{-2}. H can be expressed (e.g., Jones 1983:92) as

$$H = c_p (T_c - T_a)\, g_b \tag{7.2}$$

where c_p is the molar specific heat of dry air at constant pressure, in $J\ mol^{-1}\ K^{-1}$; T_a is the temperature of the air above the big leaf, in K; and g_b is the aerodynamic boundary layer conductance of the canopy, in mol $m^{-2}s^{-1}$. Molar units are chosen for the derivation of the expression of the latent heat flux to enable comparisons with the derivations of Jarvis and McNaughton (McNaughton and Jarvis 1983; Jarvis and McNaughton 1986). The conversion of conductances from molar units to ms^{-1} is achieved by multiplying by RT_a/P (0.0244 m^3 mol^{-1} at 20°C, at sea level), where P is the atmospheric pressure (10^5 Pa) and R is the gas constant (8.3144 Pa m^3 mol^{-1} K^{-1}).

It will prove useful to introduce the isothermal net radiation, R_{ni}, which can be defined as "the net radiation that would be received in the same environment by an identical surface at air temperature" (Monteith 1973:163):

$$R_{ni} = S_n + L_a - \varepsilon_c \sigma T_a^4. \quad (7.3)$$

From Equations 7.1 and 7.3, the apparent (long-wave) radiative loss from the big leaf at temperature T_c to the surrounding air at temperature T_a can be calculated:

$$R_n - R_{ni} = -\varepsilon_c \sigma (T_c^4 - T_a^4). \quad (7.4)$$

Assuming that $T_c - T_a$ is small, Equation 7.4 can be approximated to the first order as

$$R_n - R_{ni} \simeq -c_p (T_c - T_a) g_r, \quad (7.5)$$

in a manner which reminds us of the expression for the sensible heat flux (Eqn. 7.2), and where g_r is a (long-wave) radiative transfer "conductance" of the canopy, in mol $m^{-2}s^{-1}$, defined as

$$g_r = 4\varepsilon_c \sigma T_a^3 \Lambda / c_p,$$

where Λ is the leaf area index, in m^2m^{-2} (for $\varepsilon_c = 1$ and $\Lambda = 1$, $g_r = 4.69 \times 10^{-3}$ ms^{-1} at 20°C). Equation 7.5 shows the correction needed to account for slight differences in canopy and air temperature. The net radiation available to the canopy is somewhat less than isothermal net radiation if the canopy is hotter than the ambient air.

It will also prove useful for the derivation of the latent heat flux to express the temperature gradient as a function of the isothermal net radiation and the latent heat flux. Combining Equations 7.1, 7.2, and 7.5, and rearranging, gives

$$T_c - T_a = \frac{(R_{ni} - G) - \lambda E}{c_p (g_b + g_r)}. \quad (7.6)$$

Latent Heat

The latent heat flux (i.e., the energy equivalent of the evaporation flux) can be expressed as being proportional to the water vapor gradient between the air within the stomatal cavities at canopy temperature and the air above the canopy. It is quite standard to formulate it as a function of the vapor pressure deficit between the canopy and the atmosphere since the air in the stomatal cavities is close to saturation (e.g., Jarvis and McNaughton 1986:38):

$$\lambda E = (c_p/\gamma)\,[e_s(T_c) - e_a]\,g_w \tag{7.7}$$

where γ is the psychrometric constant at big-leaf temperature ($\gamma = 66.1$ at 22°C), in PaK^{-1}; e_s is the saturation vapor pressure, also at big-leaf temperature, in Pa; e_a is the ambient vapor pressure, in Pa; and g_w is the total conductance for water vapor, calculated as $g_w^{-1} = g_c^{-1} + g_b^{-1}$, where g_c is the canopy conductance, in mol $m^{-2}s^{-1}$.

The determination of the latent heat flux using Equation 7.7 requires that canopy temperature be measured. To circumvent this requirement, Equation 7.7 can be combined with Equation 7.6 and an approximation of the saturation vapor pressure. Saturation vapor pressure can be approximated to the n-th order using a Taylor series expansion around $T = T_a$, as

$$e_s(T_c) = e_s(T_a) + \sum_{j=1}^{n} \frac{1}{j!}\, e_s^{(j)}(T_a)\,(T_c - T_a)^j,$$

where $e_s^{(j)}(T_a)$ is the j-th derivative of saturation vapor pressure with respect to temperature at air temperature. This approach was originally suggested by Penman (1948:125), who approximated saturation vapor pressure linearly (i.e., $n = 1$). More recently, Paw U and Gao (1988) successively fitted a polynomial of degree 4 to e_s (p. 123), which is roughly equivalent to making a fourth-order approximation, i.e., $n = 4$. They used a second-order approximation, i.e., $n = 2$ (p. 125), to study the error resulting from the first-order approximation as the temperature gradient between the air and the canopy increases. Paw U and Gao show that in some instances significant errors occur. This is discussed later.

Here the first-order (Penman) approximation is used because of its simplicity. Equation 7.7 can now be rewritten as

$$\lambda E = (c_p/\gamma)\,[s\,(T_c - T_a) + D_a]\,g_w \tag{7.8}$$

where D_a is the saturation vapor pressure deficit of the ambient air, in Pa; and s is the derivative of the saturation vapor pressure with respect to temperature at air temperature ($s = e_s^{(1)}(T_a) = de_s(T_a)/dT$), in PaK^{-1} ($s = 145$ at 20°C). Substituting Equation 7.6 in Equation 7.8 yields a Penman-Monteith formulation of the latent heat flux from a big-leaf equivalent to the one used by Jones (1976:605).

$$\lambda E = \frac{\varepsilon(R_{ni} - G) + c_p D_a\,(g_b + g_r)/\gamma}{\varepsilon + (1 + g_r/g_b)\,(1 + g_b/g_c)}, \tag{7.9}$$

where ε is the dimensionless ratio of the increase of latent heat content to the increase of sensible heat content of saturated air ($\varepsilon = s(T_a)/\gamma(T_c)$). Since γ is weakly dependent on temperature, it is reasonable to assume that ε is a function of air temperature alone ($\varepsilon = 2.20$ at 20°C). Equation 7.9 is a generalization of the result derived for a symmetrical amphistomatous leaf, i.e., a leaf with stomata on both sides, by Jarvis and McNaughton (1986:40, their Eqn. A9), who assumed that radiative losses are negligible, i.e., that $g_r = 0$. It should also be noted that Equation 7.9 can be formally obtained from their Equation A17 (p. 42) for a hypostomatous leaf, i.e., a leaf with stomata on its lower side only, by translating g_{b1} to g_b, g_{b2} to g_r, and g_{s1} to g_c. To deal with changes over significant vertical distances and to allow for pressure changes, Jarvis and McNaughton (1986:43) use the concept of a "potential saturation deficit," D. In this case, their Equation A24 for canopy evaporation cannot be derived from Equation 7.9.

Vegetation Control of the Evaporative Flux

Using the concept of isothermal net radiation, the latent heat can be reformulated in a manner which eliminates temperature effects on radiation. It is important to know whether or not this alternative formulation affects the evaluation of control of evaporation by vegetation.

Evaporation Control without Radiative Losses

In their analysis, which omits radiative losses, i.e., $g_r = 0$, Jarvis and McNaughton consider two extreme cases, namely, the complete decoupling of the canopy from and the perfect coupling of the canopy to the atmosphere.

The evaporative behavior of a canopy which is decoupled from the atmosphere by very thick boundary layers ($g_b \rightarrow 0$) and therefore driven by available energy (here, net radiation) is similar to that of a wet surface in the absence of advection. As "time increases without limit" (McNaughton and Jarvis 1983:40), with R_n, G, T_a, T_c, g_c, and g_b remaining constant, the evaporation rate is said to approach an "equilibrium evaporation" rate, E_{eq} (e.g., Brutsaert 1982:218). The corresponding latent heat flux can be written as

$$\lambda E_{eq} = \frac{\varepsilon(R_n - G)}{\varepsilon + 1}. \tag{7.10}$$

Conversely, extreme ventilation corresponds to a situation where the canopy has negligible boundary layers ($g_b \rightarrow \infty$). When the drying power of the air rather than available energy drives evaporation, the evaporation rate tends to the "imposed transpiration rate," E_{imp} (Jarvis and McNaughton 1986:41, Eqn. A13), which can be expressed as

$$\lambda E_{imp} = \frac{c_p}{\gamma} g_c D_a. \tag{7.11}$$

On the basis of these two results, Jarvis and McNaughton (1986:41, Eqn. A14) rewrite the evaporative flux as

$$E = \Omega_c E_{eq} + (1 - \Omega_c) E_{imp}, \tag{7.12}$$

where Ω_c is a "decoupling factor that describes how closely the saturation deficit at the . . . surface is linked to that of the air outside the boundary layer" (Jarvis and McNaughton 1986:14), and which can be expressed as

$$\Omega_c = \frac{\varepsilon + 1}{\varepsilon + 1 + g_b/g_c}. \tag{7.13}$$

Clearly, Ω_c is between 0 and 1. Since E_{eq} is independent of vegetation and E_{imp} is a function of canopy conductance, the Ω_c factor defined in Equation 7.13 determines the extent to which vegetation controls evaporation. As summarized by Equation 7.12, when the source of water vapor provided by the vegetation is perfectly coupled aerodynamically to the atmospheric sink for water vapor ($g_b \rightarrow \infty$) or when the stomata are completely closed ($g_c \rightarrow 0$), the vapor pressure deficit between the canopy and the air determines the evaporative demand ($E \propto D_a$) and the vegetation controls the evaporative flux ($\Omega_c \rightarrow 0$). When vegetation is completely decoupled aerodynamically from the atmosphere ($g_b \rightarrow 0$) or (hypothetically) when water diffuses freely through the stomata ($g_c \rightarrow \infty$), vegetation exerts no control over transpiration ($\Omega \rightarrow 1$), and evaporation is driven by available energy ($E \propto R_n - G$). As a reminder, it should be pointed out that the aerodynamic conductance increases as vegetation height and/or wind increase and decreases as the characteristic dimension of the leaves increases. (The characteristic dimension of the leaves is roughly proportional to the square root of the leaf area.)

If vegetation is completely uncoupled from the atmosphere, it has no control over evaporation, which is instead determined by surface moisture; conversely, if it is perfectly coupled, it has total control because virtually all the available water is transpired. Hence, within the framework of the definitions used here, vegetation/atmosphere *coupling* and vegetation *control* of evaporation are synonymous. Therefore, an explicit way to evaluate the degree of coupling between the atmosphere and vegetation is to compute the elasticity of evaporation with respect to canopy conductance, because it provides a measure of the control of vegetation over evaporation. (The elasticity can also be thought of as the logarithmic derivative of evaporation with respect to canopy conductance.) Taking $g_r = 0$, differentiating Eqn. 7.9, substituting Eqns. 7.9 and 7.13, and rearranging the terms, yields the elasticity derived by Jarvis and McNaughton (1986:14, Eqn. 13):

$$\frac{dE/E}{dg_c/g_c} = 1 - \Omega_c. \tag{7.14}$$

(For $\Omega_c = 0$, a change in conductance causes a proportional change in evaporation; for $\Omega_c > 0$, the change is less than proportional.)

Evaporation Control with Radiative Losses

The results presented above do not take into account the effects of slight differences between canopy and air temperature. Using the same formulation for the "imposed transpiration rate" (Eqn. 7.11) and similar ones for the "equilibrium evaporation rate" and the "omega decoupling factor," and rearranging Eqn. 7.9, yields

$$E = \tilde{\Omega}_c \tilde{E}_{eq} + (1 - \tilde{\Omega}_c) E_{imp} \tag{7.15}$$

where

$$\lambda \tilde{E}_{eq} = \frac{\varepsilon(R_{ni} - G)}{\varepsilon + (1 + g_r/g_b)} \tag{7.16}$$

and

$$\tilde{\Omega}_c = \frac{\varepsilon + 1 + g_r/g_b}{\varepsilon + (1 + g_r/g_b)(1 + g_b/g_c)}. \tag{7.17}$$

By differentiating Equation 7.9, substituting Equations 7.9 and 7.17, and rearranging the terms, an elasticity equation similar to Equation 7.14 can be derived:

$$\frac{dE/E}{dg_c/g_c} = 1 - \tilde{\Omega}_c. \tag{7.18}$$

As can be seen by taking $g_r = 0$, Equations 7.15–7.18 are generalizations of Jarvis and McNaughton's Equations A14, A10, A16, and A22, respectively. As previously, Equations 7.15–7.18 can be formally obtained from Jarvis and McNaughton's (1986:42) Equations A18, A19, A20, and A22 for a hypostomatous leaf by translating g_{b1} to g_b, g_{b2} to g_r, and g_{s1} to g_c.

$\tilde{E}_{eq}$ is now a function of isothermal net radiation and the ratio of the radiative to the aerodynamic conductance, g_r/g_b. $\tilde{\Omega}_c$ describes how closely the vegetation is coupled convectively and radiatively to the air around it. $\tilde{\Omega}_c$ is between 0 and 1. When the big leaf is in radiative equilibrium with the air around it, i.e., when it is well coupled radiatively ($g_r \rightarrow 0$), the behavior of the decoupling factor with radiative coupling converges toward the behavior of the decoupling factor without radiative coupling ($\tilde{\Omega}_c \rightarrow \Omega_c$). When the apparent (long-wave) radiative loss from the big leaf is large, i.e., when the big leaf is poorly coupled radiatively to the air around it ($g_r \rightarrow \infty$), vegetation control is solely determined by the relative importance of physiological control as compared with aerodynamic control ($\tilde{\Omega}_c \to g_c/(g_b + g_c)$). In both cases, the limits of $\tilde{\Omega}_c$ and Ω_c for $g_b \to$, $g_b \to \infty$, $g_c \to 0$, and $g_c \to \infty$, respectively, are identical. Having commented on the role of the radiative conductance, it is useful to recall that g_r is proportional to the leaf area index, the emissivity, and the third power of the air temperature.

From Equation 7.17 and the above comments, it appears that the strength of the coupling between the atmosphere and vegetation is determined by the aerodynamic, physiological, and radiative characteristics of the vegetation, as well as the ambient air temperature.

Theoretical and Practical Implications

The expressions for the equilibrium evaporation flux and the decoupling factor presented in the previous section will now be compared analytically and the consequences of their differences assessed.

Analytical Comparisons

The behavior of evaporation, E, resulting from the new formulation of the decoupling factor, $\tilde{\Omega}_c$, agrees qualitatively with that predicted by McNaughton (1976:188–189) and Paw U and Gao (1988:135). In particular, it can be shown from Equations 7.15, 7.16, and 7.17 that zero is now the limiting value for λE, instead of $\varepsilon (R_n - G)/(\varepsilon + 1)$, when g_b goes to zero (McNaughton and Jarvis 1983:40). Also, $\tilde{\Omega}_c$ is relatively insensitive to the approximation chosen for the saturation vapor pressure, as can be inferred from Paw U and Gao (1988:136, Fig. 6).

From Equations 7.2, 7.5, 7.10, and 7.16,

$$\frac{E_{eq}}{\tilde{E}_{eq}} = \left(1 + \frac{1}{1+\varepsilon} g_r/g_b\right) \Big/ \left(1 + \frac{H}{R_n - G} g_r/g_b\right).$$

Hence, in most cases, E_{eq} will be larger than $\tilde{E}_{eq}$.

Comparing $\tilde{\Omega}_c$ and Ω_c (Equations 7.13 and 7.17), it can be shown analytically that $\tilde{\Omega}_c \leq \Omega_c$. Therefore, Jarvis and McNaughton overestimate the Ω factor in the case of a canopy. As a result (Eqns. 7.14 and 7.18), canopy control of transpiration is underestimated.

Practical Implications for Vegetation Control

In the previous subsection, it was shown analytically that Jarvis and McNaughton underestimate vegetation control of transpiration. In the present section, the magnitude of the difference between the two derivations is estimated by using empirical data.

Let ξ be defined as the ratio of the elasticities computed by the two methods:

$$\xi = \left(\frac{dE(\tilde{\Omega}_c)/E(\tilde{\Omega}_c)}{dg_c/g_c}\right) \bigg/ \left(\frac{dE(\Omega_c)/E(\Omega_c)}{dg_c/g_c}\right) = \frac{1 - \tilde{\Omega}_c}{1 - \Omega_c}.$$

Where Ω_c is large, i.e., when vegetation has little control over its transpiration, a small correction results in large changes in ξ, e.g., with $\Omega_c = 0.9$ (a value typical for grasslands) and $\tilde{\Omega}_c/\Omega_c = 0.9$ (i.e., a 10% error), $\xi \simeq 2$. In other words, the value of the corrected vegation control is twice the uncorrected one. So, one can expect that this correction will significantly affect the vegetation with the least control over its transpiration.

Calculations based on empirical observations confirm this. As shown in Martin (1989:51, Tab. 2), the difference between the uncorrected and corrected Ωs can cause significant differences in $\xi (1.2 \leq \xi \leq 1.55)$. From the arithmetic in the previous paragraph, vegetation types characterized by a low canopy roughness (i.e., low aerodynamic conductance), such as grasslands, are more affected by the correction than vegetation with rougher canopies. Taking into account temperature gradients amounts to including an additional energy transfer mechanism, increasing the coupling between the vegetation and the atmosphere, and hence the control of vegetation over evaporation. Vegetation with low canopy roughness weakly controls its own evaporation since it is essentially driven by available energy. Thus, the relative increase in control is larger for this type of vegetation than for vegetation with high canopy roughness.

Practical Implications for Evaporation

To assess the importance of the radiative coupling for evaporation, a series of sensitivity studies was performed. To focus on the role of vegetation rather than aerodynamic characteristics, canopy resistance, vegetation height, and leaf area index were varied and the other variables set at constant values (T_a = 15°C, T_c = 20°C, $(R_i - G)$ = 1000 W m^{-2}, e_a = 500 Pa). Simple formulations of the boundary-layer resistance around the leaf and the aerodynamic resistance above the canopy were implemented (e.g., Jones 1983:53, 57).

From the examination of Figures 7.1–7.6, several conclusions may be drawn. The $\tilde{\Omega}_c$ decoupling factor increases nonlinearly and not always monotonously depending on the height as canopy conductance increases. $\tilde{\Omega}_c$ increases as height increases and as leaf area index decreases. As vegetation height increases (and the aerodynamic resistance decreases), the relative difference between the latent heat fluxes computed using the two formulations decreases. At high leaf area indices, however, the error increases to a finite value as height reaches its maximum (chosen here to be 100 m). As leaf area index increases, the relative difference increases nonlinearly and the relationship between the canopy resistance and the decoupling factor becomes increasingly nonlinear.

Additional studies show that, in general, the relative difference between $\lambda E(\tilde{\Omega}_c)$

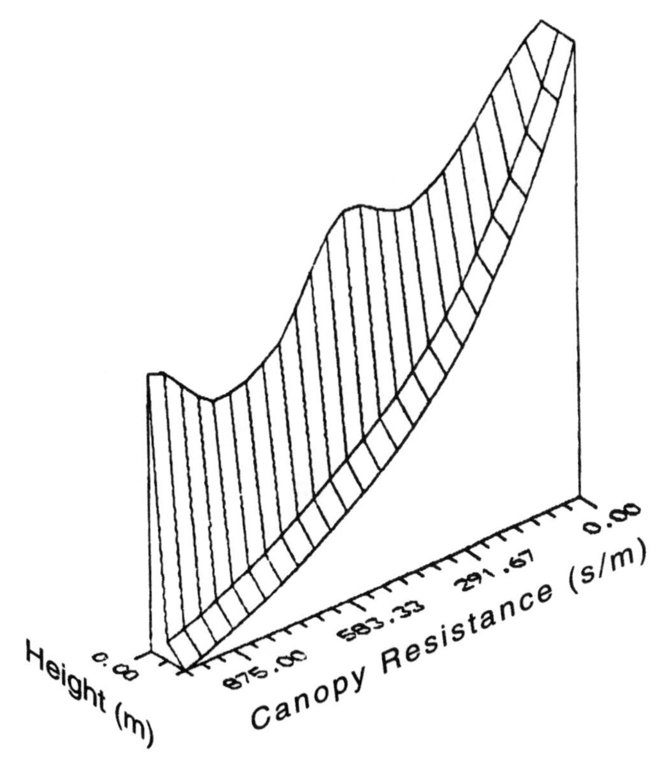

Figure 7.1. Absolute sensitivity of Omega to height and canopy resistance (LAI = 0.1).

and $\lambda E(\Omega_c)$ increases as roughness decreases, as air temperature increases, and as the temperature gradient between the air and the canopy increases.

Discussion and Conclusion

An analytical expression is derived for the elasticity of evaporation with respect to canopy conductance (i.e., the fractional change in evaporative flux, dE/E, caused by a fractional change in canopy conductance, dg_c/g_c) which takes into account the changes in latent heat flux and net radiation caused by stomatal closure. The simplicity of the derivations presented here makes a detailed discussion of the various processes at work possible, which was not the case in the context of the treatment by Paw U and Gao (1988).

Comparisons with the computations of Paw U and Gao (1988) show that the

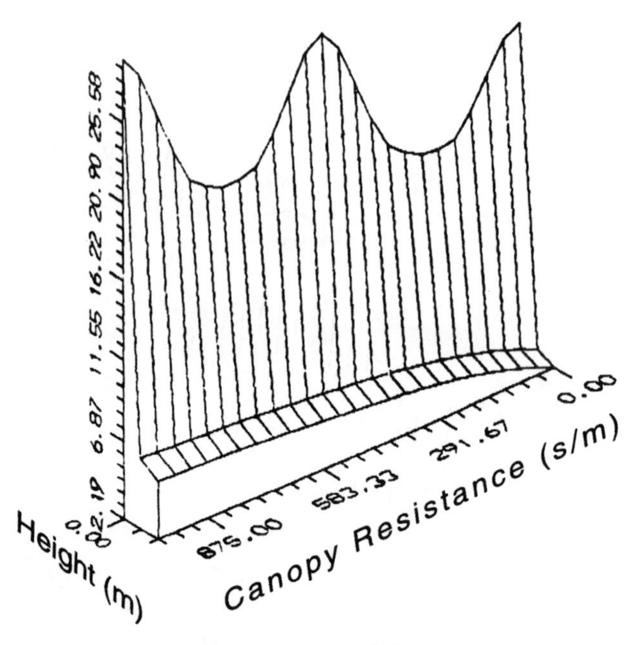

Figure 7.2. Relative sensitivity of Omega to height and canopy resistance (LAI = 0.1).

value of the decoupling factor is relatively insensitive to the type of approximation made in the calculation of the saturation vapor pressure. Because a big-leaf approach is used, the treatment of radiative transfers within the canopy is excluded. This may alter the exact formulation of the decoupling factor, but it is unlikely to affect its qualitative behavior.

The general conclusion reached by Jarvis and McNaughton (1986), that tall vegetation has a greater control over its transpiration than shorter vegetation, still holds. However, the formulations for equilibrium evaporation and the decoupling factor which take into account temperature gradients between the canopy and the air are different from the ones they obtained (Jarvis and McNaughton 1986). In particular, the new expression for equilibrium evapotranspiration yields a correct behavior for zero aerodynamic conductance and zero vapor pressure deficit. Finally, computations of the latent heat flux can be significantly affected by the inclusion of a radiative exchange between the vegetation and the atmosphere.

Stepping back and looking now at the "big picture," when present, vegetation has some control over the partitioning of energy at the surface of the Earth. This means that vegetation affects the local hydrology and atmospheric motion. Omitting its root system, which is also potentially important, vegetation may be

Omega (LAI = 1.0)

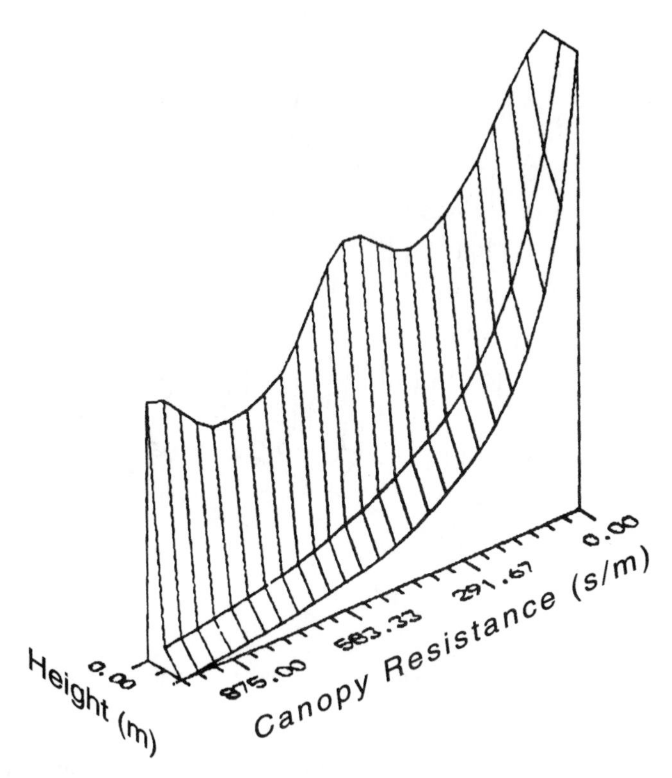

Figure 7.3. Absolute sensitivity of Omega to height and canopy resistance (LAI = 1.0).

characterized by its height, its leaf area, its stomatal conductance, and the emissivity of its foliage. As was exemplified in Figures 7.1, 7.3, and 7.6, the effects of changes in any of these four parameters have a highly nonlinear effect on the relative importance of the control of vegetation over atmosphere/vegetation interactions. In addition, the strength of the coupling between the atmosphere and vegetation decreases as scale increases. Hence, changes in vegetation characteristics present challenges both in terms of their nature and the way in which their effects are felt at increasingly larger scales.

The single most important point to be made here is that a detailed knowledge of the aerodynamic, physiological, and radiative characteristics of the vegetation, as well as the ambient environmental conditions, is essential to model the interaction between the atmosphere and vegetation, and therefore to evaluate its strength. Such information requirements have direct implications for future investigations into global vegetation dynamics and their climatic consequences. Because of the

Relative Diff. (LAI = 1.0)

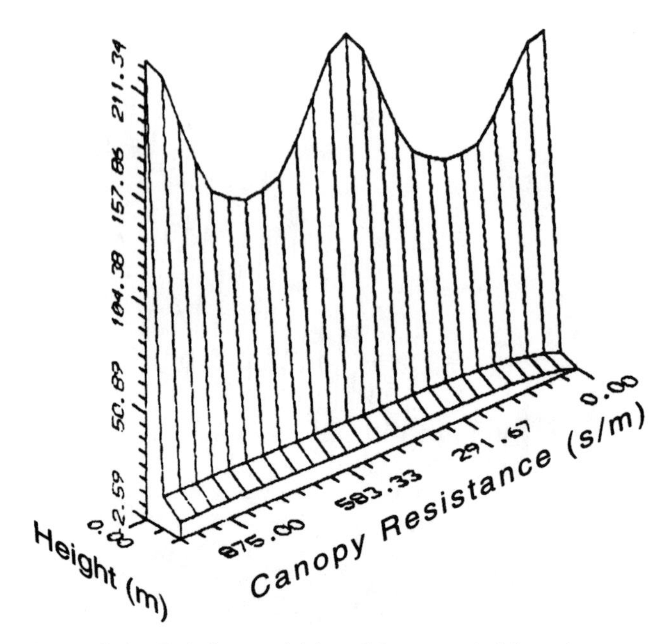

Figure 7.4. Relative sensitivity of Omega to height and canopy resistance (LAI = 1.0).

nature of the phenomena which underlie them, climate/vegetation studies will most likely have to incorporate a variety of data sources and, for instance, combine output from atmospheric general circulation models with surface, airborne, and satellite-based observations.

Acknowledgments

The careful reading of and constructive comments on an earlier draft from Bruce Kimball, Bernard Pinty, Michel Verstraete, and Erica Schwarz are gratefully acknowledged.

References

Brutsaert, W. (1982). *Evaporation into the Atmosphere: Theory, History, and Applications*. Dordrecht: D. Reidel.

Calder, I. R. (1977). A model of transpiration and interception loss from spruce forest in Plynlimon, Central Wales. *Journal of Hydrology*, **33**, 247–65.

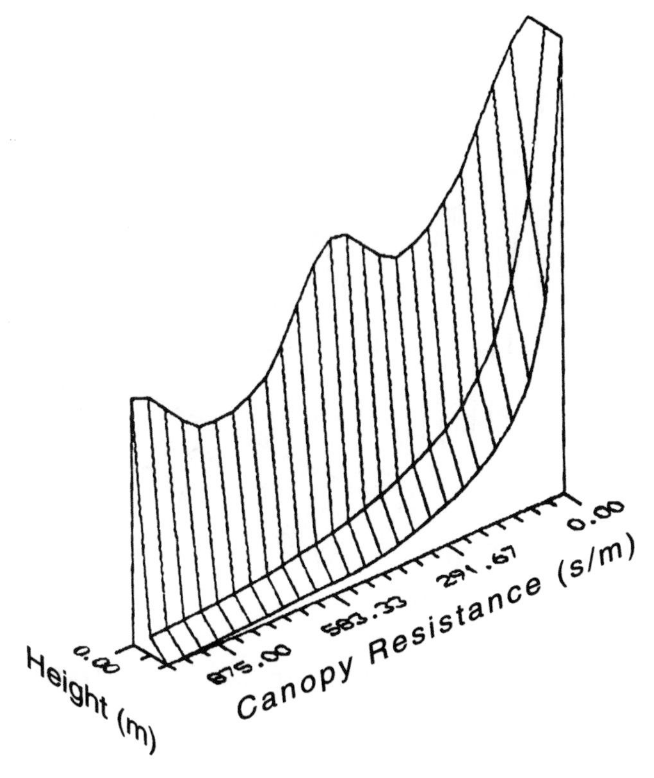

Figure 7.5. Absolute sensitivity of Omega to height and canopy resistance (LAI = 6.0).

Deardorff, J. (1978). Efficient prediction of ground temperature and moisture with inclusion of a layer of vegetation. *Journal of Geophysical Research,* **83,** 1889–903.

Dickinson, R. E. (1984). Modeling evapotranspiration for three-dimensional global climate models. In *Climate Processes and Climate Sensitivity,* ed. J. E. Hansen and T. Takahashi, pp. 58–72. Geophysical Monograph 29. American Geophysical Union, Washington, D.C.

Dickinson, R. E. and Hanson, B. (1984). Vegetation-albedo feedbacks. In *Climate Processes and Climate Sensitivity,* ed. J. E. Hansen and T. Takahashi, pp. 180–6. Geophysical Monograph 29, American Geophysical Union, Washington, D.C.

Gates, D. M. and Papain, L. E. (1971). *Atlas of Energy Budgets of Plant Leaves.* New York: Academic Press.

Hall, W. A. (1971). Biological hydrological systems. In *Biological Effects in the Hydrological Cycle—Terrestrial Phase,* ed. E. J. Monke, pp. 1–7. Proceedings of The Third International Seminar for Hydrology Professors, UNESCO, Paris, France.

Relative Diff. (LAI = 6.0)

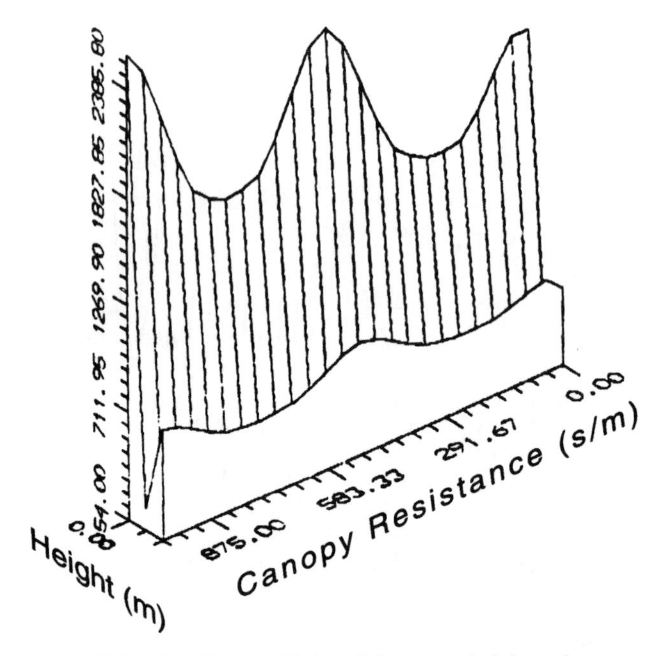

Figure 7.6. Relative sensitivity of Omega to height and canopy resistance (LAI = 6.0).

Halldin, S. and Lindroth, A. (1986). Pine forest microclimate simulation using different diffusivities. *Boundary-Layer Meteorology,* **35,** 103–23.

Jarvis, P. G. and McNaughton, K. G. (1986). Stomatal control of transpiration: Scaling up from leaf to region. *Advances in Ecological Research,* **15,** 1–49.

Jones, H. G. (1976). Crop characteristics and the ratio between assimilation and transpiration. *Journal of Applied Ecology,* **13,** 605–22.

Jones, H. G. (1983). *Plants and Microclimate: A Quantitative Approach to Environmental Plant Physiology.* Cambridge: Cambridge University Press.

Martin, Ph. (1989). The significance of radiative coupling between vegetation and the atmosphere. *Agricultural and Forest Meteorology,* **49,** 45–53.

McNaughton, K. G. (1976). Evaporation and advection. I. Evaporation from extensive homogeneous surfaces. *Quarterly Journal of the Royal Meteorological Society,* **102,** 181–191.

McNaughton, K. G. and Jarvis, P. G. (1983). Predicting effects of vegetation changes on transpiration and evaporation. In *Water Deficit and Plant Growth,* ed. T. T. Kozlowski, pp. 1–47. New York: Academic Press.

Miller, P. C. (1971). Sampling to estimate mean leaf temperatures and transpiration rates in vegetation canopies. *Ecology,* **52,** 885–9.

Monteith, J. L. (1973). *Principles of Environmental Physics*. London: Edward Arnold.

Paw U, K. T. and Gao, W. (1988). Applications of solutions to non-linear energy budget equations. *Agricultural and Forest Meteorology,* **43,** 121–45.

Penman, H. L. (1948). Natural evaporation from open water, bare soil, and grass. *Proceedings of the Royal Society A,* **193,** 120–45.

Rosenberg, N. J., Blad, B. L. and Verma, S. B. (1983). *Microclimate: The Biological Environment*. 2nd edition. New York: John Wiley & Sons.

Rutter, A. J. (1975). The hydrological cycle in vegetation. In *Vegetation and the Atmosphere, Volume 1, Principles,* ed. J. L. Montieth, pp. 111–54. London: Academic Press.

Shukla, J. and Mintz, Y. (1982). Influences of land-surface evapotranspiration on the earth's climate. *Science,* **215,** 1498–500.

Slabbers, P. J. (1977). Surface roughness of crops and potential evapotranspiration. *Journal of Hydrology,* **34,** 181–91.

Stewart, J. B. and de Bruin, H.A.R. (1985). Preliminary study of the dependence of surface conductance of Thetford Forest on environmental conditions. In *The Forest-Atmosphere Interactions,* ed. B.A. Hutchison and B.B. Hicks, pp. 91–104. Dordrecht, The Netherlands: Reidel.

Verstraete, M.M. and Dickinson, R.E. (1986). Modeling surface processes in atmospheric general circulation models. *Annales Geophysicae,* **4,** B, $, 357–64.

Walter, H. (1973). *Vegetation of the Earth in Relation to Climate and the Eco-Physiological Conditions*. New York: Springer-Verlag.

Wilson, M.F., Henderson-Sellers, A., Dickinson, R.E. and Kennedy, P.J. (1987). Sensitivity of the Biosphere-Atmosphere Transfer Scheme (BATS) to the inclusion of variable soil characteristics. *Journal of Climate and Applied Meteorology,* **26,** 341–62.

Woodward, F.I. (1987). *Climate and Plant Distribution*. Cambridge: Cambridge University Press.

PART III

Measuring Global Vegetation Change

8

Monitoring Vegetation Change Using Satellite Data

B.N. Rock, D.L. Skole, and B.J. Choudhury

Native vegetation will respond in predictable ways to a variety of stresses. Depending on the intensity and duration of the stress, the response may be at the cellular level, the morphological/macroscopic level, or the community level. Figure 8.1 presents examples of some typical vegetation responses to stress. All of the examples of change at different levels given in Figure 8.1 have diagnostic spectral characteristics which can be detected by using various types of remote sensing systems. Because of the sensitivity of native vegetation to stress factors associated with environmental change (moisture levels, nutrient levels, temperature, anthropogenic factors, etc.), the ability to remotely detect subtle levels of change (response to stress) in the vegetation may prove to be a very useful indicator of environmental change. Remote sensing techniques employing satellite multispectral data provide an accurate means of detecting, quantifying, mapping, and monitoring change in vegetation on local, regional, and global scales. Change at different scales in both vegetation kind (vegetation types, species associations, etc.) and vegetation condition (state of health, degree of deforestation, seasonal stage of growth, etc.) can be studied by using various sensor systems and image processing techniques. Table 8.1 summarizes some of the available satellite data currently in use for change detection purposes.

If current and future satellite sensor systems are to be used, either as a source of potential input data for vegetation change modeling efforts or as a means of assessing and/or validating model products, an understanding of the types, scales, uses, and limitations of remotely sensed data is needed. Toward this goal, this chapter provides some examples of current applications of satellite and aircraft data for the purpose of assessing and monitoring various types of vegetation change.

Local-Scale State-of-Health Studies

The use of satellite and aircraft multispectral data to assess forest decline (Waldsterben) damage provides an example of a local-scale study of vegetation change

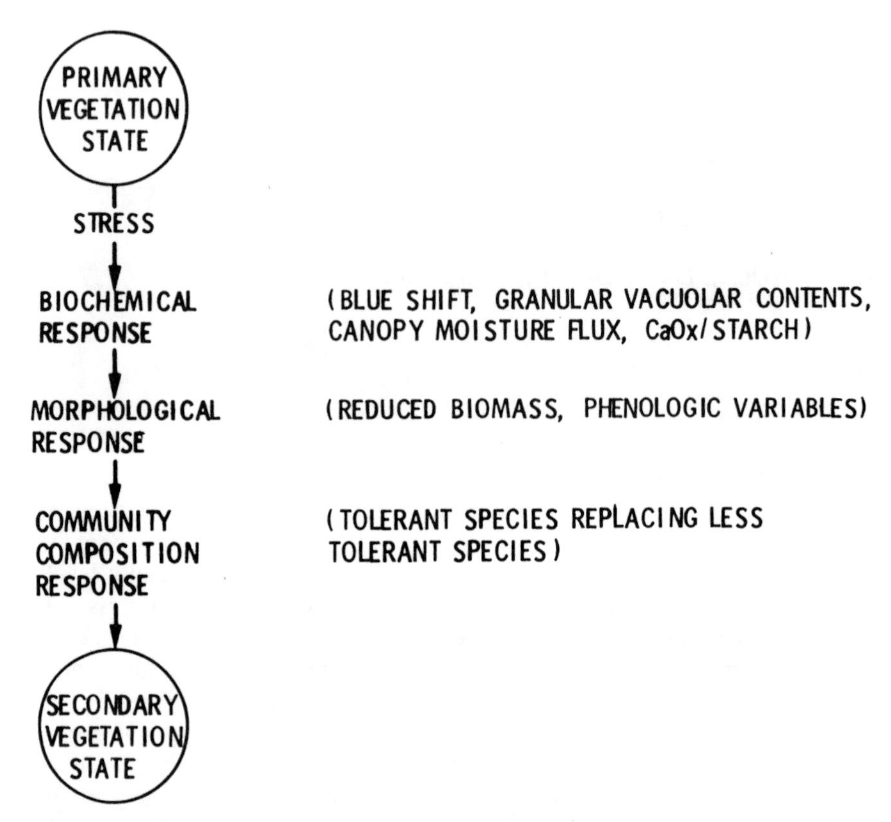

Fig. 8.1. Schematic representation of vegetation response to varying degrees of stress. Initial exposure to nonlethal levels of stress will result in previsual cellular-level change. If stress is chronic or intense, visual (morphological) damage will result. If stress continues, community-level changes will occur. Examples of specific types of cellular, morphological, and community changes are given to the right of each. All of these levels of response may be detected by using remote sensing tools.

across an area and over time. The conifer forests of the northeastern United States and central Europe have experienced a marked decline in growth and vigor over the last several decades (Johnson and Siccama 1984; Schuett and Cowling 1985; Vogelmann *et al.* 1985). Although the causes of this general decline have not been determined with certainty, it is likely that atmospheric pollutants (acid deposition, ozone, SO_2, toxic metals, etc.) have played a major role in this change in state of forest health (Klein and Perkins 1987).

Landsat Thematic Mapper (TM) data have been successfully used to detect, quantify, and map areas of forest-decline in the northeastern U.S. (Rock *et al.* 1986, 1987; Vogelmann and Rock 1986, 1988). An example of a forest decline damage assessment image, produced by using TM data, is presented in Figure

Table 8.1. Summary of current satellite sensor systems.

Satellite sensor system	Spatial resolution	Spectral coverage	Repeat cycle	Time period covered	Spatial domain
SPOT (Système Probatoire d'Observation de la Terre)	10 m pancromatic 20 m multispectral	(0.51–0.73 μm) (0.50–0.89 μm)	26 days (at nadir[a])	1986–present	Local to regional
Landsat TM (Thematic Mapper)	30 m reflected 120 m thermal	(0.4–2.2 μm) (8.0–12.0 μm)	16 days	1982–present	Local to regional
Landsat MSS (Multispectral Scanner)	80 m reflected	(0.5–1.1 μm)	16 days	1972–present	Local to regional
NOAA-7, 9 AVHRR (Advanced Very High Resolution Radiometer)	LAC (1 km) GAC (4 km) Resampled 7–15 km	(0.6–1.1 μm)	Daily, composited monthly	1982–present	Regional to global
Nimbus-7 SMMR (Scanning Multichannel Microwave Radiometer)	25 km	8 mm (37 GHz)	Weekly, composited monthly	1979–1985	Global

[a]More frequently off nadir due to pointing capability of SPOT.

8.2 (see color section). Note that the heaviest damage (portrayed as red) occurs on the west-facing slopes of the mountains in the center of the image. Spectral reflectance data of most use in portraying damage as assessed by the TM are from the near infrared (NIR—TM band 4) and short wave infrared (SWIR—TM band 5). A ratio of TM5/TM4 was used to emphasize differences seen in these spectral regions, and the resulting images (Fig. 8.2) are thought to relate to changes in canopy biomass and structure as determined by TM band 4, as well as canopy moisture status as determined by TM band 5. Both types of change are associated with forest decline damage in spruce and fir in the United States and Europe. Based on extensive ground assessment for selected mountains seen in Figure 8.2, this image is considered to be an accurate depiction of forest damage conditions (Vogelmann and Rock 1986). The extended spectral coverage in the short-wave reflected infrared provided by Landsat TM data accounts for this high degree of accuracy in the detection of forest damage (Rock and Vogelmann 1989; Zawila-Niedzwiecki 1989; Vogelmann 1990).

Landsat TM data have also been used to assess the relative amounts of forest damage detected on different mountain ranges in the northeastern United States (Rock *et al.* 1987; Vogelmann and Rock 1988). A damage-rating scale was developed by using the TM5/TM4 ratio to assess damage levels (the higher the ratio, the higher the damage) in the Adirondack Mountains in New York, the Green Mountains in Vermont, and the White Mountains in New Hampshire. From a summary of damage levels as determined by using this method (Tab. 8.2), it is apparent that there is a trend of decreasing damage from the most severe in the westernmost range (the Adirondacks) to light damage in the easternmost range (the White Mountains). It is also evident from the satellite data, as well as from field studies, that factors such as elevation, slope, and aspect alone do not account for the differences seen in relative damage levels.

The differences seen in the reflectance data acquired by the TM suggest that factors coming from the west may influence the levels of forest decline damage. Airborne pollutants are carried across the New York–Vermont–New Hampshire region from west to east and concentrations of pollutants are greater in the Adirondacks than to the east. These data suggest that a relationship may exist between the spatial patterns of pollutant exposure and the spatial patterns and levels of damage on both a local and a regional scale.

In addition to using satellite datasets to assess forest damage across a region (spatial patterns of change), the long-term databases provided by the Landsat satellite series (acquired continuously since 1972) allow for assessment of temporal patterns of change. In a multitemporal comparison of Landsat Multispectral Scanner (MSS) data acquired on August 29, 1973, and again on August 21, 1984, for the central Green Mountain Range of Vermont, the greatest levels of reflectance change (decrease in NIR reflectance, which may be due to decrease in biomass) are seen in the red spruce zone on the west-facing slopes of each mountain (Vogelmann 1988). Other vegetation zones on the mountains, such as

Table 8.2. Conifer damage in the Adirondack Mountains, Green Mountains, and White Mountains.

Site	Elevation (m)	% Low damage pixels	% Medium damage pixels	% High damage pixels	Damage rating
Adirondacks					
Whiteface	1484	8.6	12.7	78.7	85.1
High Peaks Area	1268	2.3	6.8	90.4	94.3
Green Mountains					
Camels Hump	1244	26.5	20.7	52.8	63.2
Mt. Abraham	1260	25.6	25.6	48.9	61.7
Breadloaf Mt.	1165	37.9	24.6	37.5	49.8
White Mountains					
Mt. Moosilauke	1464	72.3	16.8	10.9	19.3
Lafayette Mt.	1585	63.9	19.0	17.0	26.6

balsam fir and hardwoods, appear to have undergone less change in NIR reflectance, suggesting that certain species (red spruce) have undergone greater change than others. These results are in agreement with ground assessment studies of one of the mountains (Camels Hump Mountain) in the range (Vogelmann *et al.* 1988).

Advanced sensor systems known as imaging spectrometers have been shown to provide more detailed spectral information than broad-band sensors such as the Landsat TM. This additional spectral information may prove useful and diagnostic regarding specific type of damage and canopy effects in both the United States and the Federal Republic of Germany (Herrmann *et al.* 1988; Hoshizaki *et al.* 1988; Peterson *et al.* 1988; Rock *et al.* 1988b). Details of spectral reflectance properties associated with variations in canopy chemistry (Peterson *et al.* 1988; Wessman *et al.* 1988), cellular-level state of health (Hoshizaki *et al.* 1988; Rock *et al.* 1988b), and moisture deficits (Hunt *et al.* 1987a,b; Rock *et al.* 1988a) may be assessed by using airborne imaging spectrometers such as the AIS (Airborne Imaging Spectrometer), AVIRIS (Airborne Visible/Infrared Imaging Spectrometer), and FLI (Fluorescence Line Imager). Figure 8.3 (see color section) compares images of the same area of forest decline (Camels Hump, Vermont) acquired with an airborne Thematic Mapper Simulator (TMS) and the FLI. The FLI image is a more accurate indication of the actual location of forest decline damage in red spruce, while the TMS image shows where more generic damage in both spruce (forest decline damage) and fir (wind and frost damage) occurs. The spatial pattern of damage detected with the FLI (Fig. 8.3) agrees well with the area of greatest change between 1973 and 1984 on Camels Hump detected with Landsat MSS data as noted above.

Imaging spectrometers such as the AIS, AVIRIS, and FLI are currently available only as aircraft systems, but similar data will be available from the HIRIS

(High Resolution Imaging Spectrometer), one of several sensor systems scheduled to be placed on polar-orbiting platforms as part of the NASA/ESA (European Space Agency) Earth Observing System (EOS). The first EOS platform is scheduled to be launched in the late 1990s. The high-spectral-resolution datasets acquired by HIRIS will provide scientists with details, to be available on a global scale, regarding canopy chemistry and pigment content (chlorophylls, carotenes, xanthophyll, tannins, anthocyanins, etc.), water status, and cellular conditions not currently obtainable by using broad-band sensors such as the Landsat Thematic Mapper. The assessment of vegetation condition on a global scale will be possible in the future, and such remotely sensed data may be used as local to regional input in global-scale vegetation change models being developed.

Regional-Scale Deforestation Studies

Vegetation change of the type described above is subtle—often detected spectrally before it can be observed in the field. Spatially, it is a scattered, fine-scale phenomenon. Remote detection of such change requires the use of high spatial- and spectral-resolution sensors targeted to specific locations or species. The observed spectral signatures reflect subtle chemical and physiological changes (Fig. 8.1) occurring in the needles or leaves of individual trees and local-scale canopies (Figs. 8.2 and 8.3). Tropical deforestation, on the other hand, is a kind of vegetation change which is quite direct and conspicuous. It can be observed at very coarse spatial scales, from 1 to 4 km in some regions (Justice *et al.* 1985; Malingreau and Tucker 1988), and is clearly detected with even the simplest set of spectral-band combinations (Nelson and Holben 1986; Woodwell *et al.* 1986).

Deforestation creates significant alterations of basic features of vegetation and community structure, including primary productivity, standing biomass, and soil organic matter (Vitousek 1983), and thus must be considered part of any vegetation change model. Deforestation has already been shown to be an important component of climate models and models of global biogeochemical cycles (Houghton *et al.* 1983; Henderson-Sellers and Gornitz 1984; Houghton and Skole 1990).

Such vegetation change is significant and globally widespread, yet the precise rate and geographic distribution of deforestation are poorly known. The best global deforestation estimates to date have been compiled from national-level surveys and have frequently been derived from changes in various land-use categories rather than from changes in forest cover directly (Houghton 1986). Moreover, since they are often compiled for one or a few inventory years, these reports provide little insight into long-term trends. Recent studies now show that deforestation rates can be obtained from satellite data, thereby providing an objective portrait of changes in forest cover directly and a capability for initiating monitoring efforts (Tucker *et al.* 1984; Tucker *et al.* 1986b; Woodwell *et al.* 1986; Nelson *et al.* 1987; Malingreau and Tucker 1988).

Regional- to global-scale monitoring of deforestation using fine-spatial-resolution imagery (20–80-m pixels) would be time consuming, computationally difficult, and costly. Hence, many now advocate the use of coarse-resolution sensors (250-m–4-km pixels) for large-area inventories. For instance, the AVHRR (Advanced Very High Resolution Radiometer) onboard the NOAA-7 and NOAA-9 satellites provides frequent (daily) coverage on a worldwide basis, permitting the compositing of cloud-free imagery on a monthly basis. The higher-resolution Landsat and SPOT (Système Probatoire d'Observation de la Terre) sensors cannot readily provide such frequent and global coverage (16-day repeat cycle for Landsat, while the SPOT satellite greatly increases its repeat frequency by utilizing off-nadir viewing). Thus, spatial resolution is often sacrificed for frequent, global coverage. Indeed, there always exists a trade-off between resolution and coverage, but there has been little recognition of the importance of choosing the appropriate spatial scale for monitoring deforestation. (See Townshend and Justice 1988 for a review.)

We have been analyzing the importance of spatial scale in deforestation monitoring by using high-spatial-resolution multispectral imagery (20-m resolution) from the SPOT satellite. Figure 8.4a (see color section) presents a false color composite of a multispectral SPOT scene acquired on August 23, 1986, over an area in the state of Rondônia, Brazil. The total area covered measures approximately 60 km $\times$ 120 km (3000 $\times$ 6000 pixels). The pattern of deforestation observed in the imagery is typical for the region (Fearnside 1982, 1986). Clearly visible in the southern (bottom) portion of the scene is the regular, symmetrical pattern of deforestation which results from planned settlements. Sparsely scattered clearings are visible ahead of the advancing front of settlement. The Cuiaba–Porto Velho highway, BR364, and the Jamari River cut diagonally across the northwest corner of the imagery and provide both a conduit and a focal point for deforestation and development activities in the region. Figure 8.4b (see color section) presents a portion of the SPOT scene at full 20-m spatial resolution.

To compare the effect of different sensor systems, the imagery was degraded to lower levels of resolution by using a technique in which a low-pass filter was used in conjunction with a convolution algorithm. The resultant image was rectified to a specific resolution by resampling. Sensor resolutions simulated in this way reflect Landsat TM (30 m), Landsat MSS (80 m), the MODIS (Moderate Resolution Imaging Spectrometer; 500 m) proposed under the NASA Earth Observing System, AVHRR LAC (Local Area Coverage; 1 km), and AVHRR GAC (Global Area Coverage; 4 km). Only results from the simulation of Landsat MSS and AVHRR LAC spatial resolutions are reported here. Figure 8.5a (see color section) shows an example of the full SPOT scene degraded to simulate the AVHRR LAC (1 km) pixel resolution. (Compare with Fig. 8.4a.) Figure 8.5b (see color section) shows a portion of the SPOT scene degraded to simulate Landsat MSS (80 m) pixel resolution. (Compare with Fig. 8.4b.)

The degraded imagery and the original SPOT reference imagery were then classified for forest and nonforest areas and compared. Preliminary results portray a complex picture, highly dependent on geometry. Overall, coarse resolutions (AVHRR/MSS) yield overestimates of area of deforestation. The simulated Landsat MSS overestimated deforestation by 10% when compared to the original SPOT reference image, while the simulated AVHRR LAC overestimated deforestation by 18%. This is not unreasonable since the classification of deforestation, represented as very bright areas, may be significantly biased by small clearings within individual pixels. This bias is in part an edge effect. It is minimized in large clearings with high area/perimeter ratios, and maximized in highly sinuous or variegated clearings with low area/perimeter ratios.

A 10% overestimate provided by Landsat MSS is probably acceptable for most analyses of deforestation, but at this resolution a stratified global sampling scheme should be developed since the amount of imagery for complete global coverage would be excessive. Woodwell *et al.* (1986) proposed that such a scheme could be developed if it were possible to identify key areas of deforestation activity. It is possible to develop a multifaceted approach, where "fronts of deforestation" are identified based on two criteria: (1) areas where deforestation rates are high and rapidly increasing and (2) areas which are significant from the perspective of model parameterization, such as being high in biomass. A sampling scheme for global change studies could be developed by first identifying fronts of deforestation by using a combination of geographic mapping in a geographic information system (GIS) and a coarse resolution sensor. At identified locations, higher-resolution imagery could be employed to obtain detailed information.

An 18% overestimate from the AVHRR LAC simulation would be acceptable if it were evenly distributed spatially; a coefficient could be applied to ratio the data to the appropriate value. Conceivably it would also be possible to identify deforested areas with a coarse-spatial-resolution sensor (250–500 m) having high spectral resolution. Band-ratio techniques described above could be employed at coarse spatial scales to delineate deforestation, even if it were not clearly visible on the image. Likewise, the simultaneous use of multiple sensors of varying degrees of spatial and spectral resolution might be as effective. Such a design is being planned for the MODIS sensor on NASA's Earth Observing System.

The identification of change requires more spatial detail than the delineation of deforested areas, but we have not yet explored this aspect of the problem. It is possible that change detection using a moderate level of spatial resolution (e.g., 500 m) and broad spectral coverage (e.g., 0.4–2.2 μm) would provide data on rates of change with sufficient accuracy for most modeling applications.

Identifying fronts of deforestation is an important priority for global-change analyses. In particular, it is an important yet highly uncertain variable in global terrestrial carbon models. Significant improvements could be made to these models if it were possible to better quantify the rate and geographic distribution of two types of human-induced vegetation change: (1) direct deforestation of the kind discussed above and (2) indirect modification or degradation of vegetation.

Unlike the more obvious, outright vegetation change associated with deforestation, forest degradation is subtle and not readily visible in imagery. It involves a less significant loss of biomass per unit area than deforestation and can occur within an intact forest canopy. Some examples, in addition to the previously cited forest decline, are fuelwood culling from forests, selective timber harvest, and forest livestock grazing—all land uses found in regions also experiencing direct deforestation. Natural forests might also be degraded by subtle changes in microclimatic conditions brought about by changes in canopy cover and evapotranspiration in regions where there is large-scale deforestation.

Optical sensor systems covering only a limited spectral region may not be capable of delineating these kinds of vegetation changes, although the aforementioned work on forest decline suggests that broad-spectral-coverage datasets can be employed by using band-ratio techniques. Of the satellite sensors currently available, the broad spectral coverage of the Landsat Thematic Mapper appears to provide the optimal datasets for assessing subtle changes in forest condition, as well as adequate spatial resolution (30 m) for detecting and quantifying fronts of deforestation.

An approach which combines optical reflectance data with narrow-beam radar altimetry may also be possible. The optical system would detect changes in parameters related to amounts of biomass, "greenness," or other canopy conditions, while the radar would provide canopy height and density data. It may be possible to estimate canopy height and tree density by analyzing the waveform pattern of the return power curve from an active radar altimeter system (Topographic Science Working Group 1988). Pulse broadening of radar echoes will occur in areas where there are variations in canopy height or areas of low vegetation density. Multifrequency, multipolarization, Synthetic Aperture Radar (SAR) systems may also provide canopy height and density parameters. These microwave datasets would permit the quantification of important canopy parameters such as total biomass, and the delineation over time of the distribution of degraded and deforested environments. These technologies and methods have yet to be tested and proven. However, approaches of this kind, coupled with existing multispectral optical techniques and in situ data, could enable us to map biomass degradation and could provide insights into its role in terrestrial carbon cycling.

Global-Scale Vegetation Studies

The use of various vegetation indices to study the spatial and temporal variations of vegetation is based upon short-wave reflectances and microwave emission data. The normalized difference vegetation index (NDVI) is based upon reflectance in the visible channel (Ch1) and near-infrared channel (Ch2) observed by the AVHRR onboard NOAA satellites, and is calculated as follows:

$$\text{NDVI} = \frac{\text{Ch2} - \text{Ch1}}{\text{Ch2} + \text{Ch1}}.$$

Field studies and radiative transfer modeling (Choudhury 1987) show that for uniform stands of agricultural crops, NDVI increases with increasing green leaf area index, and it provides a quantitative indicator of the fraction of incident photosynthetically active radiation intercepted or absorbed by vegetation (Asrar *et al.* 1984). One has to be careful in extrapolating these implications of NDVI based upon field data to the NDVI derived from satellite observations because of atmospheric effects, spatial heterogeneities within the field of view, and the aggregating/compositing procedures used in synthesizing the global datasets.

The spatial resolution of the AVHRR LAC data is 1.1 km for nadir observations. However, the global observations are achieved at 4-km nadir resolution by aggregating four 1.1-km pixels in a 5 × 3 array of scanline observations. Thus, these GAC data at 4-km resolution do not represent a true average of a 4-km × 4-km area but only a fraction of this area. These 4-km data are further composited temporally (1-week–1-month period) to minimize the effects of cloud and atmospheric opacity variations as discussed in detail by Holben (1986). The spatial resolution is further degraded in this temporal compositing by choosing one 4-km GAC data point for 7 × 7 to 15 × 15 km spatial-resolution cells (varying from equator to pole).

A detailed analysis of the NDVI global dataset has been performed by Justice *et al.* (1985). The temporal patterns of NDVI have unique distinguishing features for different plant communities of the world. A more thorough analysis of the NDVI data for monitoring grasslands may be found in Justice (1986). Temporal integration of NDVI has been shown to provide a good indicator of the productivity of grasslands (Justice 1986) and, more generally, of different biomes in North and South America (Goward *et al.* 1987). Additionally, the temporal variation of NDVI has been shown to be correlated with the temporal variation of atmospheric CO_2 concentration (Tucker *et al.* 1986a). This correlation is expected because the atmospheric CO_2 concentration is depleted by growing vegetation due to photosynthesis, while decomposition of vegetation increases the CO_2 concentration.

A second global dataset for vegetation change assessment is based upon the difference of vertically and horizontally polarized brightness temperature (ΔT) observed by the 37-GHz (wavelength about 8 mm) channel of the scanning multichannel microwave radiometer (SMMR) onboard the Nimbus-7 satellite. Field studies and radiative transfer modeling for agricultural crops have shown that ΔT is affected by soil wetness, soil surface roughness, and vegetation water content (Choudhury 1989). The ΔT values range between 25 and 30 K over dry bare soils, and this value increases to about 35 K when the soil gets moderately wet. With increasing vegetation, the ΔT value decreases due to scattering and absorption of microwave radiation emitted from the soil and also because vegeta-

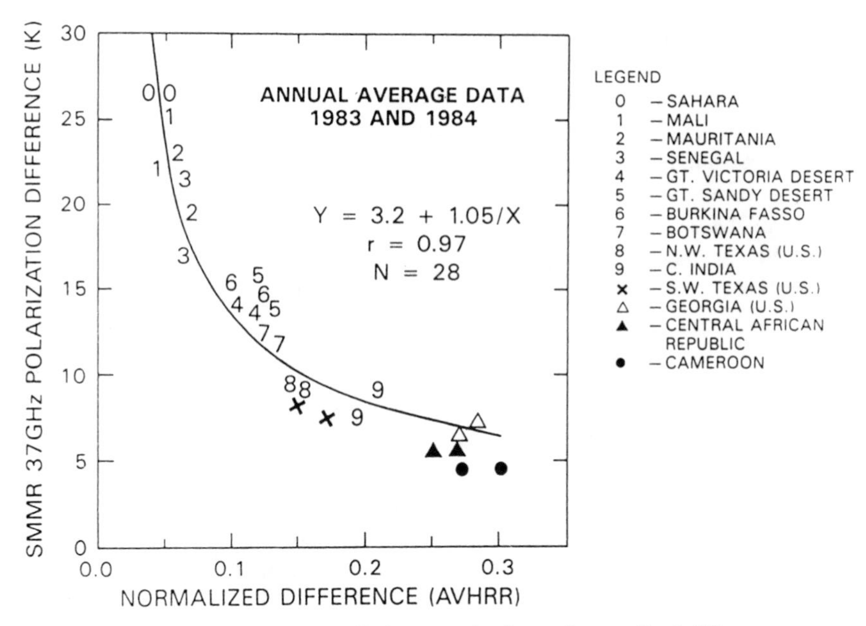

Fig. 8.7. Relation between annually integrated values of normalized difference vegetation index from NOAA-7 satellite and the 37-GHz polarization difference from the Nimbus-7 satellite. The date values are from 14 locations and, for each location, for 1983 and 1984.

tion emits largely depolarized radiation (Choudhury *et al.* 1987). Field studies show that ΔT over dense agricultural crops is about 3 K. Thus, the temporal pattern of ΔT would reflect the temporal pattern of vegetation growth and decay if the effect of temporal variations of soil moisture on ΔT could be minimized.

The ΔT data from SMMR have a spatial resolution of $0.25^\circ \times 0.25^\circ$, and at any location there are about four observations per month. These data have been screened to minimize the effect of soil moisture and gridded globally on a Mercator projection at full resolution from January 1979 to December 1985 (Choudhury and Tucker 1987). The integrated average of these 84 months of data is shown in Figure 8.6 (see color section). The color bar on top of the figure gives the value of ΔT in degrees Kelvin.

The ΔT values are generally greater than 20 K over regions of sparse vegetation (for example, Sahara, Saudi Arabia, and Gobi deserts) and less than 5 K over densely vegetated areas (such as rainforests in Cameroon and Brazil). The ΔT value exceeds 40 K over open water. Because of $0.25^\circ \times 0.25^\circ$ spatial resolution, one could see the Amazon and the Congo Rivers as color anomalies resulting from mixed pixels of high ΔT values for water and low ΔT values for the surrounding vegetation. The seasonal inundation of rivers and swamps of South America has been quantified using this satellite data (Giddings and Choudhury 1989).

The relation between annually averaged values of NDVI from AVHRR data and ΔT from SMMR data for 1983 and 1984 is shown in Figure 8.7 for 14 globally distributed locations. These two satellite-observed vegetation indices are seen to be highly correlated. By a sensitivity analysis of this relation, Becker and Choudhury (unpublished) have concluded that the ΔT data would be highly valuable for studying interannual variation of vegetation over arid and semiarid regions of the world and for studying desertification. A detailed analysis of the global ΔT data as they relate to the atmospheric CO_2 concentration, land-surface evaporation, and primary productivity has been carried out by Choudhury (1988).

Summary

This chapter presents several examples of how current and future satellite datasets may be used to assess change in both type and condition of vegetation. The studies vary in scale and kind of information derived. Remotely sensed data may be acquired which relate to state of health and damage, species/community distribution patterns, deforestation, and seasonal development of vegetation at scales ranging from local to global. Not all of the information would be needed on a continuous basis, but rather, different kinds of remotely sensed data, acquired at different temporal and spatial scales, will prove to be of great value as input in the development of vegetation models as well as for validation of model predictions. Such remote sensing input will be an important component in developing strategies for the study of global change.

Acknowledgments

This chapter describes research carried out at the University of New Hampshire, the NASA/Goddard Space Flight Center, and the Jet Propulsion Laboratory. Numerous agencies have funded portions of this work, including NASA, NOAA, and the USDA Forest Service.

References

Asrar, G., Fuchs, M., Kanemasu, E.T. and Hatfield, J.L. (1984). Estimating absorbed photosynthetic radiation and leaf area index from spectral reflectance in wheat. *Agronomy Journal*, **76**, 300–6.

Choudhury, B.J. (1987). Relationships between vegetation indices, radiation absorption, and net photosynthesis evaluated by a sensitivity analysis. *Remote Sensing of Environment*, **22**, 209–33.

Choudhury, B.J. (1988). Relating Nimbus-7 37 GHz data to global land-surface evaporation, primary productivity and the atmospheric CO_2 concentration. *International Journal of Remote Sensing*, **9**, 169–76.

Choudhury, B.J. (1989). Monitoring global land surface using Nimbus-7 37 GHz data: Theory and examples. *International Journal of Remote Sensing*, **10**, 1579–1605.

Choudhury, B.J. and Tucker, C.J. (1987). Monitoring global vegetation using Nimbus-7 37 GHz data: Some empirical relations. *International Journal of Remote Sensing*, **8**, 1085–90.

Choudhury, B.J., Tucker, C.J., Golus, R.E. and Newcomb, W.W. (1987). Monitoring vegetation using Nimbus-7 scanning multichannel microwave radiometer's data. *International Journal of Remote Sensing*, **8**, 533–8.

Fearnside, P.M. (1982). Deforestation in the Brazilian Amazon: How fast is it occurring? *Intercienca*, **7**, 82–8.

Fearnside, P.M. (1986). Spatial concentration of deforestation in the Brazilian Amazon. *Ambio*, **15**(2), 74–81.

Giddings, L. and Choudhury, B.J. (1989). Observation of hydrologic features with Nimbus-7 37 GHz data, applied to South America. *International Journal of Remote Sensing*, **10**, 1673–86.

Goward, S.N., Dye, D., Kerber, A. and Kalb, V. (1987). Comparison of North and South American biomes from AVHRR observations. *Geocarto International*, **1**, 27–39.

Henderson-Sellers, A. and Gornitz, V. (1984). Possible climatic impacts of land cover transformations, with particular emphasis on tropical deforestation. *Climatic Change*, **6**(3), 231–57.

Herrmann, K., Rock, B.N., Ammer, U. and Paley, H.N. (1988). Preliminary assessment of Airborne Imaging Spectrometer and Airborne Thematic Mapper data acquired for forest decline areas in the Federal Republic of Germany. *Remote Sensing of Environment*, **24**, 129–49.

Holben, B.N. (1986). Characteristics of maximum-value composite images from temporal AVHRR data. *International Journal of Remote Sensing*, **7**, 1417–34.

Hoshizaki, T., Rock, B.N. and Wong, S.K. (1988). Pigment analysis and spectral assessment of spruce trees undergoing forest decline in the United States and Germany. *GeoJournal*, **17**, 173–8.

Houghton, R.A. (1986). Estimating changes in the carbon content of terrestrial ecosystems from historical data. In *The Changing Carbon Cycle: A Global Analysis*, ed. J.R. Trabalka and D.E. Reichle, pp. 175–93. New York: Springer-Verlag.

Houghton, R.A., Hobbie, J.E., Melillo, J.M., Moore, B., Peterson, B.J., Shaver, G.R. and Woodwell, G.M. (1983). Changes in the carbon content of terrestrial biota and soils between 1860 and 1980: A net release of CO_2 to the atmosphere. *Ecological Monographs*, **53**, 235–62.

Houghton, R.A. and Skole, D.L. (1990). Changes in the global carbon cycle, 1700–1980. In *The Earth as Transformed by Human Action*, ed. B.L. Turner, pp. 393–408. New York: Cambridge University Press.

Hunt, E.R., Jr., Rock, B.N. and Nobel, P.S. (1987a). Measurement of leaf relative water content by infrared reflectance. *Remote Sensing of Environment*, **22**, 429–35.

Hunt, E.R., Jr., Wong, S.K. and Rock, B.N. (1987b). Relative water content of spruce needles determined by the leaf water content index. In *Proceedings of the Twenty-First International Symposium on Remote Sensing of the Environment*, Ann Arbor, Michigan, 26–30 October 1987, pp. 1093–110.

Johnson, A.H. and Siccama, T.C. (1984). Decline of red spruce in the northern Appalachians: Assessing the possible role of acid deposition. *Tappi Journal*, **67**, 68–72.

Justice, C.O. ed. (1986). Monitoring the grasslands of semi-arid Africa using NOAA-AVHRR data. *International Journal of Remote Sensing*, **7**, 1383–622.

Justice, C.O., Townshend, J.R.G., Holben, B.N. and Tucker, C.J. (1985). Analysis of the phenology of global vegetation using meteorological satellite data. *International Journal of Remote Sensing*, **6**, 1271–318.

Klein, R.M. and Perkins, T.D. (1987). Cascades of causes and effects of forest decline. *Ambio*, **16**, 86–93.

Malingreau, J.P. and Tucker, C.J. (1988). Large-scale deforestation in the southeastern Amazon Basin of Brazil. *Ambio*, **17**(1), 49–55.

Nelson, R. and Holben, B. (1986). Identifying deforestation in Brazil using multiresolution satellite data. *International Journal of Remote Sensing*, **7**(3), 429–48.

Nelson, R., Horning, N. and Stone, T.A. (1987). Determining the rate of forest conversion in Mato Grosso, Brazil, using Landsat MSS and AVHRR data. *International Journal of Remote Sensing*, **8**(12), 1767–84.

Peterson, D.L., Aber, J.D., Matson, P.A., Card, D.H., Swanberg, N., Wessman, C. and Spanner, M. (1988). Remote sensing of forest canopy and leaf biochemical contents. *Remote Sensing of Environment*, **24**, 85–108.

Rock, B.N., Defeo, N.J. and Vogelmann, J.E. (1987). *Vegetation survey pilot study: Detection and quantification of forest decline damage using remote sensing techniques*. Final Report to the USDA Forest Service, Jet Propulsion Laboratory Document D-4669, Pasadena, California.

Rock, B.N., Elvidge, C.D. and Defeo, N.J. (1988a). Assessment of AVIRIS data from vegetated sites in the Owens Valley, California. In P*roceedings of the AVIRIS Performance Evaluation Workshop*, Jet Propulsion Laboratory, Pasadena, California, 6–8 June 1988, JPL Publication 88-38, pp. 88–96.

Rock, B.N., Hoshizaki, T. and Miller, J.R. (1988b). Comparison of in situ and airborne spectral measurements of the blue shift associated with forest decline. *Remote Sensing of Environment*, **24**, 109–27.

Rock, B.N. and Vogelmann, J.E. (1989). The use of remote sensing for the study of forest damage. In *Proceedings of the International Conference and Workshop—Global Natural Resource Monitoring and Assessments: Preparing for the 21st Century*, Venice, Italy, 24–30 September 1989, pp. 453–67.

Rock, B.N., Vogelmann, J.E., Williams, D.L., Vogelmann, A.F. and Hoshizaki, T. (1986). Remote detection of forest damage. *BioScience*, **36**, 439–45.

Schuett, P. and Cowling, E.B. (1985). Waldsterben, a general decline: Symptoms, development. *Plant Disease*, **69**, 548–58.

Topographic Science Working Group. (1988). *Topographic Science Working Group Report to the Land Processes Branch, Earth Science and Applications Division, NASA Headquarter*. Lunar and Planetary Institute, Houston.

Townshend, J.R.G. and Justice, C.O. (1988). Selecting the spatial resolution of satellite sensors required for global monitoring of land transformations. *International Journal of Remote Sensing*, **9**(2), 187–236.

Tucker, C.J., Holben, B.N. and Goff, T.E. (1984). Intensive forest clearing in Rondonia, Brazil, as detected by satellite remote sensing. *Remote Sensing of Environment*, **15**, 255–61.

Tucker, C.J., Fung, I.Y., Keeling, C.D. and Gammon, R.H. (1986a). Relationship between atmospheric CO_2 variation and a satellite-derived vegetation index. *Nature*, **319**, 195–9.

Tucker, C.J., Townshend, J.R.G., Goff, T.E. and Holben, B.N. (1986b). Continental and global scale remote sensing of land cover. In *The Changing Carbon Cycle: A Global Analysis*, ed. J.R. Trabalka and D.E. Reichle, pp. 221–41. New York: Springer-Verlag.

Vitousek, P.M. (1983). The effects of deforestation on air, soil, and water. In *The Major Biogeochemical Cycles and Their Interactions*, ed. B. Bolin and E. Cook, pp. 223–45. SCOPE 21. New York: John Wiley and Sons.

Vogelmann, J.E. (1988). Detection of forest change in the Green Mountains of Vermont using Multispectral Scanner data. *International Journal of Remote Sensing*, **9**, 1187–200.

Vogelmann, J.E. (1990). Comparison between two vegetation indices for measuring different types of forest damage in the northeastern United States. *International Journal of Remote Sensing*, **11/12**, 2281–97.

Vogelmann, H.W., Bliss, M., Badger, G. and Klein, R.M. (1985). Forest decline on Camels Hump, Vermont. *Bulletin of the Torrey Botanical Club*, **112**, 274–87.

Vogelmann, J.E. and Rock, B.N. (1986). Assessing forest decline in coniferous forests of Vermont using NS-001 Thematic Mapper Simulator data. *International Journal of Remote Sensing*, **7**, 1303–21.

Vogelmann, J.E. and Rock, B.N. (1988). Assessing forest damage in high elevation coniferous forests in Vermont and New Hampshire using Thematic Mapper data. *Remote Sensing of Environment*, **24**, 227–46.

Vogelmann, H.W., Perkins, T., Badger, G. and Klein, R.M. (1988). A 21-year record of forest decline on Camels Hump, Vermont. *European Journal of Forest Pathology*, **18**, 240–9.

Wessman, C.A., Aber, J.D., Peterson, D.L. and Melillo, J. (1988). Remote sensing of canopy chemistry and nitrogen cycling in temperate forest ecosystems. *Nature*, **335**, 154–6.

Woodwell, G.M., Houghton, R.A., Stone, T.A. and Park, A.B. (1986). Changes in the area of forests in Rondonia, Amazon Basin, measured by satellite imagery. In *The Changing Carbon Cycle: A Global Analysis*, ed. J.R. Trabalka and D.E. Reichle, pp. 242–57. New York: Springer-Verlag.

Zawila-Niedzwiecki, T. (1989). Satellite images for forest decline assessment. In *Proceedings of the International Conference and Workshop—Global Natural Resource Monitoring and Assessment: Preparing for the 21st Century*, Venice, Italy, 24–30 September 1989, pp. 473–8.

9

Global Geographic Information Systems and Databases for Vegetation Change Studies

David L. Skole, Berrien Moore III, and Walter H. Chomentowski

Introduction

A description of the geographical distribution of the world's vegetation is a key ingredient of global-vegetation-change studies. One central question is: how is the current distribution of vegetation influenced by climatic factors and human disturbance, and how will it be influenced in the future? Finding the answer will depend largely on the development of models which consider the dynamic interplay between climatic factors, biogeochemistry, and human disturbance. Because the global landscape is spatially heterogeneous and complex, these models will need to depict the geographical distribution of various primary attributes. This necessitates the development of geographically referenced datasets and information systems (National Research Council 1986; IGBP 1988).

The global distribution of vegetation reflects large-scale variations in the distribution of temperature, precipitation, and various other environmental factors. (See Holdridge 1947.) Potentially, this results in a heterogeneous but somewhat predictable global biogeography. However, these factors alone do not account for the rich and changing set of vegetation types observed today. Against a dynamic phenological and successional backdrop characterizing natural communities, humans have continually modified the natural landscape. Thus, the contemporary state of the world's vegetation must be viewed as a constantly shifting mosaic of land-cover types determined by both the physical environment and human activities.

To be sure, human activities have transformed the natural landscape over the past 200 or more years in a variety of ways and to varying degrees. At one end of a gradient of change is deforestation, an obvious and direct form of human-induced vegetation change. At the other end is forest degradation and decline, a subtle and indirect result of human activity. One result of such vegetation change has been a reduction in the amount of carbon stored on land, hence shifting the distribution of carbon from the biota to the atmosphere. While some land transformations increase the amount of carbon stored on land, the overall historical trend has been one of forest conversion to agriculture and other human uses that reduce the amount of carbon stored in the biota, resulting in a net flux of carbon to the atmosphere (Moore *et al.* 1981; Houghton *et al.* 1983). Human alterations to native vegetation have also influenced other atmospheric constituents and biogeochemical cycles (Crutzen *et al.* 1979; Khalil and Rasmussen 1983; Banin *et al.* 1984; Greenberg *et al.* 1984).

Because of the number and complexity of calculations, the development of global models has been, and continues to be, an important part of efforts to understand the dynamic relationship between vegetation and biogeochemical cycles. (See, for instance, Emanuel *et al.* 1984 and Moore *et al.* 1981.) These models have been invaluable for defining and refining basic questions at the global scale but are not rich in geographical detail. Improving the geographical specifications of global models is one way to reduce the uncertainty of current results. For instance, the global net flux of carbon from land-cover changes was between 0.9 and 2.5 Gt in 1980 (Houghton *et al.* 1985a, 1987). The wide range in the estimated global net flux results, in part, from uncertainties concerning the kind of vegetation converted to human uses and the rates of deforestation. Moreover, this flux of carbon, when considered with the release from fossil fuels, appears too large to be accommodated in a balanced budget when linked to existing geochemical and oceanic models of the global carbon cycle. In addition, the long-term net flux of carbon from land-cover change is different from estimates derived by deconvolution of CO_2 concentrations in air trapped in the bubbles of deep ice-cores (Siegenthaler and Oeschger 1987; Houghton and Skole 1990).

These discrepancies might be resolved, or be more precisely defined, by developing models which consider the geographic distribution of ecosystems and land-cover change (Bolin 1984; Emanuel *et al.* 1984, 1985a,b; Houghton *et al.* 1985b). Such an approach would facilitate:

(1) The coregistration of maps of deforestation and ecosystem types which would allow the delineation of the amount of biomass which burns, decays, or is removed, and the timing of these processes
(2) The linkage of terrestrial carbon models to general circulation models (GCMs), thereby providing the GCMs with better information on fluxes and surface boundary conditions
(3) The comparison of estimated terrestrial fluxes with direct measurements of atmospheric carbon dioxide and results from three-dimensional tracer models of atmospheric circulation, which would provide independent cross-checks on global carbon balances

Adding Geographical Detail to Global Ecosystem Models

In recent years a number of authors have advocated the development of global ecosystem models with enhanced geographical detail. Bolin (1984) presents one such strategy. He argues for developing a global database in a format easily compatible with models based on a half-degree-resolution global grid. This gridded database would consist of important variables which capture the geographic and temporal heterogeneity of natural and disturbed ecosystems. The data would be derived from a suite of in situ measurements, cartographic products (e.g., maps), and remote sensing datasets. A similar approach is advocated by Emanuel *et al.* (1985a) for global models. Moore *et al.* (1989) also present such a strategy for modeling hydrology and biogeochemical cycles at the scale of whole water basins. In the approach taken by Moore *et al.* (1989), the geographical framework is based on both a gridded network of environmental factors and polygons delineating basin boundaries.

Two important themes run through these discussions. First, the best way to add geographical detail is to exploit a wide variety of existing data. This requires an organized and systematic way to interchangeably utilize many data types and data formats at different scales and over a range of levels of resolution. Second, methods and protocols for handling the large volume of data inherent in geographically referenced models need to be developed. The volume alone is problematic enough to thwart many new initiatives. For example, Kahn and Leidecker (1989) note that the data stream expected from the planned NASA Earth Observing System (Eos) is likely to be on the order of a terabit per day. The average scientist might be expected to utilize 4.1 gigabytes for a typical model and might also need the extended ability to extract this primary dataset from an initial dataset of between 20 and 40 gigabytes.

The encoding, processing, and utilization of large volumes of variously formatted data mean scientists have to adopt new methodologies, procedures, and approaches to database and model development. Clearly, new global-change research initiatives will need to be coupled to efforts to develop special geo-based information systems. This priority has been part of several national and international fora (National Research Council 1986; Earth Systems Science Committee 1988; IGBP 1988; NASA 1988).

Developing Integrated GEO-Based Information Systems for Global Ecosystem Models

In a general sense, an information system can be defined as a set of procedures or a system which facilitates the flow and processing of data through a sequence of steps from its original acquisition to its integration into analysis (Calkins and Tomlinson 1977). It is useful to think of two types of information system. The first kind is the extensive data-processing and distribution system, such as the proposed Earth Observing System Data and Information System. (See Kahn and Leidecker 1989; Dutton 1989; Chase 1989.) The second is the research-team information system, developed to suit the needs of investigators at their home institutions. The latter would complement the former, and presumably both would exchange high-level data products. Researchers have recently discussed the important role of research-team information systems (Chase 1989; Dutton 1989), but to date there has been little experience in actually developing them for global-scale modeling. In the discussion below, we focus our remarks on the attributes of this second kind of information system.

The quality of being geo-based is not necessarily explicit in the aforementioned definition of information systems. Ten years ago, geographic information systems (GISs) were not in widespread use, and few scientists knew anything about them. Today, there are as many definitions and descriptions of geographic information systems as there are applications and users. Definitions range from simple characterizations as software for computerized cartography to robust characterizations as integrated decision-support technologies. A discussion of the fine distinction between various definitions is beyond the scope of this chapter. We do wish to make one clarification, however. We make a distinction between traditional GISs and integrated geo-based information systems (IGISs). We will use the term GIS specifically in reference to its more traditional, and somewhat more restricted, meaning. (See Smith *et al.* 1987 for a review of traditional GIS.)

Berry (1987) provides a simple but functional definition of GISs as a special class of information system which has three characteristics: (1) they process and manipulate geographically referenced, or spatially coherent, data; (2) they are automated; and (3) they internally relate geographical location and attribute data. These characteristics imply that geographic information systems utilize spatial datasets, such as maps and remote sensing imagery, in digital formats suitable for computational or statistical analyses in both numeric and geographic space. For a review of various definitions and descriptions of GISs, see Parker (1988) and Cowen (1988).

We view integrated geo-based information systems, on the other hand, in a similar but broader context. Architecturally, they are multiple-platform, distributed networks of hardware and software specifically configured for encoding, processing, and analyzing spatial data of many types and formats, including remote sensing data, cartographic data, tabular statistics, and in situ data. In this view, a locally networked research team IGIS would be linked to a larger data and information network. The research team IGIS would also contain a suite of datasets, methodologies, models, and protocols tailored to the team's specific research questions. Thus, the implementation of an integrated geo-based information system transcends engineering and is a crucial element of a research strategy.

Operationally, a research team IGIS would have five components: (1) a data automation component, which permits the digitizing of cartographic products or the importation of other digital data; (2) a geo-referencing component which permits the management of all coordinate, topological, and projection information with capabilities for editing and updating this information; (3) an attribute or database management component, which permits the organization of raw data and data attributes, with capabilities for relational querying of various datasets linked by common attribute and coordinate information; (4) a data analysis component, which would consist of a tool box of utilities for the quantitative and topological analysis of datasets and for cartographic modeling; and (5) an output component which permits the display and drawing of cartographic products or the exporting of digital data (presumably in some standard data-interchange format).

This kind of research-facility information system would have four critical features of system architecture: (1) it would be a hierarchical, networked, distributed processing configuration having links to other networks and data systems; (2) it would be capable of supporting geographically referenced numeric models at the global scale; (3) it would be capable of the development, organization, and distribution of large databases for global-scale research; and (4) it would integrate remote sensing and traditional GIS technology. This last feature is crucial. For global modeling, traditional GIS data sources (e.g., maps, digital elevation models, tabular data) are significantly improved by the use of remotely sensed data.

A Prototype Integrated GEO-Based Information System

This section provides a brief description of a research-team IGIS which we have been developing to support models of global biogeochemistry. Our work has had three basic objectives: (1) to formulate the hardware and software environments for the system, with a particular focus on linking remote sensing and traditional GIS environments; (2) to develop a working compilation of geographically referenced global datasets; and (3) to use the system and datasets to derive geographic descriptions of land-cover change.

Developing the Basic System

Figure 9.1 shows the general architecture of the prototype system. There are four basic components: (1) a geographic information subsystem; (2) a remote sensing subsystem; (3) a "toolbox"; and (4) a model driver. We have utilized standard commercial products whenever possible. In some cases, though, these products have been enhanced or tailored to our needs by the incorporation of our software. The geographic information subsystem manages most of the geographical datasets. Since remote sensing plays a crucial role as a source of geographical data, we have also included a structure to manipulate remote sensing imagery and combine it with other data in the GIS. We have taken particular care to ensure that the remote sensing component is fully compatible with the GIS component. This means that classified imagery can be brought into the GIS as another data

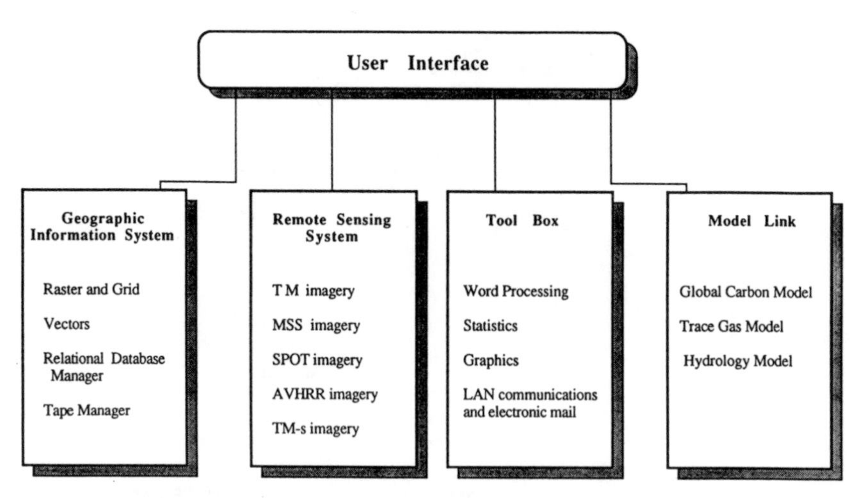

Fig. 9.1. The general structure of a prototype research-team-integrated geo-based information system (IGIS) for global-vegetation-change studies.

layer, or vector datasets can be interactively drawn over imagery in the image processor. The toolbox is a collection of utilities. Included are utilities for Local Area Networking on an EtherNet (TCP/IP) medium, gateways to public data networks and electronic mail, word processing, statistics, and graphics.

The Hardware Environment

Figure 9.2 shows the general hardware configuration of the information system. The IGIS is comprised of several computing platforms linked together by a communications network. We took this approach primarily because model building and data management have different hardware requirements. The former is a central processing unit (CPU)-intensive task which is best handled by a high-end bulk processor, while the latter requires interactive access to data. No single computer could meet both requirements simultaneously. The optimal configuration is a networked system of distributed processors combining the number-crunching capabilities of a mainframe computer with the interactive and imaging capabilities of a workstation.

Data Sources and Data Structures

Generally, we utilize four sources of geographical data: (1) maps and atlases; (2) tabular summaries and documentary statistics; (3) remotely sensed data and imagery; and (4) in situ studies or point measurements. These sources provide information which is both thematic and numerical. Thematic data represent attributes which describe a land surface characteristic according to some naming convention. Typical thematic datasets depict the geographic distribution of types

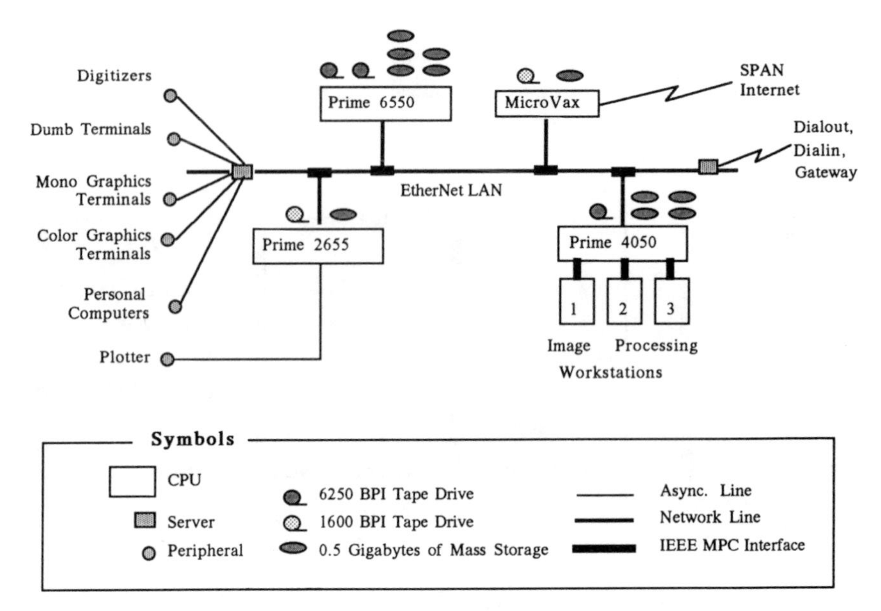

Fig. 9.2. The general layout of the hardware environment developed for a prototype research-team-integrated geo-based information system (IGIS). The figure shows a distributed network configuration of several CPU platforms linked through an EtherNet, with gateways to connect the research-team IGIS with larger data-distribution information systems.

of soil and vegetation and may not provide any kind of quantitative description. Numerical datasets provide geographic distributions of parametric and quantitative information.

Whatever the source, these data are encoded into a computer in some kind of structure or data model. Generally, we employ two broad classes of data models. The first class is the tessellated data model. The grid (a regular square mesh) is the most often used tessellated structure in the system, but other models are employed depending on the specific purpose of the dataset and application. Figure 9.3 shows several kinds of tessellated data structures. Our efforts have focused on implementing the grid and the triangulated irregular network (TIN), but there is much recent literature to support increased utilization of hexagonal structures (Crettez 1980; Gibson and Lucas 1982; van Roessel 1988) and quad-tree structures (Samet 1984). The latter is likely to be an important tessellated structure in the future, since it can effectively capture detailed boundaries between map features in an efficient manner. (See Peuquet 1984 for a good discussion of various geographical data models.)

The second class of data model used in the system is the vector model. Vector structures are often called polygon structures. Figure 9.4 presents various types

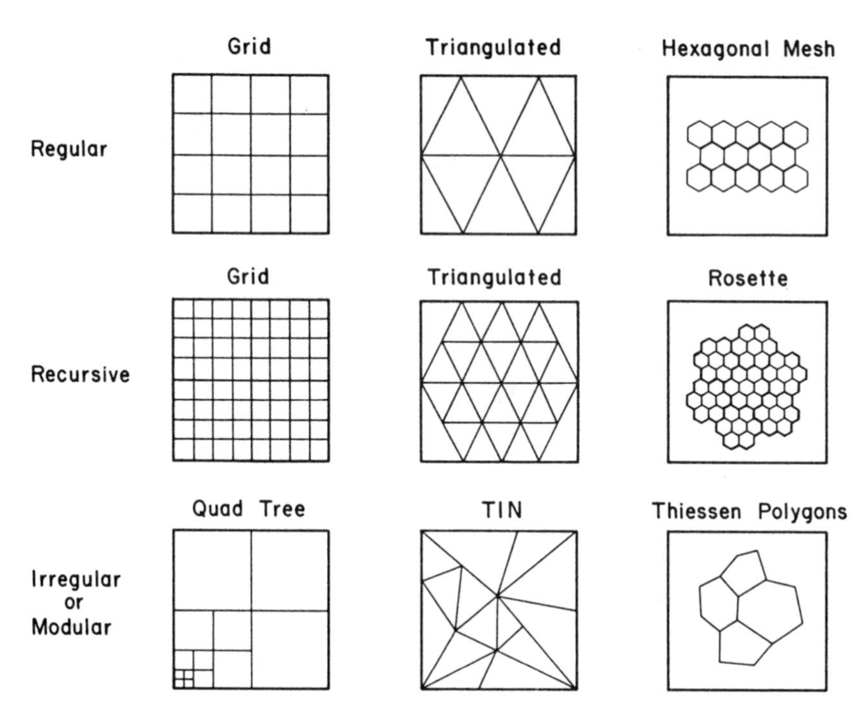

Fig. 9.3. Some common tessellated data structures. The grid, triangulated, and hexagonal formats constitute three subgroups. Each general subgroup can be divided into finer spatial units in a recursive or irregular manner.

of vector data structures employed in the information system, while Figure 9.5 shows how each is encoded. Note that polygon, line, and point data structures are actually different types of vector data. Point data is represented as a zero-order line with a single x-y coordinate. Several points are then connected to make up line data, which can be either branched networks such as river systems or irregular networks such as roadways. Polygon data are comprised of line data which close on themselves. This model is the most effective one for accurately computerizing source data. It is relatively easy to convert polygon data to grid data. It is thus possible to encode the original source without information loss and to create coarse grid datasets which retain subcell inventories of the polygon attributes. This is valuable since typical global models utilize large grid cells on the order of 2500 square kilometers.

These data models organize a dataset into discrete geographical units (cells or polygons) which have associated with them coordinate information and descriptor

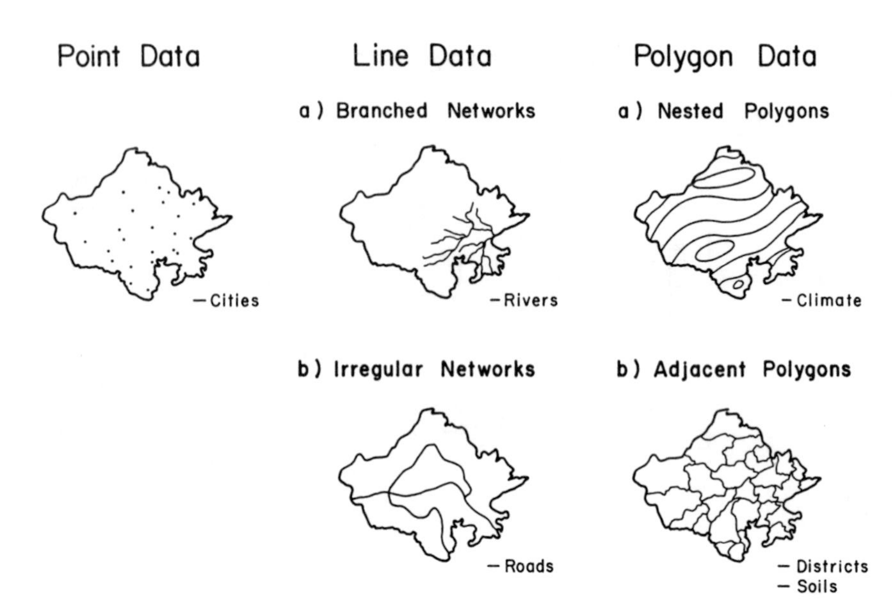

Fig. 9.4. Some common representations of vector data structures. The point, line, and polygon form three general classes of vector data.

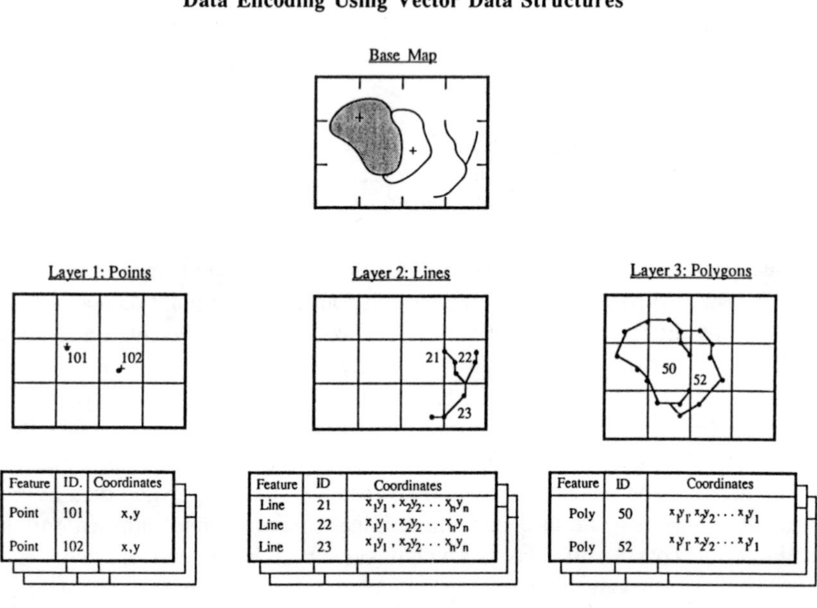

Fig. 9.5. The basic internal structure of the three general classes of vector data. It is useful to conceive of each class of vector data as being organized as separate layers of data.

information. All datasets are divided into two basic components: (1) a cartographic component and (2) a thematic, or feature, component. The cartographic component provides location and coordinate information for each geographical unit (i.e., administrative district, stream segment, meteorological station for polygons, lines, and points, respectively). The thematic component provides the basic information for each geographic unit of the cartographic component (e.g., district name, stream name and order, station name), often with associated numeric datasets (e.g., population, monthly flow rates, monthly precipitation rates).

Datasets

Table 9.1 lists most of the global datasets being developed under this project. The datasets are primarily grid cell data, but as we encode datasets in the future, more will have vector formats. Although the grid cell format is optimal for most modeling, it is better to encode the original data in vector form and convert it to grid cell format for linking to models.

After datasets have been encoded they are coregistered. For this purpose we developed a standard global land surface template to which all datasets conform. This template was developed by using the World Data Bank II coastline and political border data. It is a vector-digitized dataset compiled at an approximately 1:1 000 000 scale. Vector datasets are "rubber-sheeted" to this template.

A grid cell conversion of this template was made for all grid cell datasets. This grid cell template defines which grid cells are to be included in each dataset at several resolutions: 1.0°, 0.5°, and 15 km (for remote sensing data). The grid cell conversion conforms in area to the original vector data to within 1% (at the 0.5° resolution). For each global gridded dataset, cells are added or subtracted so that all datasets match grid cell to grid cell.

All datasets are archived on computer-compatible tape at four levels:

Level 0: The Received Data. The raw dataset as received or encoded in its original form and format.

Level 0A: Backup Copy. The same dataset in a backup format, written to tape in a format consistent with our in-house computer backup system.

Level 1: Model Compatible Copy. This dataset is the original dataset converted to 0.5° grid cells. For instance, a 10-minute grid cell dataset is represented as XLongitude, XLatitude, (Code (i), i=1,9). Point data is represented as XLongitude, XLatitude, (Points(i), i=1, n).

Level 2: Vector-Encoded Copy/Display Copy. This is a digital map which can be plotted or displayed with a legend, annotation, and projection and scale markers.

Level 3: Synthetic/Corrected Dataset. These are datasets which have been derived from a combination of other datasets, produced as model output, or are modified versions of earlier datasets.

Table 9.1. A partial listing of global and regional digital datasets encoded in the prototype research team IGIS. All datasets can be mapped and projected in either vector or grid cell formats and contain both cartographic information (Earth coordinates) and thematic information.

Map description and sources	Nominal resolution and format
Base maps:	
Global land surface template[a, d]	1:1 million vector format
Physical features datasets:	
Global topographic elevation[e]	5 arc-minute grid format
Global mode elevation[f]	10 arc-minute grid format
Global minimum elevation[f]	10 arc-minute grid format
Global maximum elevation[f]	10 arc-minute grid format
Global terrain characteristics[f]	10 arc-minute grid format
Soil datasets:	
FAO World Soils I.[a]	30 arc-minute grid format
FAO World Soils II.[b]	2 arc-minute grid format
USDA 7th Approximation[c]	30 arc-minute grid format
World Organic Soil Carbon[g]	Point format
World Organic Soil Nitrogen[g]	Point format
Soil Texture and Phase[a]	30 arc-minute grid format
Vegetation datasets:	
Holdridge Life Zones[h]	30 arc-minute grid format
World Ecosystem Complexes[i]	30 arc-minute grid format
Global Vegetation and Albedo[j]	60 arc-minute grid format
Primary and Secondary Land Cover[k]	60 arc-minute grid format
Land Cover and Land Use[a]	1:15 million vector format
Hydrology and climatology datasets:	
World Rivers and Lakes[a, d]	1:1 million vector format
World Monthly Surface Climatology[f]	Met station point data format
Terrestrial Global Water Budget[m]	60 arc-minute grid format
Global Cloud Cover[n]	30 arc-minute grid format
Global Irradiance[a]	30 arc-minute grid format

[a]Dataset developed/encoded by the authors and colleagues.

[b]Dataset developed/encoded by UNEP-GRID.

[c]Dataset developed/encoded by W.R. Emanuel, Oak Ridge National Lab.

[d]Central Intelligence Agency (1987).

[e]National Oceanic and Atmospheric Administration (1988).

[f]U.S. Navy (1984).

[g]Zinke *et al.* (1984); see also Post *et al.* (1982, 1984).

[h]Emanuel *et al.* (1985b).

[i]Olson *et al.* (1985); see also Olson *et al.* (1983).

[j]Matthews (1983).

[k]Wilson and Henderson-Sellers (1985).

[l]Spangler and Jenne (1984).

[m]Willmott *et al.* (1985).

[n]Hahn *et al.* (1988); see also Hahn *et al.* (1982, 1984).

An Application: Mapping Land-Cover Change in Brazil

The information system was used to develop a detailed account of the geographic distribution of land-cover change in Brazil from 1960 to 1985. Brazil offers an important test for several reasons. The country is itself continental in scale and provides a good location for a global-scale pilot study. It encompasses the largest tropical-forest biome in the world. It also encompasses a diversity of ecosystems other than forests, including woodlands, grasslands, and wetlands. It is a region which has recently undergone, and is still undergoing, rapid land transformation (Lanly 1982; Fearnside 1986). Finally, it is a region which has many ongoing ground-based and satellite-based studies and a large amount of ancillary data (IBGE 1936–80; Nelson and Holben 1986; Malingreau and Tucker 1988).

Our intention was to develop an approach to mapping land-cover change over large areas and over time. We tried to develop a database which was internally consistent both temporally and geographically and which could provide detailed estimates of the specific ecosystems disturbed at particular points in time. The research-team IGIS described above permitted incorporation of several types and formats of data into a single analysis. The analysis thereby lends insights into some of the major forcing agents of land-cover change and yields improved quantification of a phenomenon which is poorly documented and understood.

We focused the data collection efforts on mapping the geographical distribution of agricultural expansion. Previous work indicates that globally, and in Latin America, the most significant human agent of vegetation change is agricultural expansion (Houghton *et al.* 1983, 1990). A dataset of vegetation conversion was computed indirectly from land-use data. Data for each of the 3970 municipios (towns) in Brazil were obtained from Brazilian census documents (IBGE 1970–80). These data were geo-referenced onto a digital base map of political districts. The data were adjusted to account for the fraction of agricultural areas in natural ecosystems. By organizing the data in a GIS framework it was then possible to make a relational analysis between maps of ecosystems and areas of land-cover change (Fig. 9.6).

Figures 9.7 and 9.8 present two maps which portray changes in land cover due to agricultural expansion between 1970 and 1980. For each date they show the total area converted displayed in five broad percentage classes. Darker areas represent greater amounts of land-cover conversion; lighter areas (e.g., the majority of the Amazon River Basin) represent lesser amounts of land conversion. The total area converted in Brazil rose from 113×10^6 hectares as of 1970 to 143.9×10^6 hectares as of 1980, a 27% increase (Tab. 9.2). The total area of natural ecosystems converted to agricultural uses by 1980 amounted to 17.1% of Brazil's land area.

Several interesting patterns emerge. Two significant "fronts of deforestation" are visible. One occurs along the Belém to Brasília highway, in the vicinity of label "A" on the maps. The other extends west from the Mato Grosso grasslands

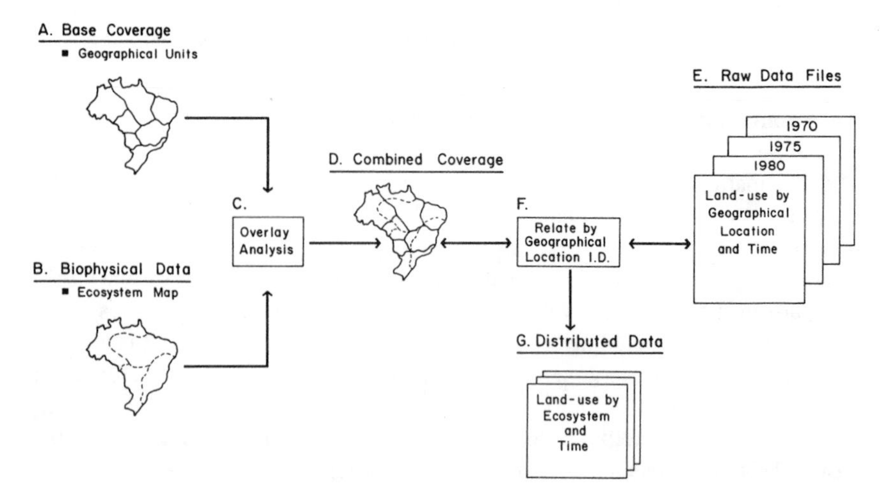

Fig. 9.6. The general approach used in developing an analysis of the ecosystem-specific geographic distribution of vegetation conversion in Brazil. Three separate datasets are interrelated. The base map consists of geographically referenced maps of Brazilian municipios, which constitute the basic cartographic modeling unit in vector format. A second map provides data on the distribution of natural vegetation in a tessellated data structure. Finally, a time-series of land-use data in tabular format is combined with the first two. Remote sensing imagery in raster format is also combined with the other data structures to cross-check the results in selected geographical regions (not shown).

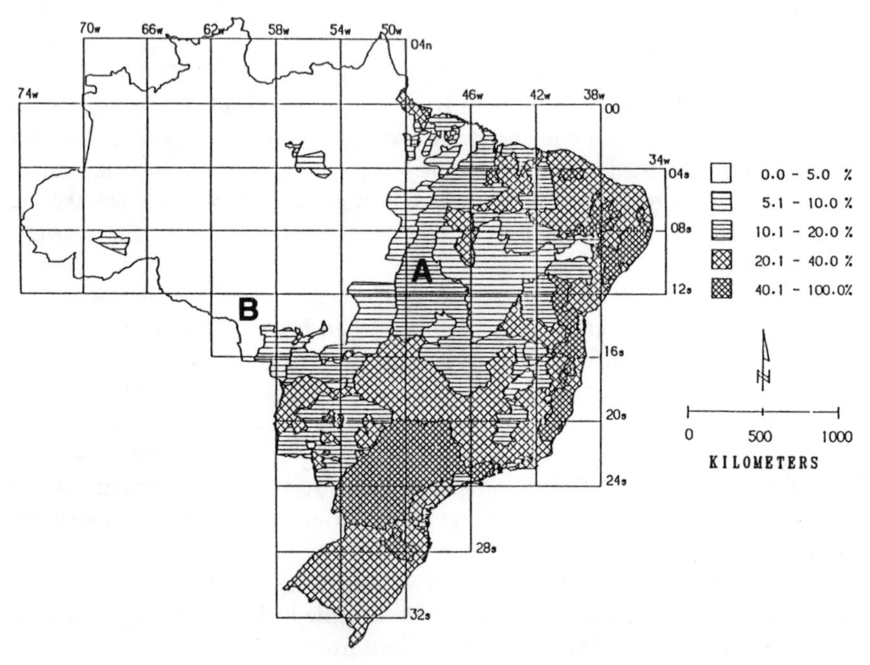

Fig. 9.7. The geographical distribution of natural vegetation conversion by 1970, arranged into classes based on the intensity of disturbance. The shades represent the fraction of a given area converted to agriculture as of 1970.

Fig. 8.2. Landsat Thematic Mapper image of the central Green Mountains of Vermont (USA), produced using TM band 2 (blue), TM band 5 (green), and TM bands 5/4 (red). Green areas on the image represent healthy forests, while red areas represent damage in forests. Camels Hump Mountain is in the top center, south of the Winooski River seen at the very top of the image (north is at the top). Note that areas of heavy damage are located on the west-facing slopes of the mountains. Approximate dimensions of this image are 30 km × 38 km. Image courtesy of Dr. James Vogelmann.

Fig. 8.3. Comparison of high-spectral-resolution Fluorescence Line Imager (FLI) damage assessment image with a broad-band Thematic Mapper Simulator (TMS) damage assessment image for Camels Hump Mountain, Vermont (USA). In both images, red depicts areas of heavy damage in conifers. The spectral data used in the FLI dataset are an indication of variations in chlorophyll absorption in the red/near-infrared (0.65–0.75 m) region, while the TMS spectral dataset includes the short-wave infrared (TM band 5). The FLI image depicts less-extensive damage and is a more accurate portrayal of where forest decline damage is known to occur; the TMS image depicts both areas of forest decline damage in spruce and frost and wind damage in fir. North is at the bottom in each image.

Fig. 8.4. a. A false color-composite SPOT image of a scene acquired on August 23, 1986 in Rondônia, Brazil. The entire scene covers an area of approximately 60 km × 60 km. Most of the deforestation shown in this image is the result of government sponsored settlement projects. b. A full-resolution color-composite image of a portion of the above scene. (Note arrow.) Details of deforestation and land-use patterns are readily visible. A small town can be seen in the upper right. The bright, unpaved roads are also clearly apparent. At this resolution, individual farm plots can be seen, as well as individual farm settlements and structures. The forest canopy surface structure and the shadows of standing trees adjacent to clearings are visible.

Fig. 8.5. a. The entire SPOT scene after having been degraded from the original SPOT reference image to simulate AVHRR LAC spatial resolution (1-km pixels). Compare with Figure 8.4a. b. A portion of the SPOT scene which has been degraded to the spatial resolution of Landsat MSS imagery (80-m pixels). Compare this scene with the same area represented in the SPOT reference image in Figure 8.4b.

Fig. 8.6. Color-coded global image of 37-GHz polarization difference from the Nimbus-7 satellite. The color scale gives the value of the polarization difference in degrees Kelvin.

Figure 10.7b

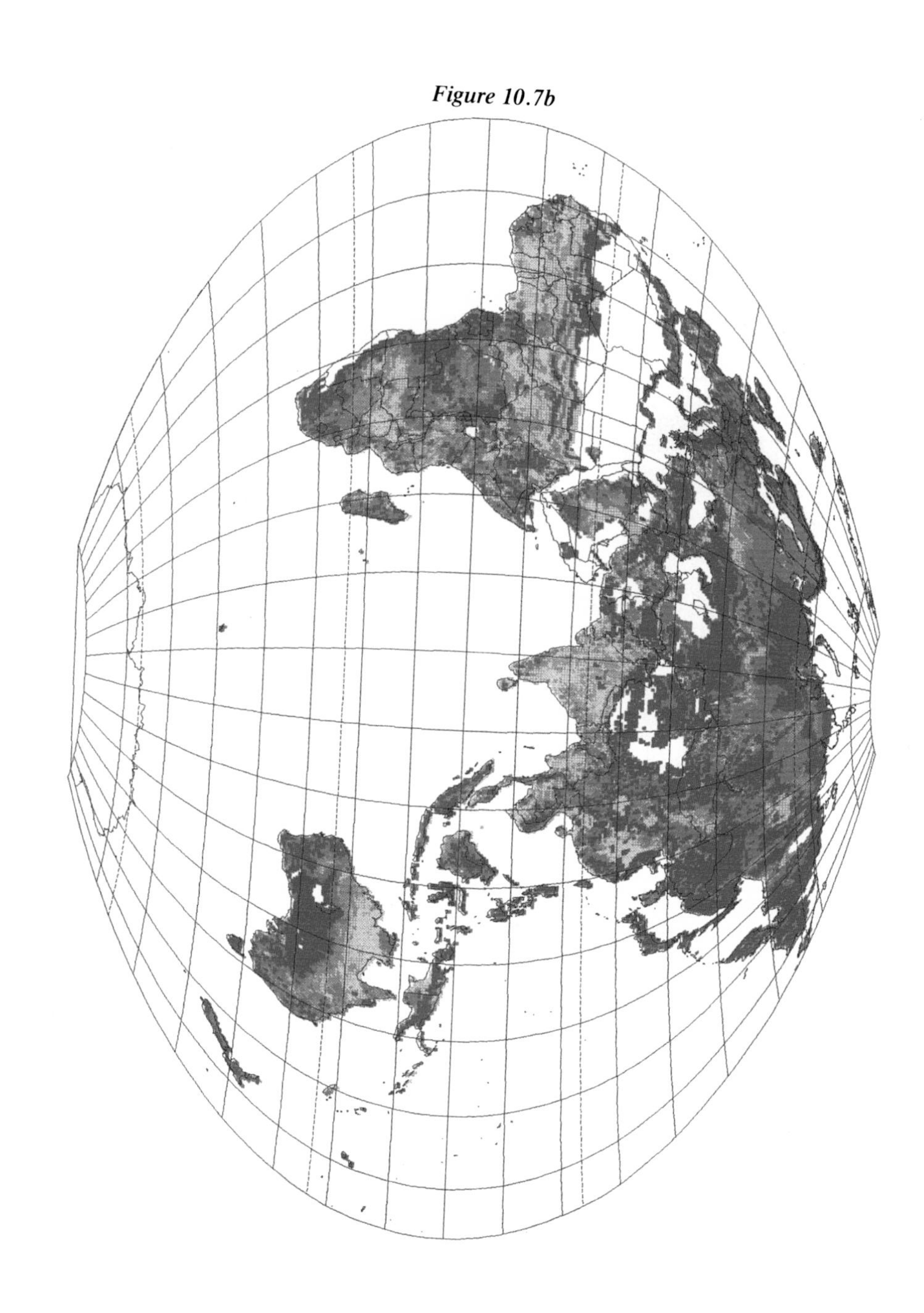

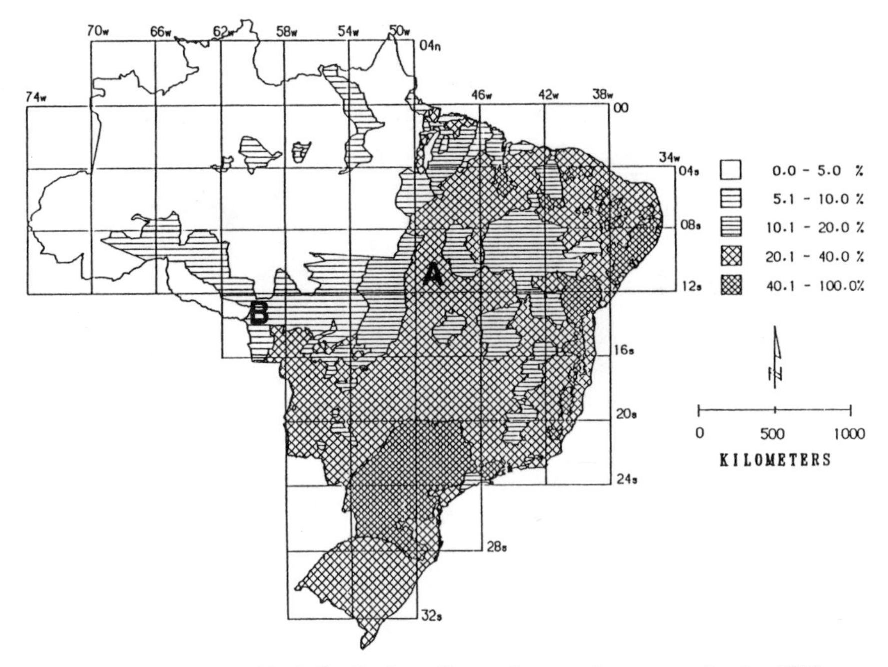

Fig. 9.8. The geographical distribution of natural vegetation conversion by 1980, arranged into classes based on the intensity of disturbance. The shades represent the fraction of a given area converted to agriculture as of 1980.

to the colonization areas in the state of Rondônia, shown by the letter "B" on the maps. This region has been the site of efforts to develop the nation's frontier economically and has experienced explosive deforestation in recent years. Roads in once-inaccessible frontiers act as demographic corridors catalyzing human migration into once-undisturbed ecosystems. Our mapped geographical patterns generally agree with recent reports from the region (Fearnside 1986).

Table 9.2. Land-cover conversion in Brazil, 1970–80. All units expressed as 10^3 hectares, except when given as percentages.

	1970	1975	1980
Total area converted	113 481	128 852	143 931
As a percent of Brazil	13.4	15.3	17.1
Net change between dates		15 371	15 074
As a percent of Brazil		1.82	1.78
As a percent of undisturbed area		2.10	2.06
Average annual change		3074	3016
As a percent of Brazil		0.36	0.36
As a percent of undisturbed area		0.42	0.41

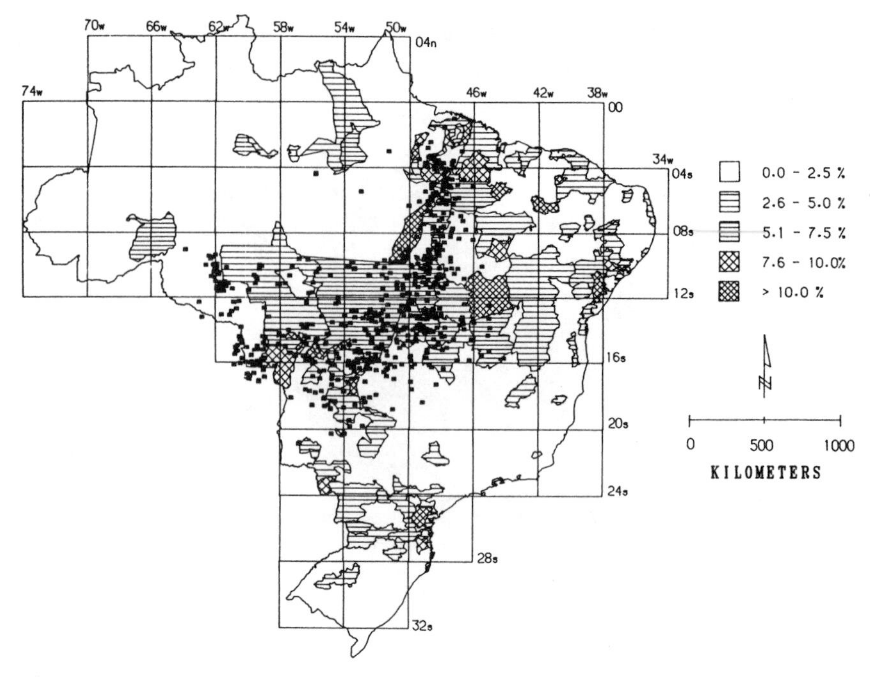

Fig. 9.9. A map showing the net change in natural vegetation cover in Brazil between 1975 and 1980. This map was produced as the difference image of maps for 1975 and 1980. Thermal anomalies (fire *hot spots*) from the thermal channel AVHRR of the NOAA-9 orbital satellite are overlaid on the derived map and shown as dark dots.

Between dates, map overlay algorithms can be used to estimate annual conversion rates and their spatial characteristics; essentially two maps are geographically and geometrically coregistered, and one is numerically subtracted from the other. We performed this change detection analysis by using three dates: 1970, 1975, and 1980. Figure 9.9 shows the change between 1975 and 1980. It is interesting that the geographical positioning of land-cover conversion in the first half of that decade occurred in southern Brazil around the states of Mato Grosso do Sul and Paraná and along a northward connecting corridor from Brasília to Belém. There was also conversion occurring in the southern state of Rio Grande do Sul. The "front" of land-cover conversion in the latter half of the decade pushed north and west into the states of Mato Grosso and Rondônia, with scattered fronts in east-central Brazil.

The net change for the whole of Brazil between 1970 and 1975 was 15.4×10^6 hectares; between 1975 and 1980 the net change was 15.1×10^6 hectares (Tab. 9.2). Averaged annually, the rate of change for all of Brazil appears to be fairly constant (Tab. 9.2). However, there are certain geographical regions which differ considerably from the national average. The results for the state of Rondônia are shown in Table 9.3. There, the total area converted more than doubled during

Table 9.3. Land-cover conversion in Rondônia, Brazil, 1970–80. All units expressed as 10^3 hectares, except when given as percentages.

	1970	1975	1980
Total area converted	479	524	968
As a percent of Rondonia	1.9	2.1	4.0
Net change between dates		45(9%)	443(91%)
As a percent of Rondonia		0.2	1.8
As a percent of undisturbed area		0.2	1.9
Average annual change		9	88
As a percent of Rondonia		0.04	0.36
As a percent of undisturbed area		0.04	0.37

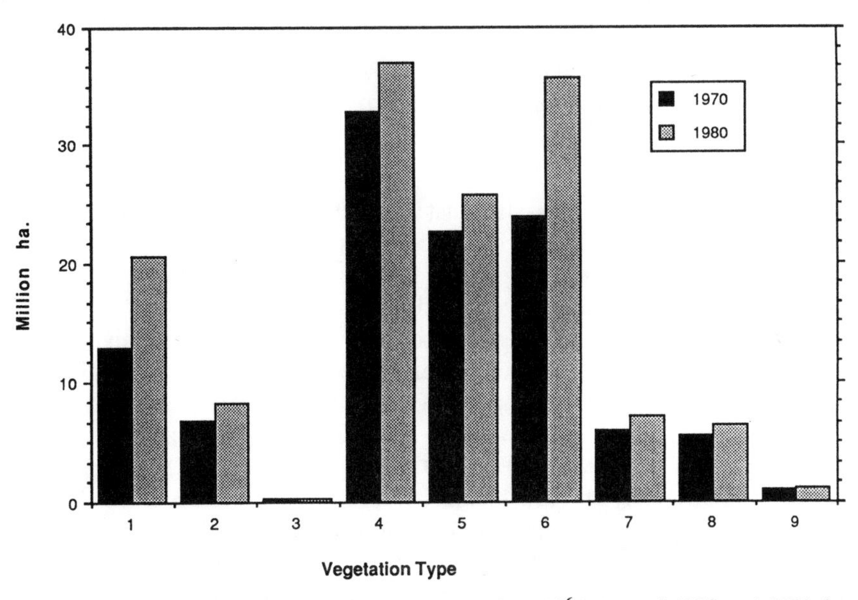

Fig. 9.10. The distribution of vegetation conversion (10^6 ha) as of 1970 and 1980 for each general vegetation type mapped in Brazil. Vegetation types are based on the UNESCO classification of Matthews (1983): (1) tropical evergreen rainforest, mangrove forest; (2) tropical/subtropical evergreen seasonal broadleafed forest; (3) subtropical evergreen rainforest; (4) tropical-subtropical drought-deciduous forest; (5) evergreen broadleafed sclerophyllous woodland; (6) woodland-grassland with 10–40% woody cover; (7) grassland-woodland with less than 10% woody cover; (8) grassland with shrub cover; and (9) tall grassland.

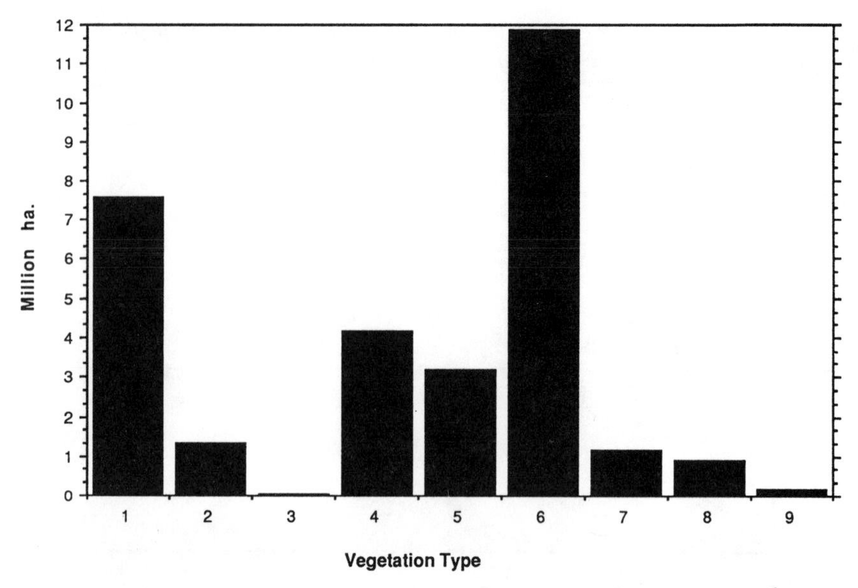

Fig. 9.11. The distribution of the net conversion between 1970 and 1980 (in 10^6 ha) for each general vegetation type mapped in Brazil. The same vegetation types labeled in Figure 9.10 are used here.

the decade 1970–80. It is interesting that 91% of the conversion which had taken place by 1980 occurred in the previous 5 years (Tab. 9.3). The average annual change rose from 9090 hectares per year in the first half of the decade to 88 560 hectares per year in the last half of the decade.

These changes in the geographical positioning of conversion activities are associated with different vegetation communities. By overlaying the digital information on land-cover conversion with a digital map of ecosystems (Matthews 1983), the geographical positioning of conversion can be related to specific ecosystem types as shown in Figure 9.10. Human activities have primarily disturbed the relatively dry, low-biomass drought-deciduous forests, woodlands, and woodland-grassland complexes. The greatest net change between 1970 and 1980 occurred in two distinct ecosystems: the woodland-grassland complexes and the tropical evergreen rainforest (Fig. 9.11), reflecting the geographical shift in land-conversion activities into Mato Grosso and Rondônia.

These data were cross-checked against other datasets derived from remote sensing analyses to obtain an estimate of the accuracy of this approach. The region of rapid deforestation in the state of Rondônia has been extensively studied using satellite imagery, mostly Landsat MSS data and AVHRR data (Tucker *et al.* 1986; Woodwell *et al.* 1986; Malingreau and Tucker 1988) for the period 1980–87.

A simple analysis using satellite data can be made by comparing the derived maps with information on the location of fires derived from the thermal channel of the AVHRR (Matson and Dozier 1981; Matson *et al.* 1984). Areas of current human activity observed from space can be compared with the derived maps for geographical similarities. Figure 9.9 is a map of land transformation discussed earlier, along with a single day's (September 1987) fire spots, or thermal anomalies, overlaid. From this comparison, one can see generally good geographical correspondence between the two datasets.

Acknowledgments

This paper describes research carried out at the Institute for the Study of Earth, Oceans, and Space, University of New Hampshire, with support from the National Aeronautics and Space Administration (NASA grant NAGW 980) and with partial support from the Department of Energy.

References

Banin, A.J., Lawless, J.G. and Whitten, R.C. (1984). Global N_2O cycles, terrestrial emissions, atmospheric accumulation, and biospheric effects. *Advances in Space Research*, **4**(12), 207–16.

Berry, J.K. (1987). Computer assisted map analysis: Potential and pitfalls. *Photogrammetric Engineering and Remote Sensing*, **53**(10), 1405–10.

Bolin, B. (1984). Global biogeochemical cycles: Studies of interaction and change—Some views on the strategy of approach. In *The Interaction of Global Biochemical Cycles*, ed. B. Moore and M. Dastoor, pp. 25–51. JPL Publication 84–21. Jet Propulsion Lab, NASA, Pasadena, California.

Calkins, H.W. and Tomlinson, R.F. (1977). *Geographic information systems, methods and equipment for land use planning*. International Geographic Union Commission on Geographic Data Sensing and Processing, Resource and Land Investigations Program, U.S. Geologic Survey, Reston, Virginia.

Central Intelligence Agency. (1987). World Data Bank II. PB87–184768. Washington, D.C.: National Technical Information Service.

Chase, R.R.P. (1989). Toward a complete Eos data and information system. *IEEE Transactions on Geoscience and Remote Sensing*, **27**(2), 125–31.

Cowen, D.J. (1988). GIS versus CAD versus DBMS: What are the differences? *Photogrammetric Engineering and Remote Sensing*, **54**(11), 1551–5.

Crettez, J. (1980). A pseudo-cosine transformation for hexagonal Tessellation with an heptarchical organization. *Proceedings of the IEEE Computer Society Conference on Pattern Recognition and Image Processing*, Miami Beach, 1980, pp. 192–4.

Crutzen, P.J., Heidt, L.E., Krasnec, J.P., Pollock, W.H. and Seiler, W. (1979). Biomass burning as a source of the atmospheric gases CO, H_2, N_2O, NO, CH_3Cl and COS. *Nature*, **282**, 253–6.

Dutton, J.A. (1989). The Eos Data and Information System: Concepts for design. *IEEE Transactions on Geoscience and Remote Sensing*, **27**(2), 109–16.

Earth Systems Science Committee. (1988). *Earth System Science, A Closer View*. Report of the NASA Advisory Council. Washington, D.C.: National Aeronautics and Space Administration.

Emanuel, W.R., Fung, I.Y.S., Killough, G.G., Moore, B. and Peng, T.H. (1985a). Modeling the global carbon cycle and changes in the atmospheric carbon dioxide levels. In *Atmospheric Carbon Dioxide and the Global Carbon Cycle*, ed. J.R. Trabalka, pp. 141–73. DOE/ER-0239. Washington, D.C.: U.S. Department of Energy.

Emanuel, W.R., Moore, B. and Shugart, H.H. (1984). Some aspects of understanding changes in the global carbon cycle. In *The Interaction of Global Biochemical Cycles*, ed. B. Moore and M. Dastoor, pp. 55–83. JPL Publication 84-21. Jet Propulsion Lab, NASA, Pasadena, California.

Emanuel, W.R., Shugart, H.H. and Stevenson, M.P. (1985b). Climatic change and the broad-scale distribution of terrestrial ecosystem complexes. *Climatic Change*, **7**, 29–43.

Fearnside, P.M. (1986). Spatial concentration of deforestation in the Brazilian Amazon. *Ambio*, **15**(2), 74–81.

Gibson, L. and Lucas, D. (1982). Spatial data processing using generalized balanced ternary. *Computer Graphics and Image Processing*, **20**, 82–9.

Greenberg, J.P., Zimmerman, P.P., Heidt, L.E. and Pollock, W. (1984). Hydrocarbon and carbon monoxide emissions from biomass burnings in Brazil. *Journal of Geophysical Research*, **98**, 1350–4.

Hahn, C.J., Warren, S.G., London, J., Chervin, R.M. and Jenne, R.L. (1982). *Atlas of simultaneous occurrence of different cloud types over the ocean*. NCAR Technical Note TN-201+STR (NTIS number PB83-152074). Boulder, Colorado: National Center for Atmospheric Research.

Hahn, C.J., Warren, S.G., London, J., Chervin, R.M. and Jenne, R.L. (1984). *Atlas of simultaneous occurrence of different cloud types over land*. NCAR Technical Note TN-241+STR (NTIS number PB88-118641/AS). Boulder, Colorado: National Center for Atmospheric Research.

Hahn, C.J., Warren, S.G., London, J., Jenne, R.L. and Chervin, R.M. (1988). *Climatological data for clouds over the globe from surface observations*. NDP026. Oak Ridge, Tennessee: Carbon Dioxide Information Analysis Center, Oak Ridge National Laboratory.

Holdridge, L.R. (1947). Determination of world plant formations from simple climatic data. *Science*, **105**, 367–8.

Houghton, R.A., Boone, R.D., Fruci, J.R., Hobbie, J.E., Melillo, J.M., Palm, C.A., Peterson, B.J., Shaver, G.R., Woodwell, G.M., Moore, B., Skole, D.L. and Myers, N. (1987). The flux of carbon from terrestrial ecosystems to the atmosphere in 1980 due to changes in land use: Geographic distribution of the global flux. *Tellus*, **39B**, 122–39.

Houghton, R.A., Boone, R.D., Melillo, J.M., Palm, C.A., Woodwell, G.M., Myers, N., Moore, B. and Skole, D.L. (1985a). Net flux of CO_2 from tropical forests in 1980. *Nature*, **316**, 617–20.

Houghton, R.A., Hobbie, J.E., Melillo, J.M., Moore, B., Peterson, B.J., Shaver, G.R. and Woodwell, G.M. (1983). Changes in the carbon content of terrestrial biota and soils between 1860 and 1980: A net release of CO_2 to the atmosphere. *Ecological Monographs*, **53**, 235–62.

Houghton, R.A., Lefkowitz, D.S. and Skole, D.L. (1991). Changes in the landscape of Latin America between 1850 and 1985. I. Progressive loss of forest. *Forest Ecology and Management*, **38**, 143–172.

Houghton, R.A., Schlesinger, W.H., Brown, S. and Richards, J.F. (1985b). Carbon dioxide exchange between the atmosphere and terrestrial ecosystems. In *Atmospheric Carbon Dioxide and the Global Carbon Cycle*, ed. J.R. Trabalka, pp. 113–40. DOE/ER-0239. Washington, D.C.: U.S. Department of Energy.

Houghton, R.A. and Skole, D.L. (1990). Changes in the global carbon cycle between 1700 and 1985. In *The Earth as Transformed by Human Action*, ed. B.L. Turner, pp. 393–408. New York: Cambridge University Press.

IBGE (Instituto Brasileiro do Geografia e Estatistica). (1936–80). *Anuario Estatistico do Brasil*. Rio de Janeiro.

IBGE (Instituto Brasileiro do Geografia e Estatistica). (1970–80). *Censo Agropecuario*. Rio de Janeiro.

IGBP (International Geosphere-Biosphere Programme). (1988). *Global Change*. Report No. 4. Special Committee for the IGBP, Stockholm, Sweden.

Kahn, R. and Leidecker, H. (1989). The crush of new data knocking at our door, or how to read "one Library of Congress" every few weeks. *Renewable Resources*, 7:8–13.

Khalil, M.A.K. and Rasmussen, R.A. (1983). Sources, sinks, and seasonal cycles of atmospheric methane. *Science*, **224**, 54–6.

Lanly, J.-P. (1982). *Tropical Forest Resources*. FAO Forestry Paper 30. Rome: Food and Agriculture Organization, United Nations.

Malingreau, J.P. and Tucker, C.J. (1988). Large scale deforestation in the southeastern Amazon Basin of Brazil. *Ambio*, **17**(1), 49–54.

Matson, M. and Dozier, J. (1981). Identification of subresolution high temperature sources using thermal IR sensors. *Photogrammetric Engineering and Remote Sensing*, **47**(9), 1311–8.

Matson, M., Schneider, S.R., Aldridge, B. and Satchwell, B. (1984). *Fore Detection Using NOAA-Series Satellite*. NOAA Technical Report NESDIS-7. Washington, D.C.

Matthews, E. (1983). Global vegetation and land use: New high resolution data bases for climate studies. *Journal of Climate and Applied Meteorology*, **22**(3), 474–87.

Moore, B., Boone, R.D., Hobbie, J.E., Houghton, R.A., Melillo, J.M., Peterson, B.J., Shaver, G.R., Vorosmarty, C.J. and Woodwell, G.M. (1981). A simple model of the role of terrestrial ecosystems in the global carbon budget. In *Carbon Cycle Modelling*, ed. B. Bolin, pp. 365–85. SCOPE 16. Chichester: John Wiley & Sons.

Moore, B., Gildea, M.P., Vorosmarty, C.J., Skole, D.L., Melillo, J.M., Peterson, B.J., Rastetter, E.B. and Steudler, P.A. (1989). Biogeochemical cycles. In *Global Ecology*, ed. M.B. Ramber, L. Margulis and R. Fester, pp. 113–41. San Diego: Academic Press.

NASA (National Aeronautics and Space Administration). (1988). *From Pattern to Process: The Strategy of the Earth Observing System*. Eos Science Steering Committee Report, Volume II. Washington, D.C.: National Aeronautics and Space Administration.

National Oceanic and Atmospheric Administration. (1988). *ETOPO 5-Minute Gridded World Elevations*. Boulder, Colorado: National Geophysical Data Center.

National Research Council. (1986). *Global Change in the Geosphere-Biosphere*. Washington, D.C.: National Academy Press.

Nelson, R. and Holben, B. (1986). Identifying deforestation in Brazil using multiresolution satellite data. *International Journal of Remote Sensing*, **7**(3), 429–48.

Olson, J.S., Watts, J.A. and Allison, L.J. (1983). *Carbon in Live Vegetation of Major World Ecosystems*. ORNL-5862. Oak Ridge, Tennessee: Oak Ridge National Laboratory.

Olson, J.S., Watts, J.A. and Allison, L.J. (1985). *Major ecosystem complexes ranked by carbon in live vegetation: A database*. Oak Ridge, Tennessee: Carbon Dioxide Information Center, Oak Ridge National Laboratory.

Parker, H.D. (1988). The unique qualities of a geographic information system: A commentary. *Photogrammetric Engineering and Remote Sensing*, **54**(11), 1547–9.

Peuquet, D.J. (1984). A conceptual framework and comparison of spatial data models. *Cartographica*, **21**(4), 66–113.

Post, W.M., Emanuel, W.R., Zinke, P.J. and Stangenberger, A.G. (1982). Soil carbon pools and world life zones. *Nature*, **298**(5870), 156–9.

Post, W.M., Pastor, J., Zinke, P.J. and Stangenberger, A.G. (1984). Global patterns of soil nitrogen storage. *Nature*, **317**(6038), 613–6.

Samet, H. (1984). The quadtree and related hierarchical data structures. *ACM Computing Surveys*, **16**(2), 187–260.

Siegenthaler, U. and Oeschger, H. (1987). Biospheric CO_2 emissions during the past 200 years reconstructed by deconvolution of ice core data. *Tellus*, **39B**, 140–54.

Smith, T.R., Menon, S., Star, J.L. and Estes, J.E. (1987). Requirements and principles for the implementation and construction of large-scale geographic information systems. *International Journal of Geographical Information Systems*, **1**(1), 13–32.

Spangler, W. and Jenne, L. (1984). *World monthly surface station climatology*. National Center for Atmospheric Research, Scientific Computing Division, (unpublished).

Tucker, C.J., Townshend, J.R.G., Goff, T.E. and Holben, B.N. (1986). Continental and global scale remote sensing of land cover. In *The Changing Carbon Cycle: A Global Analysis*, ed. J.R. Trabalka and D.E. Reichle, pp. 221–41. New York: Springer-Verlag.

U.S. Navy. (1984). *Global 10-Minute Elevation Data*. Boulder, Colorado: National Center for Atmospheric Research.

van Roessel, J.W. (1988). Conversion of Cartesian coordinates from and to generalized balanced ternary addresses. *Photogrammetric Engineering and Remote Sensing*, **54**(11), 1565–70.

Willmott, C.J., Rowe, C.M. and Mintz, Y. (1985). Climatology of the terrestrial seasonal water cycle. *Journal of Climatology*, **5**(6), 589–606.

Wilson, M.F. and Henderson-Sellers, A. (1985). A global archive of land cover and soils data for use in general circulation climate models. *Journal of Climatology*, **5**, 119–43.

Woodwell, G.M., Houghton, R.A., Stone, T.A. and Park, A.B. (1986). Changes in the area of forests in Rondonia, Amazon Basin, measured by satellite imagery. In *The Changing Carbon Cycle: A Global Analysis*, ed. J.R. Trabalka and D.E. Reichle, pp. 242–57. New York: Springer-Verlag.

Zinke, P.J., Stangenberger, A.G., Post, W.M., Emanuel, W.R. and Olson, J.S. (1984). *Worldwide Organic Soil Carbon and Nitrogen Data*. ORNL/TM-8857. Oak Ridge, Tennessee: Oak Ridge National Laboratory.

10

Assessing Impacts of Climate Change on Vegetation Using Climate Classification Systems

Wolfgang P. Cramer and Rik Leemans

Introduction

In response to the growing public and scientific concern about global environmental change and its impact on terrestrial ecosystems, a demand for spatially explicit predictions of potential vegetation under increased atmospheric CO_2 has appeared (Prentice *et al.* 1989). Maps based on such predictions are expected to serve as an indication of likely impacts of global warming on natural vegetation (Busby 1988), wildlife (Arnold 1988) and agriculture and forestry (Parry *et al.* 1988; Graetz *et al.* 1988). Other important issues which could be addressed are likely changes in biodiversity, as well as levels in and feedback mechanisms between the sources and sinks of the global carbon cycle (Prentice and Fung 1990; Adams *et al.* 1990).

In the first section of this chapter, we briefly outline the history of static models that can be used to predict future potential vegetation. We show that such models, either by explicit formulations or (more often) implicitly in other theories, form an important part of traditions in global mapping of both climate and vegetation. This large body of accumulated knowledge can now be used as a predictive tool. In the second half of the chapter, we review two such models and discuss how they are used to derive global scenarios of terrestrial ecosystem change.

Static models of climate-vegetation interactions rest upon the basic assumption that climate and vegetation are in equilibrium with each other. This view has caused controversy for some time, but many authors now agree that it is valid for moderately long time-scales and on a regional spatial scale (Davis 1981; Prentice 1986; Ritchie 1986; Webb 1986, 1987). Given this assumption, the insights developed in paleoenvironmental studies can also be expected to have a meaning for understanding issues of future global change. Maps derived from simple formulations of climate-vegetation relationships may allow for the recognition of regions with large predicted changes in potential natural vegetation. Static models can therefore also be used to determine likely shifts in anthropogenic vegetation, e.g., regions climatically suitable for agricultural or forestry crops (Blasing and Solomon 1984).

Rates of climate change currently predicted (ca. 0.3°C/decade; Houghton *et al.* 1990) exceed the rates of observed Holocene vegetation change by at least one order of magnitude (Solomon and Cramer, in this volume). Successional processes (e.g., diaspore transport, soil development, change in vegetation structure) limit the rate of large-scale biome shifts, and the equilibrium assumption might therefore become invalid (Prentice 1988). Some species may not establish in areas with a now-suitable climate because they may not reach them fast enough. Predictions derived from static models should therefore be more realistic for *declining* natural vegetation than for scenarios of species invading new areas.

Traditional Approaches to Climate-Vegetation Classification

In the absence of a dense network of weather stations, vegetation has long been recognized as an integrative indicator of regional and global terrestrial climate. During earlier phases of geography and vegetation science, this recognition was implicit in the overall purpose of most expeditions to unknown and remote regions. Travel descriptions from these days often surprise the modern reader with a strange mixture of observations about climate and vegetation. This may appear unsystematic and coincidental, because the description also included notes on land use and other cultural aspects (Linnaeus [1732] 1977; Hearne 1772).

Most explorers of the 19th century used a more organized structure to present their observations and thereby laid the ground for modern regional geography. A systematic treatment of any region consisted of a concise description of the regional bedrock and sediment geology, details of the prevailing soil types, some climatic observations, an account of vegetation and land use (these two factors were often seen as synonymous), and finally a detailed account of the human settlements (von Humboldt 1807; Schouw 1823; de Candolle 1855; Merriam 1898).

The perception that the vegetation of a given area represents a good summary of regional gradients in climate has been used indirectly for the compilation of large-scale global climate maps. Species distribution limits or vegetation gradients, both with changing altitude and with increasing distance to the sea, were used as indications about major climatic gradients. Vegetation thus became a main source of information for global climate classifications (e.g., Köppen 1884, 1936; Holdridge 1947; Thornthwaite 1948; Troll and Paffen 1964). These mapping techniques assumed both that patterns of vegetation distribution and those of climate were in equilibrium and that climate was the single determining factor for vegetation. Walter (1964) made this assumption explicit and asserted that it should be roughly valid for the global-scale distribution of biomes and vegetation types.

Not until the 20th century did global maps of vegetation and/or climate become clearly divided conceptually. A major step in this direction was Holdridge's (1947) "life zone" concept in which he attempted to clearly define regional climate types in such a way that the dominating ecosystem type of the region was predicted. The climate classification itself was based on measured variables (temperature and precipitation) only. This approach extended earlier correlative analyses of species' distributions and isotherms (reviewed by several authors, e.g., Hutchins 1947) through a multidimensional classification scheme basically involving warmth and moisture. It was recognized that standard meteorological data such as mean monthly or annual temperatures and precipitation sums were poorly related to ecosystem types. Therefore, bioclimatic variables (*biotemperature* and estimated evapotranspiration; for details see below) were derived from such input, and ecosystem types were derived empirically from critical thresholds of these. The resulting 20 major ecosystem types were termed *life zones* to account for the width of the concept, which was meant to involve not only potential natural vegetation but also fauna and possible land use. Although the *life zone approach* still ignored major environmental determinants of ecosystem function and structure, such as seasonality or soil characteristics, it laid the ground for all later multidimensional approaches based on standard weather-station data.

Holdridge's approach has several parallels with different terminology, e.g., those of Sukachev (1958, *biogeocoenosis*), Krajina (1959, *edaphotope*), or Kojima (1979, *biogeoclimatic zones*). The main difference, however, is the strictly numerical formulation of his classification, which makes it suitable for computerized applications.

Owing to the zonal concept of the Holdridge system, each locality is assigned only one unique ecosystem type. Box (1981) extended this system to a more complex concept related to species and the possibility of their coexistence. Using earlier theory about plant types and their relation to the environment (Raunkiær 1907; Boysen Jensen 1949) and current database techniques, the link was made back to the species-climate correlations of earlier phytogeographers. Also the aim was to achieve global coverage in one model. Climate is represented in this model by eight bioclimatic variables derived from measured temperature and precipitation data. These variables were developed to account better for drought stress and seasonality than the Holdridge system did. The more important result was, however, that most localities can now, as in the real world, support several plant types, while others, such as deserts or ice-sheets, can be entirely empty.

A shortcoming of this approach is that it depends entirely on the current knowledge about the species' major functional characteristics (reduced to a few types) in any given area. In contrast to information about the more general aspects of biomes (life zones), this knowledge probably does not exist. The system requires a high degree of parameterization in terms of defining distribution limits for the 90 plant types along each of the eight climatic dimensions. Box attempted this task by one of the most extensive literature surveys performed about climate-vegetation relationships. In spite of the extent of this literature, many plant-type distributions still remain poorly reflected for large regions, and "tuning" must be performed if real-world distributions are to be compared with predictions from this model (Sargent 1988; Cramer and Leemans 1990).

Global climate maps based entirely on measured climate parameters appeared still later, because the global network of weather stations had (and has) large gaps, particularly in less-populated or technologically less-advanced areas. Weather stations are often located in urbanized coastal areas, and weather records spanning one or several decades are usually only available for the industrialized countries. Reliable long weather records and the maps derived from them are therefore biased toward the temperate zones and are uncertain even for some of the largest sparsely populated areas, such as the Amazon Basin, Tibet, and Inner Mongolia. Besides, many stations are not representative for a larger surrounding area, because they are located at the bottom of valleys, near mountain summits, or on airfields. Owing to this incomplete and heterogeneous network, a global climate map based exclusively on interpolations between the averages of longer weather records will be relatively reliable in some regions and very unreliable in others. Some rather dense global climate databases are now available (e.g., Walter and Lieth 1960–67; Müller 1982; Bradley *et al.* 1985; Willmott and Rowe 1985), but still none of them covers all available data and large gaps exist.

Global predictions of vegetation distribution based on climate characteristics are entirely dependent on gridded climate data. In this chapter, we use the International Institute for Applied Systems Analysis (IIASA) gridded global climate database (Leemans and Cramer 1990) to compile maps of vegetation patterns (life zones) and potential species-type distributions. The database consists of long-term monthly means for a variety of meteorological variables, interpolated to a regular grid. We have mapped life zones and plant types for current climate as a baseline (here defined as average monthly weather of the period 1931–60) and for a climate scenario under a double-CO_2 equivalent "greenhouse" atmosphere. The scenario for double-CO_2 conditions is based on the output of a General Circulation Model (GCM) for the atmosphere (Oregon State University [OSU] model, Schlesinger and Zhao 1989).

Databases and Models

Gridded Terrestrial Climate Database

The climate database used here consists of monthly temperature and precipitation values from weather stations scattered around the world. It was first compiled at the International Institute for Applied Systems Analysis (IIASA), based on published (Weather Bureau 1959; Walter and Lieth 1960–67; Meteorological

Office 1966–83; Müller 1982) and unpublished sources.[1] The input data were carefully screened for errors, outliers, and apparently unreliable stations, and the total array of stations was checked for doublets. An observation record of at least 5 years (if possible standardized for the period 1931–60) was used for the computation of mean monthly temperature and precipitation values. The argument for using such a short minimum period is that we were giving priority to spatial coverage: given the choice between long-distance interpolation and short observation periods we decided for the latter source of uncertainty. Most stations, however, provided data for the 30-year standard period.

The weather-station data set was used to create a global database of monthly mean temperature and precipitation observations for land areas. The final gridded data had a resolution of 0.5° longitude and latitude (grid cells being rectangles of approximately 55 × 55 km at the equator, becoming more and more narrow toward the poles). Where several stations were available for the same grid cell, the most "reliable" station for that cell (judged by the record length and its source) was selected exclusively. After this selection, the database was reduced from more than 12 000 raw data records to 6279 stations with temperature data and 6088 stations with precipitation data worldwide. Each locality is characterized by longitude, latitude, and altitude above sea level. An interpolation scheme based on horizontal triangulation of the positions of available stations, followed by a smooth surface fitting (Akima 1978), was used to obtain values for all 62 483 terrestrial cells, excluding Antarctica.[2] The interpolation was made for all monthly temperature and precipitation values and resulted in a database on a 0.5° × 0.5° grid with monthly values for each of the cells.

Monthly temperature values were corrected for altitude by applying a linear lapse-rate. Free atmosphere lapse-rates are frequently around −0.6°C/100-m altitude change (moist adiabatic lapse-rate for saturated air) and up to −1.0°C/100 m (dry adiabatic lapse-rate; Strahler and Strahler 1989). Such lapse-rates cannot be used in local or short-term studies because they vary considerably over time and may even become positive in the case of temperature inversions. On a regional-to-continental basis, however, altitudinal zonations may well be explained by them, and they are used where no real weather-station data are available. We used a typical value of −0.6°C/100 m (Su 1984; Woodward 1987; Ohsawa 1990). Prior to the interpolation, the temperature values of all stations were linearly transformed to sea-level values (using their altitude and this lapse-

[1]The databases mentioned were extended with data from the following countries: Bulgaria, China, Czechoslovakia, Denmark, Japan, Norway, and The Netherlands. The database will be extended further.

[2]Owing to the long time-scale of potential ice-sheet degradation and the low reliability of climate predictions for this special case, we did not attempt any vegetation prediction for a world with ice-covered areas differing from those today. Greenland appears on the maps for technical reasons, but the predictions for this area are not part of this study. In addition, no correction for changes in shorelines due to sea-level change was made.

rate). The grid cell values resulting from the interpolation were transformed back to the modal height of that cell, using a global topography database (National Geophysical Data Center 1988).

The influence of altitude on precipitation varies within much wider ranges than it does for temperature. This is true even on a continental scale, because precipitation at any given point is strongly affected by movements of air masses in large regions, aspect, and slope. Lapse-rates for precipitation in mountainous areas (caused by orographic condensation) have been determined in local studies (e.g., Blaesdale and Chan 1972; Delijaniec 1972; Khurshid Alam 1972; Rydén 1972; Storr and Ferguson 1972; Boryczka and Stopa-Boryczka 1986). Owing to the high variability in time and space, a worldwide generalized lapse-rate would give unreliable results. Precipitation was therefore interpolated directly by using the observed monthly mean precipitation values within the triangulation of stations without any reference to altitude. Since the selection procedure was aimed at acquiring the most representative value for each variable in each grid cell, precipitation and temperature data were not always derived from the same station.

A complete description of the IIASA gridded climate database is given by Leemans and Cramer (1990).

Double-CO_2 Climate Scenario

A climate scenario for double-CO_2 in the atmosphere was derived from the results of a GCM for the atmosphere. Most recent GCMs indicate similar trends for climate change, although considerable differences occur in both magnitude and regional pattern. The current consensus among most GCM results is that larger temperature increases will occur in higher latitudes, with peaks during the winter season and relatively smaller increases during the summer season (Mitchell *et al.* 1987). The variation in monthly precipitation patterns simulated by different GCMs is much larger than that for temperature (Mitchell 1983; see Solomon and Cramer, in this volume). Details about general circulation models and their predictions for global warming, and a discussion of the reliability of their different results, are given by Bach (1988), Tucker (1988), and Harrison (1990).

For this study, we used output from the model developed at Oregon State University (OSU) by Schlesinger and Zhao (1989) for both baseline climate (control run) and double-CO_2 conditions. The global increase in temperature given by this model is ca. 3.5°C (Tab. 10.6) for the continents, which places the model on an average level among the global-warming predictions as given by other GCMs. We selected the OSU model because of its relatively high spatial resolution ($4^\circ \times 5^\circ$) as compared with all other models to which we had access. It predicts clouds by using a relative humidity scheme such that large-scale clouds form in the lower atmosphere when relative humidity exceeds 85% (95% in the upper atmosphere). The oceanic component of the model is a 60-m-deep mixed-layer ocean with a simple thermodynamic sea ice model and prescribed oceanic heat flux. (See Harrison 1990 and Houghton *et al.* 1990 for a comparison with other models.)

Patterns in global temperature are simulated relatively well with most GCMs, but there are major differences in their precipitation patterns, both between different GCMs and between any one GCM prediction and observed data. By increasing horizontal spatial resolution, a GCM should perform better in this respect (Wilson and Mitchell 1987), but no output from experiments of this kind was available to us. To predict the climate for an increase in greenhouse gases corresponding to doubled preindustrial CO_2 levels, we first calculated temperature and precipitation anomalies as differences between normal (control) and double-CO_2 runs. The coarse ($4^\circ \times 5^\circ$) output grid of these GCM data was then interpolated to our finer $0.5^\circ \times 0.5^\circ$ grid, using the same routine as for the baseline climate database. This method should estimate future conditions better than gridded double-CO_2 predictions (without baseline climate overlay) directly.

In determining the double-CO_2 climate anomaly for a GCM grid cell, temperature change was expressed as a positive or negative value to be added to the baseline climate data. Precipitation was derived as percentage change to avoid negative precipitation where GCM-predicted "normal CO_2 precipitation" had been less than the observed data. Using the differences between a control and a double-CO_2 run and overlaying this difference on baseline climate is presently the only way to down-scale climate-change scenarios. It should be noted, however, that the spatial resolution of such results may be overinterpreted, but we believe that medium-scale features such as mountain ranges or other azonal climates are preserved in a meaningful way.

Holdridge's Life Zone Classification

An attempt to express objectively major types of global climate through simple functions of frequently measured climatic variables was published in 1947 by Holdridge (revised 1967). Holdridge used three bioclimatic variables derived from standard meteorological data to express explicitly the relation of climate patterns and broad-scale vegetation formations.

The first variable is *biotemperature*, defined as the mean value of all daily mean temperatures above 0°C, divided by 365. The biotemperature concept rests upon the observation that plant growth is usually not very strongly affected by how far below freezing point the temperature is. Biotemperature therefore gives a measure of heat during the growing season that is likely to be more directly related to plant growth than simply mean temperature. Another, more common, measure of heat is the number of growing degree days, calculated as the sum of all daily mean temperatures exceeding a certain base level, e.g., 0°C or 5°C. Global patterns of biotemperature sensu Holdridge are closely related to those of growing degree days (Tuhkanen 1980; Leemans 1989).

The biotemperature concept in this model is extended by a special plant-related temperature factor that accounts for the risk of occurrence of killing frost. This is done on an empirical basis by dividing the warm temperate and subtropical zones at a biotemperature cutoff level of 16°C. The result is the division of seven life zones into a warm temperate and a subtropical subset, with each subset comprising seven separate zones (Fig. 10.1; see color section).

The demand of plants for moisture is expressed in two ways, which are not clearly functionally separated: *mean annual precipitation* and *potential evapotranspiration (PET) ratio*. PET is rarely measured directly in meteorological networks. In the Holdridge scheme, it is therefore estimated as a pseudo-empirical function of mean annual biotemperature, being this value multiplied by 58.43. This ratio is based on a limited set of observations, rather than on physical principles, and because of the restricted and undocumented data supporting this function, it is probably not empirically valid even for larger regions. Its values are low compared with Thornthwaite's method (see below) for high-latitude regions and high for the subtropical and tropical regions (Tuhkanen 1980). Different humidity provinces are defined by the ratio between PET and mean annual precipitation. The complete chart of the Holdridge Life Zone Classification is given in Figure 10.1.

The first computerized implementation (to our knowledge) of the Holdridge Life Zone Classification on a global scale was carried out by Emanuel *et al.* (1985a). Unfortunately, they did not compute biotemperature strictly according to Holdridge, but used average annual temperature only. Their climate index roughly corresponds with biotemperature only as long as temperature never drops below freezing or exceeds 30°C. Therefore, this modification has major consequences for the colder boreal and polar life zones. This was later corrected in a much more rarely quoted article (Emanuel 1985b and used also by Warrick *et al.* 1986). In both studies, Emanuel *et al.* (1985a,b) used a horizontal interpolation of weather stations (omitting some high-altitude stations). Therefore mountainous regions are poorly represented on this map, resulting in erratic life zones in central-eastern Siberia and in the Himalayas, Andes, and other mountain ranges.

We implemented the Holdridge Life Zone Classification using the IIASA climate database and following the scheme given by Holdridge (1967). Mean annual precipitation was computed as the sum of all monthly values. Biotemperature was computed for daily temperature values using smoothed pseudodaily temperature values generated by fitting a quadratic spline function through the monthly average temperature values. The specific PET ratio was computed from biotemperature and total annual precipitation. We used these three climate indices in each terrestrial cell to generate a world life zone map. To simplify this map, which had 39 life zones, several zones were combined to form a total of 14 zones (Tab. 10.1).

The simplified life-zone map (Fig. 10.2a) visually compares well with the global vegetation map of Olson *et al.* (1983). In a statistical comparison of our

Table 10.1. Holdridge life zones (Holdridge 1967) including subsections and their simplification for the present study. The life zones and subsections (39 in total) are renamed according to their unique climate characteristics. Our 14 simplified groups were defined by visual comparison with global vegetation maps, particularly the one by Olson et al. *(1983). The clusters are reflected by corresponding colors in Figure 10.1 and Figure 10.2.*

Holdridge life zone	Simplified group
Ice	Polar desert
Polar desert	Polar desert
Subpolar dry tundra	Cool forest
Subpolar moist tundra	Tundra
Subpolar wet tundra	Tundra
Subpolar rain tundra	Tundra
Boreal desert	Cool forest
Boreal dry scrub	Cool forest
Boreal moist forest	Boreal forest
Boreal wet forest	Boreal forest
Boreal rainforest	Boreal forest
Cool temperate desert	Cool desert
Cool temperate desert scrub	Cool desert
Cool temperate steppe	Steppe
Cool temperate moist forest	Cool temperate forest
Cool temperate wet forest	Cool temperate forest
Cool temperate rain forest	Cool temperate forest
Warm temperate desert	Hot desert
Warm temperate desert scrub	Hot desert
Warm temperate thorn steppe	Chaparral
Warm temperate dry forest	Chaparral
Warm temperate moist forest	Warm temperate forest
Warm temperate wet forest	Warm temperate forest
Warm temperate rain forest	Warm temperate forest
Subtropical desert	Hot desert
Subtropical desert scrub	Hot desert
Subtropical thorn woodland	Savanna
Subtropical dry forest	Dry tropical forest
Subtropical moist forest	Subtropical forest
Subtropical wet forest	Tropical rainforest
Subtropical rain forest	Tropical rainforest
Tropical desert	Hot desert
Tropical desert scrub	Hot desert
Tropical thorn woodland	Savanna
Tropical very dry forest	Savanna
Tropical dry forest	Dry tropical forest
Tropical moist forest	Tropical rain forest
Tropical wet forest	Tropical rain forest
Tropical rainforest	Tropical rain forest

map with the digitized map by Olson *et al.* (1983), Monserud (1990) found that agreement was just at the threshold between *poor* and *fair*, although Olson's database was derived with a different purpose and with a different set of classes for reflecting human land use. The map comparison statistic κ (Cohen 1960) calculated for the entire map was 0.40.[3] Four of the life zones had a *good* agreement: tundra ($\kappa = 0.62$), boreal forest ($\kappa = 0.55$), hot desert ($\kappa = 0.64$), and tropical rainforest ($\kappa = 0.55$).

The life-zone map presented here is an improvement over the one by Emanuel *et al.* (1985b) because of the improved climate databases used and our topography-related interpolation scheme.

Box's (1981) Model for Macroclimate and Plant Types

A unified global expression for the relationship between macroclimate and vegetation was made by Box (1981). Rather than describing climatic demands of entire vegetation formations, as many earlier climate mappers had done, Box defined climate limits on plant species types (termed *plant forms* in his original text). To overcome the difficulty of exceedingly large numbers of (possibly unknown) plant species potentially occurring in any region, he lumped all higher plant species (including some cryptogam life forms) into 90 plant types. (See Appendix A in Box 1981 for a detailed description.) These plant types are not meant to resemble real taxonomical units, nor are all known taxonomical units explicitly assigned to plant types. Rather they form a special, much more detailed variant of earlier authors' life-form systems (e.g., Raunkiær 1907).

Climate is represented in Box's model by a set of eight bioclimatic indices (Tab. 10.2). These indices are designed to reflect the plants' required conditions for principal climatic resources or constraints, namely, warmth, frost frequency, and moisture. It is important to note that all indices are still derived from standard weather data to achieve global coverage. This goal had to be balanced with the need to define variables which can, on the basis of physical response processes, be expected to have maximum importance in determining the geographical ranges of plants. Box's indices differ from Holdridge's not only in their total number: they also give a better estimation of drought stress and address seasonality explicitly (separately for warmth and moisture). The latter enables the model to separate subtropical drought-deciduous plant types occurring in savannas and other warm regions from evergreen plants, without conflicting with the high-latitude winter-deciduous plant types. This strategy is also illustrated by the division between boreal deciduous and evergreen conifers, which form separate types in this scheme.

---

[3]Monserud (1990) used a scale defined by Landis and Koch (1977) for the interpretation of κ (upper bound of each class in parentheses): no agreement (0.05), very poor (0.20), poor (0.40), fair (0.55), good (0.70), very good (0.85), excellent (0.99), perfect (1.00).

Table 10.2. Bioclimatic variables used by Box (1981) to determine distribution limits of plant types.

T_{max}	Mean temperature of the warmest month (°C)
T_{min}	Mean temperature of the coldest month (°C)
D_T	Range between T_{min} and T_{max} (°C)
P	Mean total annual precipitation (mm)
MI	Moisture index defined as the ration between P and annual potential evapotranspiration (PET, estimated according to Thornthwaite and Mather 1957)
P_{max}	Mean total precipitation of the wettest month (mm)
P_{min}	Mean total precipitation of the driest month (mm)
P_{Tmax}	Mean total precipitation of the warmest month (mm)

Warmth is expressed by the *mean temperature for the warmest month* (reflecting largely the latitudinal total radiation pattern) and by the *range between the warmest and the coldest month's mean temperature* (reflecting continentality). The other major temperature constraint, frost frequency, is expressed by the *coldest month's mean temperature*. Moisture stress is estimated by several measures derived directly from precipitation (*mean precipitation for the driest and the wettest month* and, notably, *precipitation for the warmest month*), but also by a moisture index (MI), which is the *ratio between annual precipitation and PET*. PET is estimated according to Thornthwaite and Mather (1957), using an empirically derived look-up table that was later modified into a set of algebraic equations. The Thornthwaite and Mather method, although extensively used, has been deemed unsatisfactory by its authors (Thornthwaite and Mather 1957) and by many later workers (e.g., Priestley and Taylor 1972). In 1981, however, when Box's model was published, this was probably the only computationally feasible approximation for a global grid and it certainly represented an improvement on Holdridge's moisture index.

Climatic "envelopes," i.e., lower and upper limits of each of the eight bioclimate indices, are defined for each of the 90 plant types. (See Tab. 7 in Box 1981.) The limits provided by Box (1981) are based on extensive literature studies for the world. The predicted plant types for each site are found by "sieving" all potentially available plant types through a mesh consisting of all climatic envelopes. As soon as any of the 16 (upper or lower) climatic limits prevents the plant type from occurring in a given cell, it is removed[4]. After selection of all plant types potentially occurring at a given locality, a simple dominance hierarchy is applied, based on two principles: (1) that certain plant types always dominate others (such as trees over grasses) and (2) that within each level of this hierarchy, plants occurring closer to their environmental limits are "less dominant" than those with all index values closer to their optimum.

[4]This *ecosieve* approach is essentially congruent with Woodward's *decision tree* concept in that a limited number of environmental envelopes determine presence or absence of certain plant types (Woodward 1989).

Primary output from the model is a list of plant types, sorted according to the final dominance hierarchy. Calibration of this list can be done by comparison with real-world localities. Weather-station data are used to prescribe a plant type assemblage which is then compared to the life-form spectrum of the surroundings. Box (1981) checked the predictions for a number of stations and found reasonable agreement. To compare plant-type distributions for large areas with real plant distributions, he implemented the model in such a way that a plant-type list is produced for each weather station in the database. It is problematic to construct global "plant-type distribution maps" from such a database unless a very large number of climate stations are used. In Box's (1981) published version of the model the maps are constructed by centering vegetation zones on the data-sites rather than by deriving plant-type lists from gridded climate data. A transition between different units must in this case result in a border drawn halfway between the stations independent of both the distance between them and possible nonlinear climatic gradients along this distance. Also, different climate indices can display different patterns between weather stations, which are not reflected by this approach.

The climatic limits for each plant type can obviously only reflect the limited current knowledge of each type's distribution and the environmental factors restricting it. This is a serious constraint for two reasons. First, the limits are assumed to be valid worldwide, although each of them is derived from a set of studies in only one or a few parts of the world. Second, it is assumed that the entire realm of higher plants around the globe can be reasonably subdivided into 90 types in the way proposed. None of these assumptions is strictly verifiable. The value of the model, however, is not restricted by this: it represents a major structural improvement on earlier attempts to synthesize widely scattered global phytogeographical knowledge and it has made this knowledge (including hypotheses derived later) available for testing. Virtually all later approaches to modeling the response of plants to climate change using functional types (e.g., Smith *et al.*, this volume) are essentially based on this concept.

Comparisons between the model's predictions and real vegetation have been made by Box and several others (Sargent 1988; Cramer and Leemans 1990), and these studies have confirmed that an exact fit cannot be achieved. This is mainly due to the generic simplicity of the model but also is due to the shortage of climate and vegetation data from large parts of the world. The general usage strategy as first developed by Sargent (1988) is to select the climatic limits for one or a few plant types for the particular purpose of each study, and then determine their distribution within the same structural framework. (See below.) Mismatch between present distribution and the prediction is then used to "tune" the limits and possibly define a new plant type, which can be used for testing climate-change predictions. Using this method, specific assumptions about certain climatic driving forces can be "tested" within entire biotic regions.

We implemented the Box model in a dedicated geographic information system

(GIS) structure based on the interpolated grid of the IIASA climate database. Box's eight climate indices (Tab. 10.2) were determined for each grid cell and a plant-type list was generated. Using the ecosieve method, all potentially occurring plant types were sorted according to decreasing dominance (increasing numerical proximity to the plant type's environmental limit) and stored in a global plant-type database. From these data, maps can be drawn for any one plant type. Also, a climate anomaly overlay as described above can be used to determine equilibrium distributions under any given climate scenario. Tools for comparison of these maps with each other in visual and statistical terms are included.

Assessing the Impact of Climate Change on Vegetation Distribution

A Short Review of Earlier Attempts

Probably the first global study of the impact of global warming on potential natural vegetation was carried out by Emanuel *et al.* (1985a,b) using the Holdridge Life Zone Classification. Their application of this classification to a double-CO_2 climate scenario demonstrated how GCM output can be used to determine impacts on broad-scale vegetation patterns. The study was based on a temperature change scenario only, with no predicted change in precipitation regime. The temperature anomaly field was derived from the quadrupled-CO_2 GCM analysis from Manabe and Wetherald (1980) by calculating the difference between control run and 4 × CO_2 for each grid cell; half of this difference was then overlaid on a baseline climate database to estimate double-CO_2 conditions. The resulting maps displayed likely principal trends of future change in life-zone patterns, indicating major poleward shifts of many zones. Extent and position of the affected areas are likely to be modified as this approach develops and as new GCM scenarios become available.

Another recent application of the Holdridge scheme to a climate-change scenario was designed to estimate changes in terrestrial carbon storage from ice-age times to the present (Adams *et al.* 1990). Adams and co-workers derived life-zone maps from both a baseline climate scenario and a reconstruction of climatic conditions 18 000 years ago. The reconstruction also involved changes in the availability of land surface due to sea-level changes and ice-sheet cover. From a comparison of these life-zone maps and estimated values of carbon storage within the different vegetation types, Adams *et al.* conclude that terrestrial carbon storage has more than doubled since the last glacial maximum.

An application of a regional variant of Box's model for a climate-change scenario was made for the Canadian boreal forest by Sargent (1988). Climatic limits for the boreal forest and each of Box's eight variables were derived from a map of forest types by using visual tuning of Box's values for "boreal/montane needle-trees." Sargent's climate database consisted of a coarse grid of 1.4° longitude and latitude. The grid values for monthly temperature and precipitation were based upon gridded data from 74 weather stations, evenly distributed throughout Canada. A double-CO_2 scenario was derived from the results of a quadrupled-CO_2 GCM simulation (Manabe and Wetherald 1980), using the same technique as Emanuel *et al.* (1985a,b). Here, however, both temperature (absolute change) and precipitation (percentage change) scenarios were used. Boreal forest distributions for baseline and future double-CO_2 climate were predicted and the two predictions were compared both visually and numerically. The result of Sargent's analysis was a dramatic change in boreal forest distribution for Canada. In terms of area, it suggested a net loss of boreal forest area of approximately 1.0×10^6 km^2. As a result of the general northward shift of climate zones, large areas would experience gain or loss of these forests (0.7×10^6 km^2 gain in the north and 1.7×10^6 km^2 loss in the south). Sargent also tested the same procedure with another GCM scenario (Hansen *et al.* 1984) and reported little or no difference in predicted change.

Dynamic models which can simulate transient vegetation change over time more explicitly have resulted in predictions for regions (Solomon 1986) or ecosystems (Overpeck *et al.* 1990). No dynamic vegetation simulator has yet been used to assess the impacts of environmental change on a global scale. Problems to be solved during the development of such models are the simulation of spatial transitions between different ecosystems (e.g., forests and grasslands), the large number of species and their environmental constraints to be modeled, the spatial and temporal heterogeneity in vegetation patterns, and the lack of appropriate data to estimate parameters for the model components. An initial attempt at delimiting and defining the domain of such a global vegetation model is described by Prentice *et al.* (1989). An introduction to this dynamic approach and its potential is given elsewhere in this volume (Chapter 12, Prentice *et al*).

Impacts of Climate Change, Derived from Static Models

If change in potentially natural vegetation is to be predicted from static models, dynamic equilibrium conditions must be assumed for climate and vegetation on a grid-cell-scale level. Static models are by definition incompatible with scenarios on transient change. This is a major limitation, because all known climate-change studies predict gradual changes, and those predicting abrupt changes do not result in any immediately occurring steady-state climate.

GCM-based predictions of future vegetation as described below rest upon the simplified scenario of climate changing abruptly from present to predicted conditions. This change is then assumed to be followed by a longer period with more stable climate. In the following, the length of that period is assumed to be sufficient to allow for complete removal of those species that will be beyond their climatic limits for establishment (at least one complete life cycle for the longest-lived plants). For most plants, this period may not be sufficient to allow for migration to newly available sites (Solomon and Cramer, in this volume). The areas given in the following impact assessments for various life zones or plant types are therefore termed "decline" areas (removal of certain plants/biomes being more certain than arrival of replacement plants), "stable" (no change likely to occur), and "available" (to indicate that the region might be climatically suitable, but that prediction of arrival is impossible).

Holdridge Life Zone Classification

Our implementation of the Holdridge Life Zone Classification for both baseline climate and double-CO_2 climate was based on the IIASA climate database with 62 483 gridded localities. The comparison between baseline and "greenhouse" conditions displays large changes (Tab. 10.3, Fig. 10.2a–c; see color section). Leemans (1989) has presented figures and maps for all 39 possible life zones, and Monserud (1990) has compared them numerically for a variety of GCM scenarios. Here, we show only the simplified life-zone classification (Fig. 10.2a for baseline and Fig. 10.2b for GCM-modified climate) which coincides best with observed broad-scale vegetation patterns. The spatial extent of the different life zones is mapped by overlaying the two classifications. The resulting map of "affected areas" (Fig. 10.2c) shows the areas which change from one life zone to another. The main trend seen on the maps is the poleward shift of the broad-scale vegetation patterns, similar to that predicted by Emanuel *et al.* (1985a,b).

Differences in the area of different life zones between our projections and those of Emanuel *et al.* (1985a,b) are explained by our use of a more recent GCM scenario and by the different ways of implementing the model. We simulate a larger area of subtropical life zones than do Emanuel *et al.* (1985a,b) because of a different definition of frost occurrence: they defined the border between warm temperate life zones and subtropical life zones at an average annual temperature of 20°C, while we use the lower biotemperature value of 16°C, as in the original Holdridge scheme (Fig. 10.1). The second main difference between the two lies in the topography-related method of temperature interpolation in our climate database. This leads to a larger relative amount of tundra or alpine life zones, which should agree better with reality.

The Emanuel *et al.* (1985b) implementation assumed no change in precipitation. Greater warmth generally leads to greater evapotranspiration and there is therefore a general trend of decreased humidity throughout the life zones. The extent of the world's forested area must therefore decrease, and the area of drier life zones increase. This is in contrast to our analysis (which accounts for changing precipitation), where the areas available for these life zones are more similar. We argue therefore that even the present rough predictions of precipitation changes should be included in impact scenarios; indeed, the assumption of stable precipitation already implies a modification of the hydrological cycle. Moreover, most GCMs now generally confirm an overall precipitation increase (Schlesinger and Mitchell 1987). On a very large scale, this change balances the life-zone changes

Table 10.3. Areas (in 10^6 km^2) occupied by modified Holdridge life zones for present climate and a 2 × CO_2 scenario (Schlesinger and Zhao 1989). "Baseline": area under present (baseline) climate; "2 × CO_2": area under GCM-derived double-CO_2 climate; "net change": change in total area; "abs. decline": area where the baseline life-zone type is replaced by another; "% decline": percentage of area under baseline climate conditions replaced by other life zones.

Life zone	Baseline	2 × CO_2	Net change	Abs. decline	% decline
Polar desert	6.6	3.0	−3.6	−3.6	−54.5
Tundra	8.6	5.1	−3.5	−6.7	−77.9
Cool forest	3.3	3.6	0.3	−1.2	−36.4
Boreal forest	15.9	13.9	−2.0	−7.8	−49.1
Cool temperate forest	10.0	11.5	1.5	−3.5	−35.0
Warm temperate forest	3.2	2.0	−1.2	−2.5	−78.1
Cool desert	4.3	3.6	−0.7	−1.8	−41.9
Steppe	7.6	9.3	1.7	−2.0	−26.3
Savanna	9.9	12.4	2.5	−2.4	−24.2
Hot desert	21.0	19.8	−1.2	−2.9	−13.8
Chaparral	5.7	5.2	−0.5	−4.3	−75.4
Dry tropical forest	15.0	15.5	0.5	−4.9	−32.7
Subtropical forest	15.3	10.5	−4.8	−9.4	−61.4
Tropical rainforest	8.6	19.6	11.0	0.0	0.0
Total	135.0	135.0	0.0	−53.0	−39.3

rather than adding to them, as is shown by our results. Note that, as far as moisture balance is concerned, all models consider only level-ground sites with well-drained soil conditions: riverine vegetation or wetlands are not reflected.

The spatial pattern of change (Fig. 10.2c) shows that the transitional areas between major vegetation zones are the most sensitive to a changing climate. This is an inherent characteristic of the model, but it should also reflect real-world conditions, because regions that are already climatically marginal for the vegetation they support should be affected first and most heavily. A mapped representation of this phenomenon, however, is strongly dependent on the exact definition of the classes used. Subdivision of one large class into two smaller ones would necessarily open up new regions of potential change where no change is predicted in the maps shown here. This implies that in reality most areas in the world are likely to become "affected" by climate change, while this map mainly demonstrates the extent and direction of latitudinal and altitudinal shifts.

Owing to the general poleward and/or altitudinal shift of all major life zones, both polar deserts and tundra life zones decrease most, because these life zones are limited by physical boundaries, such as the Arctic Sea and mountain summits. The transitions between the higher latitude zones are more sensitive in this model, for two reasons. The first is the larger temperature increases for these latitudes given by the GCM results, and the second is that, because of the logarithmic nature of the division of biotemperature in the Holdridge scheme, increases at high latitudes are likely to generate larger effects in the life-zone distributions. The changes in patterns seen in the midlatitudes seem to be more complex than just unidirectional shifts. Clear patterns of change cannot be detected at this scale, and it is not possible to draw general conclusions on the character of the projected life-zone change for these regions.

Box's Model of Plant-Type Distribution

We implemented Box's ecosieve routine using his original bioclimatic limits, except for including Sargent's (1988) modification for boreal wintergreen trees (Tab. 10.4). The resulting database, calculated for the full IIASA climate database grid, consists of a list of plant types for each of the grid cells, ordered by their dominance, for both baseline and double-CO_2 climate. This spatial vegetation database can be sampled in many different ways, including most of the traditional classification and ordination methods developed for vegetation analysis. Just like the real world's vegetation, however, it cannot be shown in one single map covering all aspects, because each locality carries a potentially different assemblage of plant types. Here we limit our analysis to the spatial changes in the distribution of a few single plant types. We also determine regional changes in a simple biodiversity index, because this issue has recently received major attention in the global-change impact debate (e.g., Wilson 1989).

We first present changes in areal extent for three plant types, although all 90 are available. Their extent under baseline and double-CO_2 climate was mapped (Tab. 10.5, Figs. 10.3–10.5). For interpretation, we also list mean changes in the bioclimatic variables after applying the GCM anomaly fields (Tab. 10.6).

Boreal wintergreen conifers (boreal/montane needle-trees sensu Box 1981) consist of tall, evergreen, short-needled conifers from genera such as *Picea*, *Abies*, and *Pinus*. They are the dominant plant types within the boreal forest. We have defined the range of this plant type according to Sargent's (1988) modification of Box's (1981) limits (Tab. 10.4). The baseline distribution (Fig. 10.3; see color section) of this plant type coincides reasonably well with the present distribution of the boreal forest. The region dominated by these trees is clearly sensitive to climate change of the order currently predicted. At large parts of the southern edge of their distribution, they are predicted to decline severely, which, for Canada, is in agreement with Sargent's (1988) results. Compared with the extent of the decline area (6.4×10^6 km^2 or 66%, Tab. 10.5), the gain of available area is relatively small (2.6×10^6 km^2) for this plant type, resulting in a large net decrease (Tab. 10.5, Fig. 10.6). As mentioned earlier, the reality behind this climatically available region is very uncertain on a medium-term time-scale because of time lags caused by successional and dispersal processes. The considerable decline, however, is mostly due to increasing winter temperatures (Tab.

Table 10.4. Bioclimatic distribution limits for selected plant types (modified after Box (1981) and Sargent (1988)).

	T_{max}	T_{min}	D_T	P	MI	P_{max}	P_{min}	P_{Tmax}
Boreal/montaine conifers	13–20	−29 to −8	10–60	100−>	0.6−>	25−>	5−>	55−>
Boreal summergreen conifers	12–22	−60 to 3	15–90	100−>	0.4−3	30−>	0−50	25−>
Tropical rainforest	15–30	8–28	0–12	700−>	0.9−>	100−>	5−>	10−>

Table 10.5. Areas (in 10^6 km^2) occupied by three plant types derived by the model by Box (1981) using the limits given in Table 10.4, for present climate and a 2 × CO_2 scenario (Schlesinger and Zhao 1989). For explanation of column headings, see Table 10.3.

Plant Type	Baseline	2 × CO_2	Net change	Abs. decline	% decline
Boreal wintergreen conifers	9.7	5.8	−3.9	−6.4	−66.0
Boreal summergreen conifers	25.1	23.0	−2.1	−7.7	−30.7
Tropical rainforest trees	19.1	16.2	−2.9	−4.9	−25.7

Table 10.6. Changes in bioclimatic variables for the terrestrial grid (62 483 cells) due to the OSU 2 × CO_2 scenario (Schlesinger and Zhao 1989). Where estimated evapotranspiration was zero, no moisture index could be determined.

	Min	Max	Mean	n
T_{min}	0.6	5.50	3.1	62 483
T_{max}	−0.2	9.20	3.5	62 483
D_T	−7.4	3.20	−0.4	62 483
P	−389.9	3265.0	148.2	62 483
MI	−3.02	110.45	0.74	60 171
P_{max}	−99.0	1941.8	33.4	62 483
P_{min}	−78.8	151.0	2.0	62 483
P_{Tmax}	−567.4	683.3	8.0	62 483

10.6) giving competitive advantage to broad-leaved trees. This scenario should have major implications, e.g., for the forest industry.

Boreal summergreen conifers ("boreal summergreen needle-trees" sensu Box 1981) consist of tall, boreal conifers with summergreen deciduous needles from genera such as *Larix* and *Pseudolarix*. They overlap with the plant type mentioned above, but they have a much broader distribution than the other boreal plant types (Fig. 10.4; see color section). These are the most frost-resistant trees, which explains their more northern and eastern Siberian distribution. Their much wider temperature range and moisture requirements also explain their apparent southern distribution in Europe. In reality, such a plant type will be limited in its abundance by other species (the vegetation of central and western Europe is composed of a relatively large number of more vigorous broad-leafed plant types), and in most of

Europe it therefore belongs to the group of less-dominant plant types. Considering relative changes, climate change influences this plant type less than the former, but the areas involved in decline and potential expansion are extensive: decline is likely to occur on an area of 7.7×10^6 km^2 or 30% (Tab. 10.5, Fig. 10.6). It mainly moves northward and loses ground both on the warm and the most continental edges of its distribution. This is also largely due to the reduced area with low enough winter temperatures (Tab. 10.6).

Tropical rain forest ("tropical rainforest trees" and "tropical montane rainforest trees" sensu Box 1981) are tall, evergreen, canopy trees of many families. This is the dominant group characterizing tropical rainforests. We present the plant type here to demonstrate the capabilities of the method in different climatic regions. The current deforestation of the world's tropical rainforest is, of course, not addressed by our map, which, like the others, only shows the shift in regions *climatically suitable* for these plants. In contrast to the Holdridge scenario, a decline of tropical rainforests is predicted, which is probably caused largely by reduced summer rainfall in the Amazon Basin and other areas. It should be noted that the GCM scenarios do *not* agree on this point, but the observation is still alarming if seen in the context of predicted reductions in precipitation following losses in forest cover and thereby reduced evapotranspiration. If these climatic limits are realistic, then this form of degradation may have dramatic consequences even for the remaining untouched rainforest areas. In our analysis, the three major areas of tropical rainforest (Southeast Asia, central Africa, and the Amazon Basin) show very different responses (Fig. 10.5; see color section). The distribution in Southeast Asia is greatly reduced, indicating increasing drought stress for these regions. However, the largest decline is in South America, particularly in southern regions, but also, remarkably, in the center. Africa shows more stable conditions. Globally, the potential area for tropical rainforests decreases by 4.9×10^6 km^2 or 26% (Tab. 10.5), (Fig. 10.6; see color section). Tropical rainforest regions are poorly represented in our climate database and characterized mainly by weather stations along their edges. This may have led to an underestimation of the total amount of precipitation, especially for Southeast Asia and the Amazon Basin. The Box model is more sensitive to changes in precipitation than the logarithmic Holdridge Life Zone Classification, and this may explain the differences in the simulated tropical forest extent of the two models.

The global Box plant-type database can also be used to estimate changes in *biodiversity* on a global scale. To our knowledge nobody has tried to determine semiquantitative changes in biodiversity under a double-CO_2 climate on a global basis. We present here a "first guess," well aware of the fact that our estimates ignore what is probably the most significant threat to global conservation of biodiversity, viz., land use and particularly rainforest destruction by direct impact. The exercise reported here serves its purpose mainly to illustrate the potential effect rapid climatic changes may have on biodiversity, especially if species really are unable to migrate to the new areas climatically suitable for them. Given the 90 species-type spectrum from the Box model, counts of the number of species declining in any given grid cell (in percent of those occurring under baseline climate) might illustrate some of these processes. This index is probably only poorly correlated with real diversity patterns, because the link between the number of taxonomical units and the number of Box plant types in a given environment need not be correlated very closely. Landscape and topographic diversity on a local scale must also have a major impact on real biodiversity and are not reflected by the model. A simple method, such as the one given here, is, however, the only way to achieve a global pattern without sampling the world's vegetation in the field with an inconceivable accuracy.

In our analysis, higher total plant-type counts were generally found near the boundaries of major vegetation types, such as between the boreal forest and the temperate forest region, and just south and north of the tropical rainforests. The plant-type count is low in high-altitude and desert areas. These observations parallel those in the real world, although the pattern is probably exaggerated because of the use of simple plant types. We estimate the decline in biodiversity as the relative disappearance of plant types under a double-CO_2 climate. Again, only decline is considered because of the time lags needed for new species to migrate to available areas.

Few areas show no significant decline in biodiversity (Fig. 10.7a&b; see color section). Only parts of those regions which had very few plant types to start with (deserts and high latitudinal or altitudinal regions) reveal no or little decline. Boreal and temperate forests (except for central Siberia), steppe, and tundra regions show generally smaller declines (up to 30% of the original plant-type count). The largest percentage decline is predicted for the edges of the equatorial zones in Africa and Australasia. India and the savanna areas south of the Sahara display especially large losses, as do northern Australia and the southeastern United States.

The geographical pattern of decline decreasing toward the poles is partly determined by the model used. The range of plant types increases toward higher latitudes, indicating higher diversity there than in tropical areas. This assumption is clearly invalid as far as taxonomical units are concerned, although it may be closer to reality regarding life-form spectra. Whatever the cause, the scenario shows medium-dry tropical regions to be most sensitive to the decline of single plant types. Repeating the analysis with a plant-type scheme more closely field-checked against real life-form spectra might help to elucidate one important aspect of biodiversity losses: degradation of community structure, e.g., from open savanna to grasslands dominated by sparsely covering annuals. Unfortunately, such data are not available. Other important mechanisms for diversity loss caused by, for example, selective removal of economically profitable tree species from rainforest areas will not be illustrated by an approach such as this.

Possible Improvements of the Static Modelling Approach

Improvements of static models currently under development can only be summarized briefly here. Much is to be gained by more realistic derivation of bioclimatic variables. Our experiments with the eight variables used by Box (1981) proved that the sensitivity of a given plant-type distribution to changes in single variables was very difficult to analyze systematically. The high complexity of the model, with respect to the eight strongly intercorrelated variables, seemed not to improve the realism of the results, as compared to other, more simple models. On the contrary, fewer well-selected variables reflecting measurable processes might yield an equally well-performing and more easily comprehensible and testable model. Experiments reported by Woodward and Williams (1987) seem to support this. We do not, however, advocate screening experiments of large numbers of different variables, as was done by Sowell (1985). This technique might result in a reasonable map for baseline climate conditions, but it is not likely to enhance understanding of physical processes determining species or life-form distributions. The correlations found may not be valid in a warmer world, and they do not give us any clue as to why species may respond differently from what was expected. The very idea of static models as described here is to allow for explicit testing of hypotheses about plant-climate relationships on a global level.

Concerning the nature of the new variables that could be used for improved static models, warmth is probably best expressed by growing degree days, derived from a day-to-day interpolation of monthly means. Frost stress could be better reflected by the average number of frost days, again derived from interpolation of monthly temperature data, or by a negative temperature sum (*frost sum*). Permafrost is not expressed by currently used bioclimatic variables, but techniques to do this exist (Bonan and Korzukhin 1989). The Thornthwaite and Mather method for estimating evapotranspiration and drought stress, which was used by Box and others, is clearly unsatisfactory for deriving any drought stress index when compared with more recent methods. Improvements can be expected from other methods such as the Priestley and Taylor (1972) equation (cf. "bucket model" implementations by Pastor and Post 1985; Cramer and Prentice 1988; Andersson 1989a,b). Such a routine, although computationally demanding, can also incorporate parameters derived from global soil databases. Possible output is various drought indices or the number of days with growth temperature and soil moisture below a specified limit.

Final Remarks

The GCM-derived climate-change database used here consists only of simulated monthly means for temperature and precipitation. So far, no good estimates seem to be available on the potential changes of variability for any of these values

(Wilson and Mitchell 1987). Little is known about predicted changes in weather variability and their meaning for vegetation dynamics. Increased frequencies of killing frosts, severe droughts, or heavy storms may substantially alter the picture given by our prediction without being reflected in any of the mean values used. Analyses of Overpeck *et al.* (1990) indicate that, under an increase of disturbance frequency, the predictability of vegetation change may decrease. Future models assessing impacts of climate change should incorporate responses to changing climate variability. There is no reason why improved static models should not include these factors.

The results of the different models discussed and presented above are strongly limited by the quality of their underlying databases. Although a major effort has been put into the creation of a global climate database, still, very few data are available for some large regions of the world, such as the Amazon Basin, Tibet and Mongolia, and most of the high-latitude and high-altitude areas, although more data may exist. The resolution of $0.5^\circ \times 0.5^\circ$ results in a smaller actual grid cell size at high latitudes and may therefore to some extent give the misleading impression of high precision there. On the other hand, in the developed regions of the world, this resolution is well supported by underlying data. In applying the results of such models to regional studies, the quality of the database should be taken into account.

One goal for our review was to show that the history of static models dates back far beyond the age of computers and that they still have major potential in forecasting and analyzing the impacts of climate change on vegetation patterns. The results clearly demonstrate the catastrophic extent of the biological consequences of man-induced global climate change. The models used in this chapter have taken into account only those changes in vegetation patterns generated by climate change which are due to the atmospheric effects resulting from an increase in greenhouse gases equivalent to a doubling of CO_2. CO_2 "fertilization" is not considered, nor are other major disturbances of global and regional ecosystems affecting vegetation distribution, such as tropical deforestation, acid precipitation, or ozone depletion. The model output therefore refers to potentially natural vegetation in equilibrium with climate only. But the techniques presented can also be applied to most agricultural or forestry crops (Parry *et al.* 1988), especially if they also involve soil properties. Such models can (and do) help in the building of short- and medium-term policy scenarios for strategies to mitigate the impact of global environmental change.

Limits to static models are all processes that involve time as a major dimension. Although competitive relationships between species types are treated in an arbitrary manner, there is no way to express successional processes in a static model. The transient nature of global environmental change and the long response times of ecosystems are critical issues of current impact scenarios, and dynamical models are therefore required. For spatially explicit global impact predictions, however, the static approach may still have a role to play until the global dynamic vegetation simulator exists. Also, it is computationally several orders of magnitude less demanding. This makes it a suitable tool for improvement of feedback processes in atmospheric circulation models—a missing link between the earth's surface and lower troposphere processes that is currently only weakly developed.

A great deal of optimism is required to assume response by the public and by policymakers to the virtually disastrous predictions of global-change impact scenarios. The first step toward amendment is more convincing scientific knowledge about the outcome of the presently running, uncontrolled, and irreversible experiment with the Earth's atmosphere.

Acknowledgments

We gratefully acknowledge guidance and comments during the analysis given by Allen M. Solomon, comments on various drafts of the manuscript by I. Colin Prentice and Allen M. Solomon, and editorial assistance by Erica Schwarz. We particularly acknowledge the discussions we had (with Solomon and Prentice) about simplifying the Holdridge life-zone system for this purpose and Robert Monserud's involvement in map comparisons. The study was made while both authors were at the International Institute for Applied Systems Analysis, Laxenburg, Austria.

References

Adams, J.M., Faure, H., Faure-Denard, L., McGlade, J.M. and Woodward, F.I. (1990). Increases in terrestrial carbon storage from the last glacial maximum to the present. *Nature*, **348**, 711–4.

Akima, H. (1978). A method of bivariate interpolation and smooth surface fitting for irregularly distributed datapoints. *ACM Transactions in Mathmatical Software*, **4**, 148–59.

Andersson, L. (1989a). Soil moisture deficits in South-Central Sweden. I. Seasonal and regional distributions. *Nordic Hydrology*, **20**, 109–22.

Andersson, L. (1989b). Soil moisture deficits in South-Central Sweden. II. Trends and fluctuations. *Nordic Hydrology*, **20**, 123–36.

Arnold, G.W. (1988). Possible effects of climatic change on wildlife in Western Australia. In *Greenhouse. Planning for Climate Change*, ed. G.I. Pearman, pp. 375–86. Leiden: E.J. Brill.

Bach, W. (1988). Development of climatic scenarios: A. From general circulation models. In *The Impact of Climatic Variations on Agriculture. Volume 1: Assessments in Cool Temperate and Cold Regions*, ed. M.L. Parry, T.R. Carter and N.T. Konijn, pp. 125–57. Dordrecht: Kluwer Academic Publishers.

Blaesdale, A. and Chan, Y.K. (1972). Orographic influences on the distributions of precipitation. In *Distribution of Precipitation in Mountainous Areas*. Symposium Geilo, Norway, July–August 1972. WMO/OMM No. 326, Geneva, Switzerland.

Blasing, T.J. and Solomon, A.M. (1984). Response of the North American Corn Belt to climatic warming. *Progress in Biometeorology*, **3**, 311–21.

Bonan, G.B. and Korzukhin, M.D. (1989). Simulation of moss and tree dynamics in the boreal forest of interior Alaska. *Vegetatio*, **84**, 31–44.

Boryczka, J. and Stopa-Boryczka, M. (1986). A mathematical model of Poland's climate. *Miscellanea Geographica*, **1986**, 55–69.

Box, E.O. (1981). *Macroclimate and Plant Forms: an Introduction to Predictive Modeling in Phytogeography*. The Hague: Dr. W. Junk Publishers.

Boysen-Jensen, P. (1949). Causal plant geography. *Det Kongelige Danske Videnskabernes Selskab, Biologiske Meddelelser*, **21**(3).

Bradley, R.S., Kelly, P.M., Jones, P.D., Diaz, H.F. and Goodess, C. (1985). *A climatic data bank for the Northern Hemisphere land areas. 1851–1980*. DOE Technical Report No. 017. U.S. Department of Energy, Carbon Dioxide Research Division, Washington, D.C.

Busby, J.R. (1988). Potential impacts of climate change on Australia's flora and fauna. In *Greenhouse. Planning for Climate Change*, ed. G.I. Pearman, pp. 387–98. Leiden: E.J. Brill.

Cohen, J. (1960). A coefficient of agreement for nominal scales. *Educational and Psychological Measurements*, **20**, 37–46.

Cramer, W. and Leemans, R. (1990). *Static vegetation models and climate change on high latitudes*. Paper presented at the symposium on "Boreal Forests: State, Dynamics and Anthropogenic Influences." Arkhangelsk, USSR, July 1990.

Cramer, W. and Prentice, I.C. (1988). Simulation of regional soil moisture deficits on a European scale. *Norsk Geografisk Tidskrift*, **42**, 149–51.

Davis, M.B. (1981). Quaternary history and the stability of forest communities. In *Forest Succession: Concepts and Application*, ed. D.C. West, H.H. Shugart and D.B. Botkin, pp. 132–54. New York: Springer-Verlag.

de Candolle, A.L. (1855). *Géographie Botanique Raisonée*. Paris: Victor Masson & Gen'eve: J. Kessmann.

Delijaniec, I. (1972). The characteristic change of the amount of precipitation in connexion with altitude in the mountainous areas of Yugoslavia. In *Distribution of Precipitation in Mountainous Areas*. Symposium Geilo, Norway, July–August 1972. WMO/OMM No. 326, Geneva, Switzerland.

Emanuel, W.R., Shugart, H.H. and Stevenson, M.P. (1985a). Climatic change and the broad-scale distribution of terrestrial ecosystem complexes. *Climatic Change*, **7**, 29–43.

Emanuel, W.R., Shugart, H.H. and Stevenson, M.P. (1985b). Response to comment: Climatic change and the broad-scale distribution of terrestrial ecosystem complexes. *Climatic Change*, **7**, 457–60.

Graetz, R.D., Walker, B.H. and Walker, P.A. (1988). The consequences of climatic change for seventy percent of Australia. In *Greenhouse. Planning for Climate Change*, ed. G.I. Pearman, pp. 399–413. Leiden: E.J. Brill.

Hansen, J., Lacis, A., Rind, D., Russell, G., Stone, P., Fung, I., Ruedy, R. and Lerner, J. (1984). Climate sensitivity: analysis of feedback mechanisms. In *Climate Processes and Climate Sensitivity*, ed. J. Hansen and T. Takahashi, pp. 130–63. American Geophysical Union, Washington, D.C.

Harrison, S.P. (1990). *An introduction to general circulation modelling experiments with raised* CO_2. WP-90-27, International Institute for Applied Systems Analysis, Laxenburg, Austria.

Hearne, S. (1772). *A Journey from Prince of Wales Fort in Hudson's Bay to the Northern Oceans in the Years 1769, 1770, 1771, and 1772*. Toronto: The Champlain Society.

Holdridge, L.R. (1947). Determination of world plant formations from simple climatic data. *Science*, **105**, 367–8.

Holdridge, L.R. (1967). *Life Zone Ecology*. San José: Tropical Science Center.

Houghton, J.T, Jenkins, G.J. and Ephraums, J.J. (1990). *CLIMATE CHANGE—The IPCC Scientific Assessment*. Cambridge: Cambridge University Press.

Hutchins, L.W. (1947). The bases for temperature zonation in geographical distribution. *Ecological Monographs*, **17**, 325–35. p73

Khurshid Alam, F.C. (1972). Distribution of precipitation in mountainous areas of west Pakistan. In *Distribution of Precipitation in Mountainous Areas*. Symposium Geilo, Norway, July–August 1972. WMO/OMM No. 326, Geneva, Switzerland.

Kojima, S. (1979). Biogeoclimatic zones of Hokkaido Island, Japan. *Journal of the College of Liberal Arts, Toyama University, Japan*, **12**, 97–141.

Köppen, W. (1884). Die Wärmezonen der Erde, nach der Dauer der heissen, gemässigten und kalten Zeit und nach der Wirkung der Wärme auf die organische Welt betrachtet. *Meteorologische Zeitschrift*, **1**, 215–26 plus map.

Köppen, W. (1936). Das geographische System der Klimate. In *Handbuch der Klimatologie*, ed. W. Köppen and R. Geiger. Berlin: Gebrüder Bornträger.

Krajina, V.J. (1959). Bioclimatic zones in British Columbia. *University of British Columbia, Botanical Series*, **1**, 1–47.

Landis, J.R. and Koch, G.G. (1977). The measurement of observer agreement for categorical data. *Biometrics*, **33**, 159–74.

Leemans, R. (1989). Possible changes in natural vegetation patterns due to a global warming. In *Der Treibhauseffekt: Das Problem—Mögliche Folgen—Erforderliche Massnahmen*, ed. A. Hackl, pp. 105–22. Laxenburg, Austria: Akademie für Umwelt und Energie.

Leemans, R. and Cramer, W. (1990). *The IIASA climate database for land areas on a grid with 0.5° resolution*. WP-90-41, International Institute for Applied Systems Analysis, Laxenburg, Austria.

Linnaeus, C. (1977). *Iter Lapponicum, Dei Grata Institutum 1732* [Lapplands Resa År 1732], ed. M. von Platen and C.-O. von Sydow. Stockholm: Wahlström & Widstrand.

Manabe, S. and Wetherald, R.T. (1980). On the distribution of climatic change resulting from an increase in CO_2 content of the atmosphere. *Journal of Atmospheric Science*, **37**, 99–118.

Merriam, C.H. (1898). Life zones and crop zones of the United States. *Bulletin US Department of Agriculture, Division Biological Survey*, **10**.

Meteorological Office (1966). *Tables of temperature, relative humidity and precipitation for the world. Part V. Asia*. HMSO, London.

Meteorological Office (1972). *Tables of temperature, relative humidity, precipitation and sunshine for the world. Part III. Europe and the Azores*. HMSO, London.

Meteorological Office (1973). *Tables of temperature, relative humidity and precipitation for the world. Part VI. Australasia and Pacific Ocean*. HMSO, London.

Meteorological Office (1978). *Tables of temperature, relative humidity and precipitation for the world. Part II. Central and South America, the West Indies and Bermuda*. HMSO, London.

Meteorological Office (1980). *Tables of temperature, relative humidity, precipitation and sunshine for the world. Part I. North America and Greenland (including Hawaii and Bermuda)*. HMSO, London.

Meteorological Office (1983). *Tables of temperature, relative humidity, precipitation and sunshine for the world. Part IV. Africa, the Atlantic Ocean South 35°* N and the Indian Ocean. HMSO, London.

Mitchell, J.F.B. (1983). The seasonal response of a general circulation model to changes in CO_2 and sea temperatures. *Quarterly Journal of the Royal Meteorological Society*, **109**, 113–52.

Mitchell, J.F.B., Wilson, C.A. and Cunnington, W.M. (1987). On CO_2 climate sensitivity and model dependence of results. *Quarterly Journal of the Royal Meteorological Society*, **113**, 293–322.

Monserud, R.A. (1990). *Methods for comparing global vegetation maps*. WP-90-40, International Institute for Applied Systems Analysis, Laxenburg, Austria.

Müller, M.J. (1982). *Selected climatic data for a global set of standard stations for vegetation science*. The Hague: Dr. W. Junk Publishers.

National Geophysical Data Center (1988). *10-minute topography database*. U.S. Department of Commerce, Washington, D.C.

Ohsawa, M. (1990). An interpretation of latitudinal patterns of forest limits in South and East Asian mountains. *Journal of Ecology*, **78**, 326–39.

Olson, J., Watts, J.A. and Allison, L.J. (1983). *Carbon in live vegetation of major world ecosystems*. Oak Ridge National Laboratory, Oak Ridge, Tennessee.

Overpeck, J.T., Rind, D. and Goldberg, R. (1990). Climate-induced changes in forest disturbance and vegetation. *Nature*, **343**, 51–3.

Parry, M.L., Carter, T.R. and Konijn, N.T. (1988). *The Impact of Climatic Variations on Agriculture. Volume 1: Assessments in Cool Temperate and Cold Regions*. Dordrecht: Kluwer Academic Publishers.

Pastor, J. and Post, W.M. (1985). *Development of a linked forest productivity–soil process model*. ORNL/TM-9519, Oak Ridge National Laboratory, Oak Ridge, Tennessee.

Prentice, I.C. (1986). Vegetation response to past climatic variation. *Vegetatio*, **67**, 131–41.

Prentice, I.C. (1988). Paleoecology and plant population dynamics. *Trends in Ecology and Evolution*, **3**, 343–5.

Prentice, I.C., Webb, R.S., Ter-Mikhaelian, M.T., Solomon, A.M., Smith, T.M., Pitovranov, S.E., Nikolov, N.T., Minin, A.A., Leemans, R., Lavorel, S., Korzukhin, M.D., Helmisaari, H.O., Hrabovszky, J.P., Harrison, S.P., Emanuel, W.R. and Bonan, G.B. (1989). *Developing a global vegetation dynamics model: results of an IIASA summer workshop*. RR-89-7, International Institute for Applied System Analysis, Laxenburg, Austria.

Prentice, K.C. and Fung, I.Y. (1990). The sensitivity of terrestrial carbon storage to climate change. *Nature*, **346**, 48–51.

Priestley, C.H.B. and Taylor, R.J. (1972). On the assessment of surface heat flux and evaporation using large-scale parameters. *Monthly Weather Review*, **100**, 81–92.

Raunkiær, C. (1907). *Planterigets Livsformer*. Copenhagen/Kristiania: Gyldalska Bokhandel & Nordisk Forlag.

Ritchie, J.C. (1986). Climate change and vegetation response. *Vegetatio*, **67**, 65–74.

Rydén, B.E. (1972). On the problem of vertical distribution of precipitation, especially in areas with great height differences. In *Distribution of Precipitation in Mountainous Areas*. Symposium Geilo, Norway, July–August 1972. WMO/OMM No. 326, Geneva, Switzerland.

Sargent, N.E. (1988). Redistribution of the Canadian boreal forest under a warmed climate. *Climatological Bulletin*, **22**, 23–34.

Schlesinger, M.E. and Mitchell, J.F.B. (1987). Climate model simulations of the equilibrium climatic response to increased carbon dioxide. *Reviews of Geophysics*, **25**, 760–98.

Schlesinger, M.E. and Zhao, Z.-C. (1989). Seasonal climatic changes induced by doubled CO_2 as simulated by the OSU atmospheric GCM/mixed-layer ocean model. *Journal of Climate*, **2**, 459–95.

Schouw, J.F. (1823). *Grundzüge einer allgemeinen Pflanzengeographie*. Berlin: Reimer.

Solomon, A.M. (1986). Transient responses of forests to CO_2-induced climate change: Simulation modeling in eastern North America. *Oecologia*, **68**, 567–79.

Sowell, J.B. (1985). A predictive model relating North American plant formations and climate. *Vegetatio*, **60**, 103–11.

Storr, D. and Ferguson, H.L. (1972). The distribution of precipitation in some mountainous Canadian watersheds. In *Distribution of Precipitation in Mountainous Areas*. Symposium Geilo, Norway, July–August 1972. WMO/OMM No. 326, Geneva, Switzerland.

Strahler, A. and Strahler, A. (1989). *Elements of Physical Geography*. New York: John Wiley & Sons.

Su, H.-J. (1984). Studies on the climate and vegetation types of the natural forests in Taiwan (I). Analysis of the variations in climatic factors. *Quarterly Journal of Chinese Forestry*, **17**, 1–14.

Sukachev, V.N. (1958). (On the principles of genetic classification in biocoenology). Condensed translation of 'O principi geneticeskoj klassifikacii v biocenologii' by F. Raney and R.F. Daubenmire. *Ecology*, **39**, 364–7.

Thornthwaite, C.W. (1948). An approach toward a rational classification of climate. *Geographical Reviews*, **38**, 55–94.

Thornthwaite, C.W. and Mather, J.R. (1957). Instructions and tables for computing potential evapotranspiration and the water balance. *Publications in Climatology*, **10**, 185–310.

Troll, C. and Paffen, K. (1964). Die Jahreszeitenklimate der Erde (Summary: The seasonal climates of the earth). *Erdkunde*, **18**, 1–28 plus map.

Tucker, G.B. (1988). Climate modelling: how does it work? In *Greenhouse. Planning for Climate Change*, ed. G.I. Pearman, pp. 22–34. Leiden: E.J. Brill.

Tuhkanen, S. (1980). Climatic parameters and indices in plant geography. *Acta Phytogeographica Suecica*, **67**.

von Humboldt, A. (1807). *Ideen zu einer Geographie der Pflanzen nebst einem Naturgemälde der Tropenländer*. Tübingen.

Walter, H. (1964). *Die Vegetation der Erde in ökophysiologischer Betrachtung. Vol. 1.* 2nd ed. Jena: VEB Gustav Fischer Verlag.

Walter, H. and Lieth, H. (1960–67). *Klimadiagramm-Weltatlas*. Stuttgart: Gustav Fischer Verlag.

Warrick, R.A., Shugart, H.H., Antonovsky, M.Ya., Tarrant, J.R. and Tucker, C.J. (1986). The effects of increased CO_2 and climatic change on terrestrial ecosystems. In *The Greenhouse Effect, Climate Change, and Ecosystems*, ed. B. Bolin, B.R. Döös, J. Jäger and R.A. Warrick, pp. 363–92. Chichester: John Wiley & Sons.

Weather Bureau (1959). *World Weather Records 1941–1950*. U.S. Department of Commerce, Washington, D.C.

Webb III, T. (1986). Is vegetation in equilibrium with climate? How to interpret late-Quaternary pollen data. *Vegetatio*, **67**, 119–30.

Webb III, T. (1987). The appearance and disappearance of major vegetational assemblages: Long-term vegetational dynamics in eastern North America. *Vegetatio*, **69**, 177–87.

Willmott, C.J. and Rowe, C.M. (1985). Climatology of the terrestrial seasonal water cycle. *Journal of Climatology*, **5**, 589–606.

Wilson, C.A. and Mitchell, J.F.B. (1987). Simulated climate and CO_2-induced climate change over western Europe. *Climatic Change*, **10**, 11–42.

Wilson, E.O. (1989). Threats to biodiversity. *Scientific American*, **261**, 60–6.

Woodward, F.I. (1987). *Climate and Plant Distribution*. Cambridge: Cambridge University Press.

Woodward, F.I. (1989). Plants in the greenhouse world. *New Scientist*, **21**, 1–4.

Woodward, F.I. and Williams, B.G. (1987). Climate and plant distribution at global and local scales. *Vegetatio*, **69**, 189–97.

11

Vegetation Diversity and Classification Systems

G. Grabherr and S. Kojima

Introduction

The first modern attempt to classify and characterize vegetation and to relate it to environment can be traced to the beginning of the 19th century, to Alexander von Humboldt. Humboldt made extensive surveys of South America and tried to explain vegetation differentiation in relation to environmental characteristics, in particular those of the climate (Humboldt 1805, 1806, 1807). Since Humboldt, numerous studies of vegetation ecology have been carried out and many different approaches have evolved. Vegetation characteristics have been described in various ways depending on the study interest and objectives. Only those which are necessary milestones for this discussion will be cited here. The main subject of this chapter is the current concern about global vegetation change. Approaches and techniques for describing and quantifying vegetation will be evaluated for their relevance to global vegetation issues, and study procedures for future research efforts to assess the climatic influences on global vegetation will be proposed.

Evaluation of Some Existing Approaches

The physiognomic approach, floristic approach, and ecosystematic approach were chosen for consideration, and the following evaluation was made.

Physiognomic Approach

Physiognomy is a morphological characteristic of vegetation. It is primarily determined by growth-form (sensu Rübel 1930) and life-form (sensu Raunkier 1934) of the dominant or codominant plants. As it is visually recognized and distinguished, it has been extensively used to characterize vegetation. Indeed, physiognomic classification is readily applicable to any kind of vegetation of the world, even in regions where the flora is not well known, or even if a researcher is not well acquainted with the flora.

The physiognomic approach has a long history in vegetation science. Indeed, it was Humboldt who first employed physiognomy to describe vegetation and introduced the term *association* as a physiognomic unit. His concept was accepted by many students in the 19th and early 20th centuries. Notable among them were Griesebach (1838, 1872), who proposed the term *formation* as a physiognomic unit (replacing Humboldt's original "association"); Rübel (1930), who proposed a growth-form categorization, and Raunkier (1934), who is well known for his life-form classification. Schimper and von Faber (1935), as well as Rübel (1930), proposed classification schemes of formation types of the world. Küchler (1967) proposed a systematic categorization of life-form and structure of plants; he made extensive surveys and mapped the vegetation of the world based on the formation type. One further derivation of the line may be Dansereau's (1957) diagrammatic schematization of vegetation structure. A meticulous classification and coding system of physiognomy was proposed by Müller-Dombois and Ellenberg (1974).

The term "formation" became widely accepted as a physiognomic unit and was extensively used in the fields of ecology, biogeography, and vegetation science. Clements (1902; Weaver and Clements 1929) adopted the term and used it to designate geographical regions of the same climax vegetation in relation to his mono-climax concept. His formation was combined with a zoogeographical notion and generated the concept and term *biome* (Clements and Shelford 1939).

In terms of causality, the formation is usually thought to be determined by macroclimate. It is in this context that formation types are often correlated with climatic types in global distribution (Holdridge 1947; Lieth 1956; Whittaker 1970). As Walter (1976) pointed out, however, not only are the morphological character and type of reaction to seasonality very important with regard to climate—ecophysiological features are important, too. As an example, the genus *Pinus,* which, if it occurs, usually determines physiognomy, consists of species adapted to the extreme boreal climates and of those which can be found in tropical regions. The formation type would be the same, i.e., pine forest or, even more roughly, coniferous forest. Furthermore, leaf-succulent plants are almost absent from the Australian flora, a result of the isolated evolution of this local flora since the Tertiary. This is by no means an outcome of a climate different from that in other arid regions. However, this problem may be solved to some extent by classifying the whole life-form composition of a particular region (Box 1981). Box (1982) demonstrated in his comprehensive correlation of growth-form distribution to climate that Mediterranean vegetation is not mainly characterized by the dominance of sclerophyllous evergreen trees but by the highest growth-form diversity of the world.

The physiognomic approach provides a fast and efficient means to describe and characterize vegetation as it does not require much *floristic detail* about the vegetation. It is especially effective in regions where the flora is not thoroughly known. In other words, the physiognomic characterization may be carried out fairly routinely if some guidelines are provided. It is, therefore, particularly useful for conducting a reconnaissance type of vegetation survey to cover large geographical areas in a limited time. The approach is not very effective in detecting spatial and temporal changes of vegetation in detail unless such changes are great enough to substantiate a shift of formation type.

Floristic Approach

The floristic approach may be represented by the phytosociology of the Zürich-Montpellier school. It developed from the physiognomic approach in the late 19th and early 20th centuries in central Europe. The approach emphasizes floristic characteristics of vegetation such as species composition and some quantitative characters (e.g. cover value, sociability, constancy, exclusiveness) of individual species in order to classify and characterize vegetation. After early works by Schröter (1894, 1926), Flahault (1901), Brockman-Jerosch (1907), and others, a comprehensive systematization of the approach was accomplished by Braun-Blanquet (1921, 1932). This system has attracted many students of vegetation science, and numerous studies have been conducted, not only in central Europe but also in other regions of the world.

The approach focuses on analysis and synthesis of the floristic composition of plant communities. A relevé or stand represents a sample of a type of vegetation which exists in the field. Relevés are grouped and classified according to their floristic similarities (Figs. 11.1 and 11.2). For this, not only small numbers of dominant or codominant species but also a whole assemblage of the constituent species are equally weighed and examined. A group of relevés which possess some species in common and more-or-less exclusively is considered to be one type of vegetation. The association as an abstract type was adopted as a basic unit of the classification (Flahault and Schröter 1910). It has a homogeneous floristic structure and represents a typified and abstracted vegetation of site-specific stand level. Species, or a combination of species, characterizing the association are called diagnostic species. The associations are hierarchically grouped into higher units: alliance, order, and class. Each of the hierarchical ranks, so-called syntaxa, is characterized by the diagnostic species for the particular rank.

The approach attempts to produce a global vegetation system based on floristic criteria. For example, Tüxen *et al.* (1972) and Ohba (1974) presented convincing accounts on mire and alpine steppe vegetation (Fig. 11.1). However, plant sociology today is far away from a floristically defined vegetation system of the world though many elaborate systems exist on a regional scale, such as that for central Europe. (See Wilmanns 1983; Ellenberg 1987; Fig. 11.2.)

Fig. 11.1. Estimated distribution of the phytosociological orders of the class Carici rupestris–Kobresietea bellardii which are alpine grasslands of the Northern Hemisphere. 1 = Kobresio-Dryadetalia BR. BL. 1948 em.; 2 = Thymo arcticae-Kobresietalia bellardii ord. nov.; 3 = Oxytropido-Kobresietalia Oberdorfer 1957; 4 = Oxytropidetalia dinarici ordo nov.; 5 = order of the Caucasus which has still to be described; 6 = Caricetalia tenuiformis Ohba 1968; 7 = order of south Siberia which has still to be described; 8 = Vikariant class to the Carici rupestris–Kobresietea bellardii (from Ohba 1974).

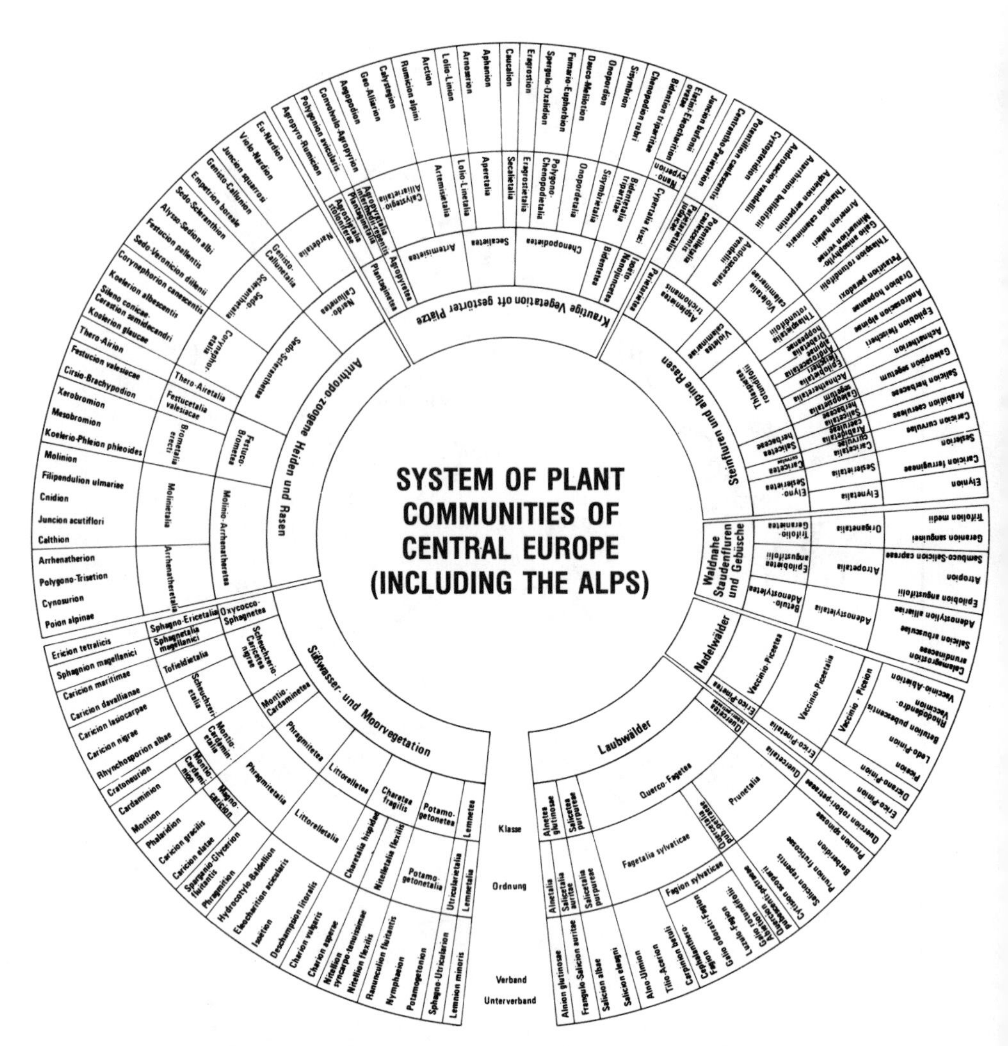

Fig. 11.2. Phytosociological vegetation units of Central Europe. Inner circle: highest level of the hierarchy which is considered here as formation type. For example: Laubwälder = deciduous forests. These are divided into four classes (Alnetea glutinosae, etc.). The class Querco-Fagetea is divided into three orders; the order Fagetalia sylvatici contains seven alliances (outer circle). The alliances are further divided into associations which are not shown here (from Kuhn 1987).

Phytosociological measures should be considered when modeling global vegetation change for the following reasons. Other vegetation classification systems rely on growth-forms, dominant species, structural features, or even ecophysiological characteristics like frost resistance as attributes. They consider convergent types of vegetation but neglect the genetic diversity of the different units. Semideserts in Patagonia show nearly the same growth-form composition as the semideserts of North Africa. The physiognomy and the composition of functional types may be similar but the floristic character is completely different. Those vegetation units which may be the best for large-scale mapping may fail when considering the genetic stock of the world, which involves the plant species diversity of 260 000 spermatophytes, 9770 pteridophytes, 20 000 bryophytes, and so on. Accordingly the question has to be posed: does it make sense to create models to forecast vegetation change without asking what happens to the genetic stock, i.e., to the evolutionary heritage of the world? Phytosociological units may overcome this problem, at least for regional purposes.

Syntaxa and Scale

Syntaxa are defined exclusively according to their content of diagnostic species; they occupy ecological niches of very different dimensions. A list of the vegetation classes of central Europe (according to Wilmanns 1983) representing space demand, alpha-diversity, and content of character species may demonstrate this (Tab. 11.1). At least 75% of classes would hardly exceed more than one or several hectares in size and therefore would never appear in vegetation mapping finer than 1:1 000 000—for example, the associations of the class Isoeto-Nanojuncetea, i.e., the plant communities of tiny puddles. These vegetation units contain twice as many character species as the large-sized units such as alpine steppe or forests.

But how can these syntaxa be subjected to vegetation modeling? One may expect that as a kind of domino effect, the small sized vegetation units would be linked to the large ones and would disappear or appear together with them. Unfortunately this does not hold true. As an example, the associations of the class Isoeto-Nanojuncetea can be found in North Africa in nearly identical composition as in central Europe, whereas the dominant and relevant class for large-scale mapping, i.e., the Querco-Fagetea, is replaced completely by the Quercetea ilicis. That means that the different classes have to be treated separately or at least as class groups. They can be stratified or ordinated according to environmental and/or phytogeographical gradients as may be demonstrated for rock communities and climax forests of Europe (Fig. 11.3). Small units can be subjected to image analysis (compare with Skole *et al.*, this volume) by displaying the distribution and change in distribution as contour maps and so on.

Syntaxa and Environmental Change

In most cases it is unlikely that a simple functional relationship would exist between vegetation described in phytosociological terms and environment, as it does for alpine steppe and soil reaction (Fig. 11.4). On the other hand, it should not be an unsolvable problem for experienced phytosociologists to figure out environmental threshold values for syntaxa, as can be done for single taxa (Fig. 11.3). Woodward's (Woodward and Williams 1987) approach using threshold values of temperature and water budget for modeling climate and plant distribution

Table 11.1. Space demand of central European syntaxa together with their contribution to floristic diversity. Syntaxa according to Wilmanns (1983; without consideration of seashore syntaxa). The figures for character species are counts of the species cited by Wilmanns, who only considered very common ones. Diversity values are rough estimates by the author.

Class	1×1	Space 10×10	Demand 100×100	(m) 1000×1000	Average alpha-diversity	Character species
Montio-Cardaminetea	x				5–8	6
Littorelletea	x				5–8	12
Utricularieta	x	x			2–4	4
Lemnetea	x	x			3–5	5
Isoeto-Nanojuncetea	x	x			5–7	14
Bidentetea	x	x			5–10	11
Polygono-Poetea annuae	x	x			5–10	5
Parietarietea	x	x			8–12	9
Artemisietea	x	x			10–20	29
Stellarietea mediae	x	x			10–20	45
Agropyretea repentis	x	x			5–10	13
Scheuchzerio-Caricetea	x	x			8–12	21
Salicetea herbaceae	x	x			5–10	18
Violetea calaminariae	x	x			5–8	5
Sedo-Scleranthetea	x	x			10–20	19
Cetrario-Loiseleurietea	x	x			5–10	4
Trifolio-Geranietea	x	x			10–20	31
Rhamno-Prunetea	x	x			5–15	20
Potamogetonetea	x	x	x		3–7	19
Thlaspietea rotundifolii	x	x	x		5–10	36
Asplenietea rupestria	x	x	x		5–10	11[a]
Epilobietea angustifolii	x	x	x		8–12	13
Phragmitetea	x	x	x		5–8	23
Oxycocco-Sphagnetea	x	x	x		5–10	7
Elynetea	x	x	x		30–40	11
Festuco-Brometea		x	x		20–50	58
Betulo-Adenostyletea		x	x		10–20	21
Salicetea purpureae		x	x		15–30	8
Sesleria variae		x	x	x	30–50	19
Caricetea curvulae		x	x	x	15–25	10
Molinio-Arrhenatheretea		x	x	x	20–30	81
Nardo-Callunetea		x	x	x	15–40	34
Alnetea glutinosae		x	x	x	10–20	8
Erico-Pinetea		x	x	x	30–40	9[b]
Vaccinio-Piceetea		x	x	x	15–25	23
Quercetea robori-petraeae		x	x	x	15–20	5
Querco-Fagetea		x	x	x	10–40	50

[a]This group especially contains many endemics in refugia of the Alps; a real estimate may increase the number to a very high value.

[b]Pine forests of the Alps are very rich in species but most of them cannot be considered as character species.

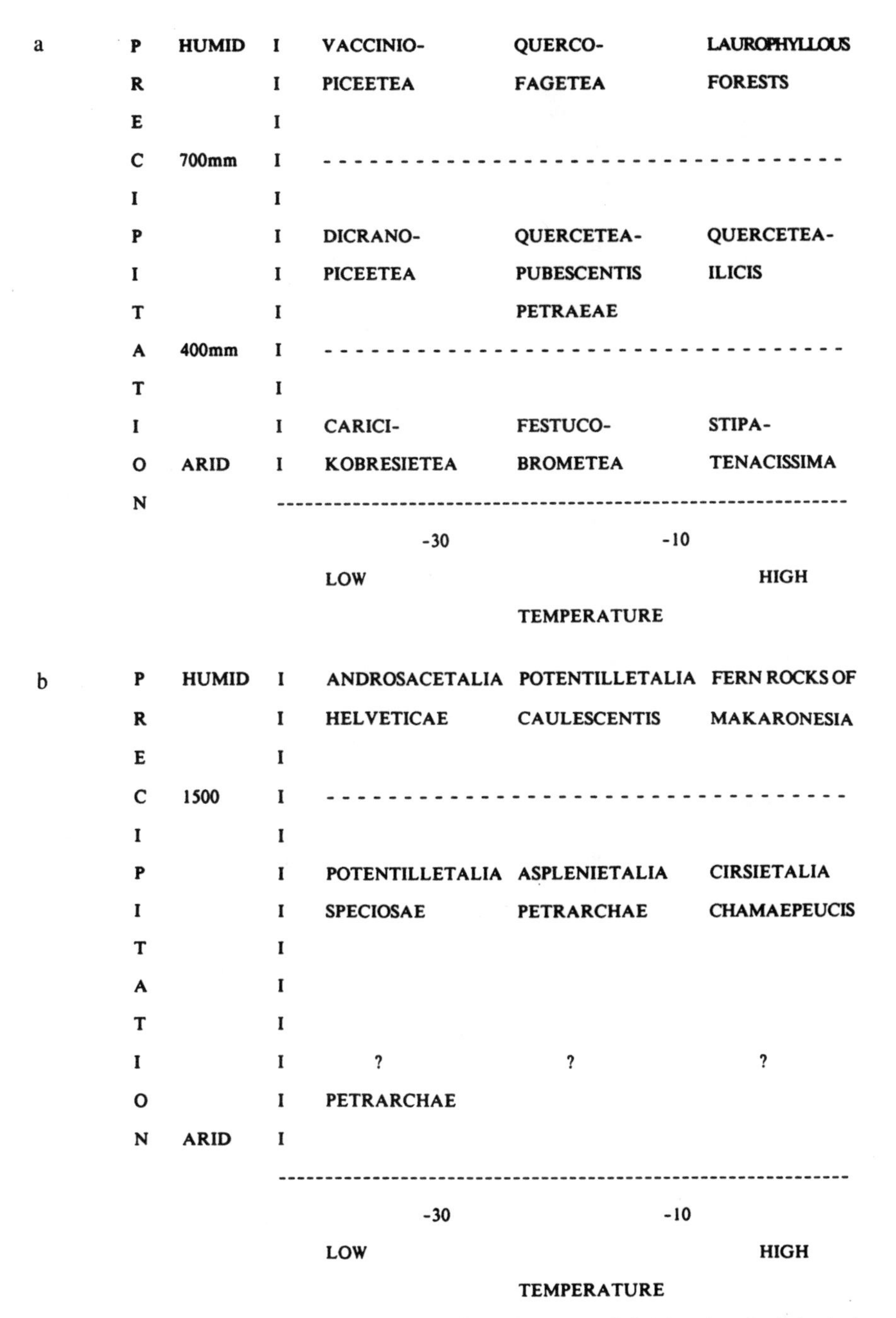

Fig. 11.3. Ecograms for (a) climax vegetation (classes) and (b) the phytosociological classes of calcareous rocks in Europe, with threshold values for annual precipitation and annual minimum temperature. The values are of course very rough estimates, but the probability is high that, for example, the *Quercus ilex* forest (Quercetea ilicis) will expand to the north into the *Quercus pubescens* forests when the winter frosts decrease in frequency.

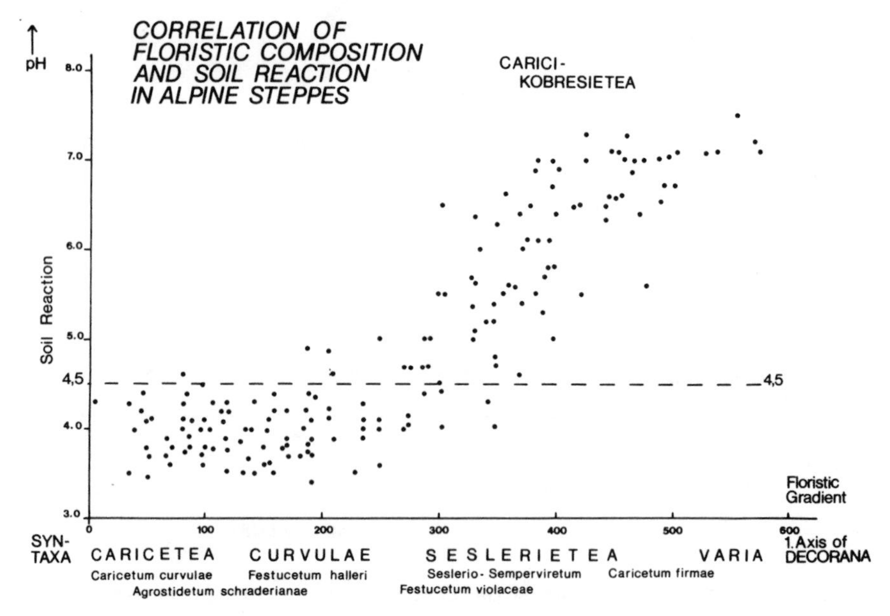

Fig. 11.4. Soil reaction of about 300 alpine steppe plots in relation to their floristic composition. The Caricetea curvulae are the communities on acid soil, and floristic variability is not affected by changes in soil reaction. Contrarily the alpine grassland on nonacid soils reacts sensitively to differences in soil reaction with a change in floristic composition. Abscissa: First axis of an ordination of 300 vegetation samples (= releves according to the Braun-Blanquet approach); ordination method: detrended correspondence analysis (DECORANA, Hill 1979).

could be extended to syntaxonomical units as well. Of course there may be some doubts as to whether a plant community will react like a single species or a species group which represents a particular growth-form. As Webb (1987) demonstrated recently, the main tree species in eastern North America changed their distribution in postglacial times individually. Species became associated which do not grow together today in the region under consideration. Contrarily, Tüxen (1974) traces some central European classes back to the Tertiary. Anyway, if the transition from one syntaxon to another is defined such that the product of vegetation change has some similarity to the unit known to occur under those circumstances today, then one's vegetation forecast would not be completely wrong.

Transition matrices may constitute another way to explore vegetation change along phytosociological dimensions. Some examples of modeling succession on the basis of Markovian models have been published during the last 10 years (Leps 1987; and others), but these were applied to very restricted datasets. Of course no major argument against an extension to syntaxonomical units can be found.

In conclusion, the principal applicability of phytosociological measures to modeling vegetation change on a global scale has been demonstrated. Their power for handling information on the genetic stock of vegetation should not be neglected; phytosociology may be the only adequate way to consider genetic diversity. However, many problems are left unsolved. Even at the highest syntax-onomical level many blank areas exist, especially for tropical regions, for which information on environmental thresholds is simply not available.

Although Braun-Blanquet presented a unifying concept which has been applied in many places in the world and to different types of vegetation, many vegetation ecologists still refuse to use it in its strict form. On the other hand, many representatives of the Braun-Blanquet school neglect modern trends, especially the use of numerical classification methods. Nevertheless, thinking about the application of phytosociology to vegetation modeling on a global scale would supply a needed stimulus for phytosociologists. In any case, in the absence of interest in precise answers, phytosociology can still contribute to the global vegetation change question, at least by supplying some pilot models valid for particular regions.

Environmental or Ecosystematic Approach

This approach may be traced back to as early as 1898, to Dokuchaev, who first proposed the concept of a soil as a functional product of climate, parent material, relief, organisms, and time. This notion was further refined by Jenny (1941) and Major (1951) to include vegetation in the dependent variables; it was assumed that vegetation and soil constitute an inseparable complex which may be termed a phytogeocoenosis. This approach emphasizes a causal relationship of vegetation (or phytogeocoensis) and environment. In North America, Clements (1916) regarded vegetation at the formation rank as a product primarily of climate and represented it with a climatic climax. His concept was inherited by Nichols (1929), Halliday (1937), and Braun (1950). In the meantime, Tansley (1935) proposed the term *ecosystem,* though preceded by some other terms, to connote a complex of natural systems. Sukachev (1945, 1960), following Morozov's original idea, proposed the term *biogeocoensis,* which is somewhat comparable to Tansley's "ecosystem." However, the fundamental difference between them seems to lie in the fact that the ecosystem is a concept applicable to any rank of generalization, whereas biogeocoensis is principally applied to an ecosystem at the community level (cf. nano-ecosystem sensu Ellenberg 1973).

Stimulated by the International Biological Program (IBP), whose activities were focused on ecosystem analysis, Ellenberg (1973) presented the most recent account on ecosystem classification of the world. Though Ellenberg tried to include attributes such as primary productivity, trophic characteristics, macroclimate, soil characters, and so on, his system is based mainly on vegetation formations. Walter (1976) criticized this approach as being a central European one extended to the whole world, suggesting, as an alternative, an ecological system of the continents based on the biome as the fundamental unit. He distinguishes zonobiomes, orobiomes (mountain systems), and pedobiomes (edaphically determined biomes such as alkali steppes). Transition zones from one zonobiome to another are called zonoecotones. Zonobiomes are mainly characterized by their macroclimate, which is considered to be uniform throughout the region. A type of zonal vegetation can be assigned to each zonobiome. Walter's ideas are best presented by an example. The Namib Desert of southwest Africa is a biome of the subzonobiome *fog-deserts* which is part of the subtropical-arid zonobiome. The particular biome consists of different biocoen complexes; the biocoens may be split such that the synusium is the smallest unit. Walter's concept might be regarded as very appropriate, especially his idea of treating mountain biomes and edaphically determined biomes separately and of basing the definition of zonobiomes on his well-known climate diagrams. Unfortunately, he did not present an elaborated system in his book, and fulfilling the biome-classification system according to his suggestion might be laborious.

A further attempt to define relationships between vegetation and environment was developed by Krajina (Krajina 1965, 1969). His *biogeoclimatic ecosystem classification* defines terrestrial ecosystems based on an integration of climate, vegetation, and soil characteristics. Vegetation is regarded as a concrete manifestation of an ecosystem. Terrestrial ecosystems are classified at two levels of integration, i.e., generalized and detailed. The former level deals with a classification of regional ecosystems which cover a broad geographical area, whereas the latter handles ecosystems at a local-site-specific level such as a plant community in a definite habitat. For the generalized level of classification, the basic unit is a biogeoclimatic zone, which is hierarchically incorporated into higher units, such as the biogeoclimatic region and the biogeoclimatic formation. The last (highest) unit is somewhat comparable to a formation of Clements (Weaver and Clements 1929) or a biome of Clements and Shelford (1939) or a zonobiome of Walter (1976).

A biogeoclimatic zone is a geographical region which is circumscribed by a uniform climate, and in practice it is determined by the same kind of climatic climax vegetation. Because a biogeoclimatic zone covers a broad geographical area, it necessarily contains a great variety of different kinds of local vegetation, i.e., site-specific ecosystems. This justifies the need for detailed classification.

For the detailed level of classification, a plant association (sensu Krajina 1960) is the basic unit and is comparable to the biogeocoenosis of Sukachev. For this classification, the phytosociological technique of the Zürich-Montpellier school is usually employed to discriminate and typify the biogeocoenoses based on their vegetation characteristics. Krajina's plant associations are also hierarchically grouped into higher units such as alliance, order, and class, each of which represents higher levels of generalization of ecosystems manifested in the form of vegetation.

With the two levels of resolution, i.e., generalized and detailed, the biogeoclimatic classification system can handle any level of ecosystem, from site-specific small ones to a vast collection of biomes on a continental scale. Since the biogeoclimatic classification emphasizes the holistic relations of terrestrial ecosystems, climate-vegetation-soil relationships are usually analyzed and synthesized. For instance, a plant association is discriminated not only on the basis of floristic composition but also on the basis of environmental characteristics. Similarly, a biogeoclimatic zone needs to be environmentally distinguished from others. In this fashion, the biogeoclimatic approach always furnishes some interpretation and analytical insight as to vegetation and environment relationships. This is one of the great advantages of this approach, which is particularly suited to dealing with vegetation change along with environmental change. On the other hand, however, this approach requires extensive and intensive studies of vegetation and environment of the region concerned beforehand. There have been some attempts to correlate broad regional vegetation at the formation level directly with environment, particularly with climate (Holdridge 1947; Lieth 1956; Whittaker 1970).

Some Recommendations for World Vegetation Characterization

In order to detect a future vegetation change, it is vitally important to collect data on the current status of the world's vegetation. Such data will serve as baseline information against which any future change can be checked. At the same time, and more importantly, correlation of the current vegetation with current environmental characteristics, especially with those of climate, needs to be analyzed.

First of all, classification and characterization of the current vegetation of the whole world should be performed. From a realistic point of view, the global vegetation classification will inevitably be fairly general. Perhaps the basic unit will be that of something comparable to a biome (sensu Clements and Shelford 1939) or to a meso-ecosystem (sensu Ellenberg 1973). It should be of a size that is mappable in 1:10 000 000 to 1:25 000 000 scale if mapping is required. For the sake of convenience, let us call it a *zone*. The entire terrestrial world should be segmented and classified into such zones. Each zone then needs to be represented by some kind of vegetation. There will be numerous kinds of vegetation reflecting diverse local environments within a zone.

Now the question is how to decide on the representative vegetation. We recommend the use of vegetation classes which may be considered as climatic climax vegetation or as zonal biogeocoenosis in the area. Climatic climax vegetation is defined by climatic influences and hence will be most sensitive to climatic change. Such vegetation will therefore be a good indicator of climatic change. Nevertheless, there will be regions where climax vegetation is not actually present. In such cases, potential climatic climax vegetation may be introduced. This is a conceptual vegetation constructed on the basis of extrapolation from nearby natural vegetation (Küchler 1967; Meeker and Merkel 1984).

Once the representative vegetation is decided, its physiognomy or formation type may be used to characterize the zone. Again realistically, vegetation should be described by a system that is readily recognizable in the field by those trained in vegetation ecology. In regions where floristic and phytosociological information is available, such information should be incorporated when the vegetation measurement needs are determined. This is especially required for naming of the representative vegetation. This should include the use of such names as a *Picea engelmannii–Abies lasiocarpa* forest region or a Vaccinio-Piceetea region instead of merely a "coniferous forest region" to describe a certain zone, because a name bearing actual plant species gives us a more realistic and accurate insight as to the kinds of ecosystems that exist.

The next requirement is to correlate such zones with environmental characteristics, especially with climate. Climatic ranges of the zones need to be circumscribed from available meteorological information. This is particularly needed to detect future vegetation and climate differences from the current status. On the other hand, if we come to know the ecological distribution pattern of the global vegetation in relation to climate, we could predict, to some extent, future global vegetation change due to climatic change as attempted by Emanuel *et al.* (1985), Bolin *et al.* (1986), and Woodward and Williams (1987), as well as demonstrated in Figure 11.1.

References

Bolin, B., Döös, B.R., Jäger, J. and Warrick, R.A. (1986). *The Greenhouse Effect, Climatic Change and Ecosystems*. SCOPE 29. New York: John Wiley & Sons.

Box, E.O. (1981). *Macroclimate and Plant Forms: An Introduction to Predictive Modeling in Phytogeography*. The Hague: Junk.

Box, E.O. (1982). Life-form composition of Mediterranean terrestrial vegetation in relation to climatic factors. *Ecologia Mediterranea*, **8**, 173–81.

Braun, E.L. (1950). *Deciduous Forests of Eastern North America*. New York: Hafner Press.

Braun-Blanquet, J. (1932). Plant sociology. NY: McGraw-Hill Book Co. Inc.

Braun-Blanquet, J. (1921). Prinzipien einer Systematik der Pflanzengesellschaften auf floristischer Grundlage. *Jahrbuch der St. Gallener Naturwissenschaftlichen Gesellschaft*, **57**, 305–51.

Brockman-Jerosch, H. (1907). *Die Flora des Puschlav und inhre Pflanzengesellschaften*. Leipzig: Engelmann.

Clements, F.E. (1902). A system of nomenclature for phytogeography. *Botanisches Jahrbuch*, **31**, 1–20.

Clements, F.E. (1916). *Plant Succession*. Publication 242, Washington, D.C.: Carnegie Institute.

Clements, F.E. and Shelford, V.E. (1939). *Bio-Ecology*. New York: John Wiley & Sons.

Dansereau, P. (1957). *Biogeography*. New York: Ronald Press.

Dokuchaev, V.V. (1898). *Study of Zones in Nature*. (Republished 1948, Moscow: USSR Academy of Sciences Press.)

Ellenberg, H. ed. (1973). *Ökosystemforschung*. Berlin: Springer-Verlag.

Ellenberg, H. (1987). *Vegetation of Central Europe and the Alps*. Stuttgart: Ulmer-Verlag.

Emanuel, W.R., Shugart, H.H. and Stevenson, M.P. (1985). Climatic change and the broad-scale distribution of terrestrial ecosystem complexes. *Climatic Change*, **7**, 29–43.

Flahault, C. (1901). A project for phytogeographic nomenclature. *Bulletin of the Torrey Botanical Club*, **28**, 391–409.

Flahault, C. and Schröter, C. (1910). *Phytogeographische Nomenklatur Extrait du compterendu: 1–29*. Bruxelles: International Botanical Congress.

Griesebach, A.R.H. (1838). Über den Einfluss des Klimas auf die Begrenzung der natürlichen Floren. *Linnaea*, **12**, 159–200.

Griesebach, A.R.H. (1872). *Die Vegetation der Erde nach ihrer klimatischen Anordnung*. Leipzig: Engelmann-Verlag.

Halliday, W.E.D. (1937). *A Forest Classification for Canada*. Forest Service Bulletin 89. Ottawa: Department of Mines and Resources.

Hill, M.O. (1979). *DECORANA—a Fortran Program for Detrended Correspondence Analysis and Reciprocal Averaging*. Ithaca, New York: Cornell University.

Holdridge, L.R. (1947). Determination of world plant formations from simple climatic data. *Science*, **105**, 367–8.

Humboldt, A. von (1805). Essai sur la geographie des plantes; accompagne d'un tableau physique des regions equinoxiales. Par A. de Humboldt et A. Bonpland, redige par A. de Humboldt.

Humboldt, A. von. (1806). *Ideen zu einer Physiognomik der Gewächse*. Tübingen: Cotta.

Humboldt, A. von (1807). *Ideen zu einer Geographie der Pflanzen nebst einem Naturgemälde der Tropenländer*. Tübingen: Cotta.

Jenny, H. (1941). *Factors of Soil Formation*. New York: McGraw-Hill.

Krajina, V.J. (1960). Can we find a common platform for the different schools of forest type classification? *Silva Fennica*, **105**, 50–5.

Krajina, V.J. (1965). Biogeoclimatic zones and biogeocoenoses of British Columbia. *Ecology of Western North America*, **1**, 1–17.

Krajina, V.J. (1969). Ecology of forest trees in British Columbia. *Ecology of Western North America*, **2**, 1–146.

Küchler, A.W. (1967). *Vegetation Mapping*. New York: Ronald Press.

Kuhn, N. (1987). Schematische Darstellung der Vegetation Mitteleuropas. *Natur und Landschaft*, **62**, 484–5.

Leps, J. (1987). Vegetation dynamics in old field succession: A quantitative approach. *Vegetatio*, **72**, 95–103.

Lieth, H. (1956). Ein Beitrag zur Frage der Korrelation zwischen mittleren Klimawerten und Vegetationsformationen. *Berlin Deutsche Botanischen Gesellschaft*, **69**, 169–76.

Major, J. (1951). A functional factorial approach to plant ecology. *Ecology*, **32**, 392–412.

Meeker, D.O., Jr. and Merkel, D.L. (1984). Climax theories and a recommendation for vegetation classification—A viewpoint. *Journal of Range Management*, **37**, 427–30.

Müller-Dombois, D. and Ellenberg, H. (1974). *Aims and Methods of Vegetation Ecology*. New York: John Wiley & Sons,

Nichols, G.E. (1929). Plant associations and their classification. *Proceedings International Congress of Plant Science. Ithaca*, **1**, 629–41.

Ohba, T. (1974). Vergleichende Studien über die alpine Vegetation Japans. 1. Carici rupestris-Kobresietea bellardii. *Phytocoenologia*, **1**, 339–401.

Raunkier, C. (1934). *The Life Forms of Plants and Statistical Plant Geography*. Oxford: Clarendon.

Rübel, E. (1930). *Pflanzengesellschaften der Erde*. Bern/Berlin: Huber.

Schimper, A.F.W. and von Faber, F.C. (1935). *Pflanzengeographie auf physiologischer Grundlage*. Jena: Fischer.

Schröter, C. (1894). Notes sur quelques associations de plantes rencontrees pendant les excursions dans la Valais. *Soc. Bot. France Bull.*, **41**, 322–5.

Schröter, C. (1926). *Das Pflanzenleben der Alpen. Eine Schilderung der Hochgebirgsflora*. Zürich: Raustein.

Sukachev, V.N. (1945). Biocoenology and phytocoenology. *Siberian Branch Soviet Academy of Sciences. USSR*, **47**, 429–31.

Sukachev, V.N. (1960). The correlation between the concept "forest ecosystem" and "forest biogeocoenose" and their importance for the classification of forests. *Silva Fennica*, **105**, 94–7.

Tansley, A. (1935). The use and abuse of vegetational concepts and terms. *Ecology*, **16**, 284–307.

Tüxen, R. (1974). Synchronologie einzelner Vegetationseinheiten in Europa. In *Vegetation Dynamics*, ed. R. Knapp, p. 364. The Hague: Junk.

Tüxen, R, Miyawaki, A. and Fujiwara, K. (1972). Eine erweiterte Gliederung der Oxycocco-Sphagnetea. In *Grundfragen und Methoden in der Pflanzensoziologie*, ed. R. Tüxen, pp. 500–520. The Hague: Junk.

Walter, H. (1976). *Der ökologischen Systeme der Kontinente (Biogeosphäre)*. Stuttgart: Fischer.

Weaver, J.F. and Clements, F.E. (1929). *Plant Ecology*. New York: McGraw-Hill.

Webb, T., III. (1987). The appearance and disappearance of major vegetation assemblages: Long-term vegetational dynamics in eastern North America. *Vegetatio*, **69**, 177–87.

Whittaker, R.H. (1970). *Communities and Ecosystems*. New York: MacMillan.

Wilmanns, O. (1983). *Ökologische Pflanzensoziologie*. Heidelberg: Quelle Meyer.

Woodward, F.I. and Williams, B.G. (1987). Climate and plant distribution at global and local scales. *Vegetatio*, **69**, 189–99.

PART IV

Modeling Global Vegetation Change

12

Modeling Large-Scale Vegetation Dynamics

I. Colin Prentice, Robert A. Monserud, Thomas M. Smith and William R. Emanuel

Introduction

The foregoing discussions have focused on information and processes needed to model global vegetation change. In this chapter, we discuss the models and modeling approaches that will be useful in analyzing responses of vegetation to large-scale environmental changes. We emphasize changing climate, but the concepts and approaches are pertinent to other global changes as well. First, we summarize the natural processes that dictate vegetation dynamics under changing environmental conditions. Models that relate natural vegetation distribution to climate are then described. These can be purely correlative—based on observed relationships between vegetation and climate—or more mechanistic, based on the physiological limits of different types of plants. To simulate transient responses, we need dynamic models that mechanistically represent community processes including competition. We summarize a class of such models that are suited to small landscape or patch analysis. We conclude by outlining a scheme for applying these models in large-scale studies by a Monte Carlo method that incorporates stochastic processes within the landscape and samples spatial variability in the environment.

Natural Processes of Vegetation Change

As plants respond to environmental change, competition for light and other resources influences reproduction, growth, and mortality to different degrees that depend on species characteristics. As a result, the dynamics of species composition, biomass, leaf area, and other community variables are more complex than the collected responses of individuals that do not interact. There may be immediate effects, for example, on annual production, but these are modified later by changes in composition and community structure. This adjustment process has similarities with secondary succession after disturbances such as fire or harvest.

Natural areas of vegetation are mosaics of patches that have been disturbed at different times by fire, wind, flood, disease, or similar events (Noble and Slatyer 1978; White 1979). Large areas of vegetation thus have disturbance regimes that are part of their normal life conditions (Gigon 1983). Such systems maintain a steady state under ordinary local disturbances but are perturbed by changes in disturbance frequency or intensity (O'Neill *et al.* 1986).

Repeated natural disturbance of patches can quicken the response of vegetated areas to climatic change (Davis and Botkin 1985), and presumably harvest of wood or crops can have the same effect. Climatic change can alter disturbance regimes. For example warmer or drier conditions can increase fire frequency, causing marginal forests to be replaced by grasslands (Grimm 1984).

The availability of propagules for local recruitment depends on the abundance of taxa in the surrounding area. At this level, climatic change can also affect vegetation composition by altering regional distributions of taxa. Such changes are well documented for the past 20 000 years (Davis 1981; Huntley and Birks 1983; Webb 1986, 1987). Trees migrated with remarkable speed in response to Late Quaternary changes (Prentice *et al.* 1991), but the time needed for large-scale spread and regional population changes may be hundreds of years (Bennett 1986; Davis *et al.*1986; Davis 1987; Woods and Davis 1989; Davis 1991) and can presumably limit the rate of vegetation response to rapid, large climatic changes.

Vegetation and soil are tightly coupled systems. Soil water-holding capacity and nutrient availability in the soil-water solution influence primary production. Although some aspects of soil formation are slow enough to be considered constant through plant community changes, nutrient turnover depends on litter characteristics, so variations in species composition can affect plant growth and further alter composition (Pastor and Post 1986).

Thus, in the face of climatic change, vegetation has a certain inertia (Smith 1965; Prentice 1986a), comprising several components: (1) the time needed for individual patches of vegetation and associated labile soil components to adjust to climatic change when the available flora is held constant (Type A Response of Webb 1986); (2) time for the spatial mosaic of vegetation to adjust to climatically induced changes in frequencies of fire and other natural hazards; and (3) time for ecesis (Type B Response of Webb 1986).

Equilibrium Relationships

The natural distribution of vegetation at a broad spatial scale is in equilibrium with climate. Plant geographers can therefore use correlation methods to relate vegetation classes to climate (e.g., Holdridge 1947; Box 1981). Bartlein *et al.* (1986) and Prentice *et al.* (1991) related regional abundances of particular taxa to climatic variables by empirical response surfaces that rely on the large-scale equilibrium between vegetation and climate. Woodward (1987) and Prentice *et al.* (1992) developed more mechanistic equilibrium models for global vegetation, based on the physiological limits of different types of plants.

Sensitivity tests of these equilibrium models suggest bounds on the impacts of climatic change on vegetation. Emanuel *et al.* (1985a,b) compared world maps of Holdridge life zones, based on climatic records from 7000 meteorological stations, with similar maps with temperature increases as simulated for a doubling of atmospheric CO_2 concentration. Lashof (1987) did analogous exercises with a vegetation-climate relationship derived from Olson's world map of ecosystem complexes (Olson *et al.* 1983) and meteorological records. Although the transient responses would be complex (Shugart *et al.* 1986; Solomon 1986), such tests indicate the sensitivity of the asymptotic distributions toward which the transient changes might eventually converge.

The experiment described by Emanuel and co-workers indicates greatest changes at high latitudes where simulated temperature increases are largest and where narrow temperature intervals define Holdridge life zones. Changes are along boundaries and are more extensive than the uncertainty in determining these boundaries within the Holdridge scheme. Such speculations indicate some of the challenges agriculture and forestry may face and bound the broad vegetation changes that might feed back on the global carbon cycle, climate, and other Earth systems.

Past vegetation patterns reflected the climates of their time, so static (equilibrium) vegetation models can be used to "predict" past vegetation from simulated or reconstructed paleoclimates (COHMAP 1988; Prentice *et al.* 1991; Prentice and Solomon, 1991). This approach assumes dynamic equilibrium (Webb 1986), i.e., vegetation changes, but stays close to equilibrium with the changing climate. Dynamic equilibrium holds when the time-scale of interest is much longer than the vegetation's response time (Prentice 1986a,b). The time-scale of interest in studies of human-induced global changes, however, is a few hundred years, comparable with the time needed for replacement of forest types. Equilibrium models are not sufficient for prediction in this context because the inertia of vegetation is too great.

For example, Emanuel and co-workers show conversion of much of today's boreal forest to temperate deciduous forest, but this result does not mean that there is likely to be a rapid or straightforward transition between the two forest types. Higher summer temperatures may at first stimulate the growth of boreal conifers, but higher winter temperatures may be unfavorable for the natural regeneration of some of these taxa at their oceanic limits (Prentice, Sykes, and Cramer 1991). Beyond a certain point, increased summer temperature will reduce growth rates again; such a warm climate would probably be suitable for temperate trees, but their recruitment would take time, so production could fall before rising again toward the higher level characteristic of temperate forests.

Community Models

Vegetation models represent transient responses from different viewpoints. Forest stand growth and succession models (Botkin *et al.* 1972; Ek and Monserud 1981; Shugart 1984; Prentice and Leemans 1990) simulate compositional and structural changes, based on species attributes and tree interactions. These models borrow many ideas from forestry yield models (Munro 1974). Nonwoody vegetation dynamics can be treated within similar modeling frameworks (Prentice *et al.* 1987; Coffin and Lauenroth 1989). Although these models represent vegetation dynamics on small patches, they simulate community phenomena with considerable generality. Environmental influences are described in terms of scalars that can be evaluated along gradients or described statistically within larger landscape units.

Forestry Yield Models

Foresters have used yield tables for nearly 200 years (Assmann 1970). These empirical models predict stem volume from stand and site attributes—usually stand basal area and height at a reference age (site index), which is a surrogate for soil moisture and nutrient status. Many European yield tables were constructed after monitoring actual stand development for one or more full rotations. North American tables are based on temporary rather than permanent plots and are thus less reliable (Spurr 1952; Monserud 1984).

Emphasis on mechanisms increased after Newnham's (1964) seminal work. Yield models now treat the growth of and interactions between individual trees or groups of similar-sized trees. The growth equations in these models are based on the following observations (Ek and Monserud 1981):

1. Tree dimensions change through time according to sigmoid growth patterns. Change in any one dimension (e.g., height) can be modeled as a sigmoid function (e.g., Richards 1959).
2. Other dimensions (e.g., stem diameter or crown width) are allometrically related to height.
3. Changes in stand density reflect thinning because of competition.
4. The heights of dominant trees of a given age and species are fairly insensitive to changes in stand density and are more closely related to potential site productivity than to other indices (Spurr and Barnes 1980).

Foresters use dominant trees as phytometers of soil moisture and nutrient status because their height is so indicative of site potential. Considerable effort has gone into methods to estimate site index directly from physical and chemical soil characteristics, but the predictive power of soil-site models has seldom met the needs of forest managers.

Thinning is sometimes modeled by the −3/2 power rule (Weller 1987a) that relates biomass increase to density decline in pure, even-aged stands. But this rule depends on an often-violated assumption of tree isometry (Weller 1987b) and cannot be applied to stands of mixed species and age. In more general models, competition occurs at the level of individual trees—the most detailed are spatially explicit and include functions that reduce the growth of each tree according to the size and proximity of competitors.

Climate has been ignored as a determinant of forest growth. As a result, most forest yield models cannot be used to analyze the consequences of climatic change. The models rely on a strong empirical base that gives accurate results under normal environmental conditions, but their performance is degraded if environmental changes occur.

Forest Succession Models

While borrowing many ideas from yield models, forest succession models, proposed by Botkin *et al.* (1972) and extended by numerous implementations and applications (Shugart 1984), pursue a different theme. Usually the same growth function is used for all trees. The growth equation is not calibrated to observations as in yield models; rather, most species-specific parameters are derived from silvical observations—for example, on maximum sizes, ages, or growth rates. The models do use empirical size relationships similar to those in yield models (Dale *et al.* 1985). Forest stand models sacrifice the precision supported by calibration in yield models for the generality obtained by using a uniform growth description and parameters that can be estimated from silvical data for a large number of species.

Unlike the yield models, forest succession models treat a plot of explicit size constrained by the requirement that resource availability be independent of horizontal position within the plot (Shugart and West 1979). With the appropriate choice of plot size, these models can produce realistic simulations of gap-phase dynamics, the species replacement sequences initiated by the death of a large tree (Watt 1947; White 1979; Shugart 1984; Smith and Urban 1988). For example, Shugart and West (1977) simulated gap-phase dynamics in an eastern Tennessee temperate forest by using an 833-m^2 (1/12 ha) plot. In boreal forests, where tree crowns are narrow, the gap created by the death of a single tree has less effect so the forest, unless disturbed, tends to a simple climax from which light-demanding trees are excluded. Leemans and Prentice (1987) found that a forest succession model with a similar plot size (1000 m^2) could also simulate this different type of dynamics. The use of plot sizes on the order of 1000 m^2 is consistent with the observation that gaps smaller than this have too low a light intensity to support regeneration of light-demanding trees (Whitmore 1982; Prentice and Leemans 1990).

In forest succession models, leaf area is conventionally directly related to stem diameter and considered to be concentrated at the top of the stem, so the total leaf area of each tree shades all trees below. Some more recent models determine leaf area from sapwood area, allowing for sapwood turnover, and vertically distribute the leaf area of each tree (Leemans and Prentice 1987; Prentice and Leemans 1990). A vertical light profile is computed from the Lambert-Beer law by assuming that shading is horizontally homogeneous.

Trees respond to the light profile according to their species-specific shade tolerance. Usually below-ground competition is modeled only in terms of crowding, but some models now include an explicit nitrogen cycle by treating litter production, decomposition, immobilization, mineralization, and growth response (Lindgren and Axelsson 1980; Pastor and Post 1986).

Random numbers of new trees are recruited annually as saplings for each species whose regeneration requirements—including light at the forest floor, mineral soil exposure, and leaf-litter depth—are met by environmental and vegetation status. The persistence of species is assured by an assumed constant propagule bath (Whittaker and Levin 1977).

The description of mortality is similar to that used in yield models (Monserud 1976). The probability that a tree dies is much higher if it is suppressed than if it is growing well, considering the optimum for trees of its species. Diameter increment is used as an index of vitality in most models, but growth efficiency in terms of stem volume increment per unit leaf area (Waring 1983; Waring and Schlesinger 1985) can be used instead. With mortality simulated as a function of individual growth, realistic thinning lines emerge because of asymmetric competition after canopy closure (Prentice and Leemans 1990).

Disturbances such as fire and wind also affect mortality and are modeled as additional stochastic processes (Mielke *et al.* 1978; Doyle 1981; Shugart and Noble 1981; Prentice, Sykes, and Cramer 1991; in press). Mortality can be immediate; trees are removed from the model stand in the year disturbance occurs. In other cases, the probability of death may be increased according to species tolerance. Disturbance can also affect growth without directly altering the probability of death.

Environmental effects enter through factors that multiply growth increment to reflect deviations from optimal conditions. Climatic conditions considered include summer warmth, winter cold, and drought (Solomon 1986). Summer warmth has usually been indexed by the annual sum of temperatures exceeding a threshold value—for example, 5°C. Maximum growth has often been assumed to occur halfway between each species' upper and lower thermal limits estimated from distribution maps. Some more recent models treat the thermal response more mechanistically in terms of separate effects on annual net assimilation, sapwood respiration, and temperature requirements for completion of the life cycle (Prentice, Sykes, and Cramer, 1991; in press). Winter cold is usually indexed by the mean temperature of the coldest month; growth is suppressed if it is too cold. Drought is described by an index such as the proportion of the growing season with subcritical soil moisture. Growth is reduced with increasing drought, reaching zero if drought index is less than or equal to the arid range limit for each species. Other environmental factors can be considered as well. Bonan's (1988) boreal forest model considers the response of each species to permafrost; Pastor and Post (1986) constrain growth by nitrogen availability.

Growth factors depend on indices (e.g., drought index) that in turn are functions of environmental variables such as temperature or soil moisture. Some of these, such as temperature, are specified directly or treated as random variables with specified distributions. Others, such as soil moisture, can be functions of additional variables, such as the status of model vegetation. These relationships are described by environmental submodels whose state variables may depend on vegetation variables. For example, Pastor and Post (1986) and Cramer and Prentice (1988) determine soil moisture from specified rainfall and temperature by considering soil water holding capacity and actual evapotranspiration. Pastor and Post (1986) incorporate a nitrogen-cycling submodel—nitrogen turnover depends on species-specific litter characteristics. Disturbance probabilities and intensities can be made joint functions of environmental variables and vegetation state variables, for example, to model the dependence of fire hazard on weather conditions and fuel load.

Solomon (1986) showed that a forest succession model treating 72 species can generate qualitatively correct forest types across eastern North America. Bonan (1989) was able to simulate successional patterns of four boreal forest sites in central Alaska. Tropical-forest dynamics have also been successfully simulated (Shugart 1984; De Sanker, Prentice, and Solomon, in press). Such results suggest that a general forest-succession-model framework can be applied anywhere simply by specifying values of basic environmental variables (temperature, rainfall, light intensity, potential evapotranspiration, and soil characteristics) and the appropriate species response parameters. The same environmental variables are then modified to analyze the responses of the forest to environmental change.

Solomon (1986) illustrated responses to climatic changes derived from general-circulation-model solutions with increased atmospheric CO_2 concentration. He found that changes in forest composition resulted in substantial changes in biomass that could be positive or negative, depending on location. This experiment considered only the effects of increased CO_2 through climatic change. Using reasonable assumptions about the direct effects of CO_2 increase on tree growth, Solomon and West (1986) showed that climatic effects are likely to predominate. Prentice, Sykes, and Cramer (in press) reached a similar conclusion. Pastor and Post (1988) report model experiments similar to Solomon's but with explicit treatment of nitrogen cycling. They find that effects of climatic change through nitrogen dynamics (e.g., effects on decomposition rate) can be as important as direct effects on plant growth.

These results point to what can be accomplished by using stand simulation models to analyze the consequences of environmental change. By developing solutions that sample landscape units or environmental gradients, we can use a collection of these models to analyze the impacts of environmental change on the natural dynamics of world forests. Human interventions of various kinds can be introduced explicitly, as are natural disturbances. Other vegetation types can be analyzed when nonwoody plants are incorporated in these or similar models.

The data requirements for estimating species parameters are not demanding compared to other modeling approaches; however, even the basic silvical data are not available for some regions, particularly tropical moist forests of South America and Africa. Huston and Smith (1987) demonstrate that reasonable growth and succession dynamics can be simulated by a stand model that treats functional plant types rather than species. Suitable functional types and their parameter values can be derived at least partly from theoretical considerations (Smith and Huston 1989). Where necessary, vegetation can be modeled at this prototypic level until data on more species are available. As theory develops, functional plant types may prove to be more practical than species for modeling purposes.

Concise Community Models

Forest models based on the Botkin formulation are uniquely successful in describing vegetation dynamics on time-scales ranging from decades to centuries. But a model that represents population changes for each species or even plant types, as opposed to treating individuals, should be satisfactory for continental- to global-scale applications—provided it captures the essentials of the processes represented in succession models. The computational requirements of population-based models can be orders of magnitude less than those of models that track individuals.

One possibility is to describe the state of the plant community on a plot by the number and/or foliage amount of individuals of each species or plant type in each of a set of vertically stacked layers of arbitrary width (Prentice *et al.* 1989; Fulton, 1991; and this volume). Again the plot is of definite size, sufficiently small that resource availability can be assumed horizontally homogeneous. Advancement through height layers can be treated as a stochastic process, with the probability of transfer to higher layers substituting for the deterministic height increments of individual plants calculated in conventional models. Size-structured population models of this type have considerable promise for large-scale modeling (see also Moore ande Noble, 1990). They also lend themselves more easily to the inclusion of nonwoody plants than do individual-based models.

Large-scale Analysis

Continental-scale simulations of vegetation dynamics can be generated by deriving very large sets of model solutions, each applying to a small plot. The region of interest is subdivided. A uniform grid in spherical coordinates is convenient, but arbitrary polygons can be used (see Skole *et al.*, this volume). Resolutions of about 0.5° are reasonable (Olson *et al.* 1983; Emanuel *et al.* 1985a)—such cells are approximately 56 km on a side at the equator. A large set of model solutions is generated for each of these landscape units, with appropriately distributed random environmental variables (Prentice *et al.* 1989). Disturbance frequency and intensity are also specified as parameters of stochastic processes. These parameters may depend on environmental characteristics such as temperature or soil moisture as well as on the status of the vegetation (Kercher and Axelrod 1984).

For landscape units small enough to support reasonable resolution, the detailed distributions of basic environmental variables, such as temperature, rainfall, soil type, and soil texture, cannot be derived from currently available observations. Instead, statistical distributions of these variables must be estimated indirectly from morphologic features, of which topography is most important; here it becomes possible to exploit the existence of very detailed topographic databases. There are models for calculating daily radiation receipts at a specified location, elevation, slope, aspect, and orbital configuration (e.g., Swift 1976; Bonan 1988; Kutzbach and Gallimore 1988). The principles are understood for other relations—for example, temperature change as a function of elevation and soil depth as a function of slope, which could also be quantified for incorporation in global vegetation models. Global relationships are not required; the aim is simply to derive the variation and covariation of environmental variables within a landscape unit from data on topography and other morphological characteristics.

Useful simulations of vegetation responses can be derived without considering spatial linkages between plots or landscape units. At least three types of connection, however, are clearly important: hydrologic transport, the dependence of disturbance frequencies on events in adjacent cells (as in the spread of fire), and the dependence of seed propagation on the abundances of species in an area and in adjacent areas (Rudis and Ek 1981). The latter two connections can probably be expressed adequately without adding complex submodels; however, the large-scale hydrologic modeling needed to include the first is only now beginning to be addressed (Eagleson 1986).

The computer implementation of this Monte Carlo scheme for analyzing large-scale vegetation responses requires geographic data management and analysis capabilities now being demanded in virtually all fields of environmental science. But some unique features are needed:

1. Values of environmental variables, such as temperature and precipitation, must be estimated for each land unit by interpolating observations

that are usually of coarser spatial resolution. In continental-scale applications, interpolation on a sphere is required, and smoothing may be necessary (Renka 1982).

2. Statistical analysis of spatial data on arbitrary landscape units of different resolutions is required both for the organization of Monte Carlo solutions by assigning joint distributions to environmental variables and to summarize and interpret model solutions.
3. Collections of Monte Carlo solutions must be analyzed to find geographic patterns in variables such as biomass and to map features that reveal the distributions of major vegetation types.

Discussion and Conclusions

Those wishing to model interactions between climate and ecosystems must reconcile the different spatial and temporal scales of atmospheric and ecological processes. The design of atmospheric general circulation models reflects the fast horizontal transport and short memory of the atmosphere—integrations are performed at intervals of less than an hour and at grid points 200 km or more apart. Vegetation models are solved at yearly intervals but on plots about 30 m across. These differences reflect the inertia of vegetation and the fine spatial scales at which vegetation processes act compared to those determining weather (Turner and Walker 1990).

There are land-surface models that allow a prescribed vegetation, in terms of height, structure, phenology, rooting depth, etc., to interact dynamically with the simulated atmosphere (Dickinson 1984; Sellers *et al.* 1986). Physical vegetation characteristics such as albedo, leaf-area index, and stomatal aperture vary diurnally and seasonally in response to atmospheric variables and in turn affect exchanges of energy, water, and momentum. The task of a global vegetation model is to simulate the slower processes by which primary vegetation characteristics—taken as constant in climate models—are transformed through time by changes in climate, including the processes of lateral spread.

We have outlined the components of a scheme for simulating global transient responses of natural vegetation to climatic change. The approach is based on a "core" simulation procedure that in our experience accurately represents the processes responsible for vegetation responses to environmental change. Substantial data are required to estimate model parameters for world-scale studies, but these needs are realistic given the importance of the task. The core model must be generalized and extended to include nonwoody plant types. To model, for example, the transitions between grasslands and forests, it must include a more explicit representation of vegetation hydrology. Most of the components needed for a generalized vegetation dynamic model exist in one or another ecosystem model; the components need to be assembled.

We are focusing on the consequences of climatic change caused by increases in atmospheric concentrations of greenhouse gases, but there are direct effects of atmospheric CO_2 concentration on photosynthesis that may complicate vegetation responses. The equations that describe plant growth need to allow coupling to process descriptions that explicitly treat CO_2 and light in photosynthesis. This is a difficult task because of the large differences in scale that must be accommodated to join our understanding of cell- and leaf-level processes with whole-plant growth (King *et al.* 1989).

As a practical matter, organization according to functional types (see Smith *et al.*, this volume) is important because the number of major plant species is overwhelming, and in many regions, data on species growth characteristics are not available now and likely will not be available in the immediate future. But there are also important conceptual benefits to the functional types approach. Large-scale analyses can clarify the role of different functional types in determining vegetation responses and the potential of various compositions of functional types under alternative climatic conditions. This kind of understanding is probably more important as a basis for planning for climatic change than are attempts to forecast changes in species composition that in actuality will be strongly influenced by human intervention.

Acknowledgments

The authors thank Dr. Herman H. Shugart, Jr., University of Virginia, Dr. Wilfred M. Post, Oak Ridge National Laboratory, and Dr. Michael Huston, Oak Ridge National Laboratory, for reviewing drafts of this chapter. W.R. Emanuel was supported by the National Science Foundation's Ecosystem Studies Program under Interagency Agreement BSR–8417923 with the U.S. Department of Energy through Contract No. DE-AC05–84OR21400 with Martin Marietta Energy Systems, Inc.

Research partially sponsored by the U.S. National Science Foundation's Ecosystem Studies Program under Interagency Agreement BSR–8417923 with the U.S. Department of Energy under Contract No. DE-AC05–84OR21400 with Martin Marietta Energy Systems, Inc., and by Swedish National Research Council grants to the project titled "Simulation modelling of natural forest dynamics."

References

Assmann, E. (1970). *The Principles of Forest Yield Study*. New York: Pergamon Press.

Bartlein, P.J., Prentice, I.C. and Webb, T. III. (1986). Climatic response surfaces from pollen data for some eastern North American taxa. *Journal of Biogeography*, **13**, 35–57.

Bennett, K.D. (1986). The rate of spread and population increase of forest trees during the postglacial. *Philosophical Transactions of the Royal Society of London*, **B314**, 523–31.

Bonan, G.B. (1989). Environmental factors and ecological processes controlling vegetation patterns in boreal forests. *Landscape Ecology,* **3,** 111–130.

Botkin, D.B., Janak, J.F. and Wallis, J.R. (1972). Some ecological consequences of a computer model of forest growth. *Journal of Ecology*, **60**, 849–73.

Box, E.O. (1981). *Macroclimate and Plant Form*. The Hague: Junk Publishers.

Coffin, D.P. and Lauenroth, W.K. (1989). A gap dynamics simulation model of succession in a semi-arid grassland. *Ecological Modelling*, **49**, 229–36.

COHMAP Project Members. (1988). Climatic changes of the last 18,000 years: Observations and model simulations. *Science*, **241**, 1043–52.

Cramer, W. and Prentice, I.C. (1988). Simulation of regional soil moisture deficits on a European scale. *Norsk Geografisk Tidsskrift*, **42**, 149–51.

Dale, V.H., Doyle, T.W. and Shugart, H.H. (1985). A comparison of tree growth models. *Ecological Modelling*, **29**, 145–69.

Davis, M.B. (1981). Quaternary history and the stability of forest communities. In *Forest Succession: Concepts and Application*, ed. D.C. West, H.H. Shugart and D.B. Botkin, pp. 132–53. New York: Springer-Verlag.

Davis, M.B. (1987). Invasion of forest communities during the Holocene: Beech and hemlock in the Great Lakes region. In *Colonization, Succession, and Stability*, ed. A.J. Gray, M.J. Crawley and P.J. Edwards, pp. 373–94. Oxford: Blackwell.

Davis, M.B. (1991). Research questions posed by the paleological record of global change. In: Bradley, R.S. (ed.) *Global Changes of the Past,* pp. 385–395. Boulder, Colorado: UCAR/OIES.

Davis, M.B. and Botkin, D.B. (1985). Sensitivity of cool-temperate forests and their fossil pollen record to rapid temperature change. *Quaternary Research*, **23**, 327–40.

Davis, M.B., Woods, K.D., Webb, S.L. and Futyma, R.B. (1986). Dispersal versus climate: Expansion of *Fagus* and *Tsuga* into the upper Great Lakes region. *Vegetatio*, **67**, 93–103.

Desanker, P.V., I.C. Prentice and A.M. Solomon. (1992). MIOMBO—A vegetation dynamics model for MIOMBO woodlands of the Zambezian region of Africa. Forest Ecology and Management (in press).

Dickinson, R.E. (1984). Modeling evapotranspiration for three-dimensional global climate models. In *Climate Processes and Climate Sensitivity*, ed. J.E. Hansen and T. Takahashi, pp. 58–72. Geophysical Monograph 29. Washington, D.C.: American Geophysical Union.

Doyle, T.W. (1981). The role of disturbance in the gap dynamics of a montane rain forest: An application of a tropical forest succession model. In *Forest Succession: Concepts and Application*, ed. D.C. West, H.H. Shugart and D.B. Botkin, pp. 56–73. New York: Springer-Verlag.

Eagleson, P.S. (1986). The emergence of global-scale hydrology. *Water Resources Research*, **22**, 6S–14S.

Ek, A.R. and Monserud, R.A. (1981). Methodology for modeling forest stand dynamics. In *Dynamic Properties of Forest Ecosystems*, ed. D.E. Reichle, pp. 177–84. Cambridge: Cambridge University Press.

Emanuel, W.R., Shugart, H.H. and Stevenson, M.P. (1985a). Climatic change and the broad-scale distribution of terrestrial ecosystem complexes. *Climatic Change*, **7**, 29–43.

Emanuel, W.R., Shugart, H.H. and Stevenson, M.P. (1985b). Response to comment: Climatic change and the broad-scale distribution of terrestrial ecosystem complexes. *Climatic Change*, **7**, 457–60.

Fulton, M. (1991). A computationally efficient forest succession model: Design and initial tests. *Forest Ecology and Management*, 42, 23–34.

Gigon, A. (1983). Typology and principles of ecological stability and instability. *Mountain Research and Development*, **3**, 95–102.

Grimm, E.C. (1984). Fire and other factors controlling the Big Woods vegetation of Minnesota in the mid-nineteenth century. *Ecological Monographs*, **54**, 291–311.

Holdridge, L.R. (1947). Determination of world plant formations from simple climatic data. *Science*, **105**, 367–8.

Huntley, B. and Birks, H.J.B. (1983). *An Atlas of Past and Present Pollen Maps for Europe: 0–13,000 Years Ago*. Cambridge: Cambridge University Press.

Huston, M. and Smith, T. (1987). Plant succession: Life history and competition. *American Naturalist*, **130**, 168–98.

Kercher, J.R. and Axelrod, M.C. (1984). A process model of fires, ecology, and succession in a mixed-conifer forest. *Ecology*, **65**, 1725–42.

King, A.W., Emanuel, W.R. and O'Neill, R.V. (1990). Linking mechanistic models of tree physiology with models of forest dynamics: Problems of temporal scale. pp. 241–248 In *Process Modeling of Forest Growth Responses to Environmental Stress*. ed. R.K. Dixon, R.S. Meldahl, G.S. Ruark and W.G. Warren. Auburn, Alabama: Timber Press, Auburn University.

Kutzbach, J.E. and Gallimore, R.G. (1988). Sensitivity of a coupled atmosphere/mixed layer ocean model to changes in orbital forcing at 9000 years B.P. *Journal of Geophysical Research*, **93**, 803–21.

Lashof, D.A. (1987). *The Role of the Biosphere in the Global Carbon Cycle: Evaluation Through Biospheric Modeling and Atmospheric Measurement*. PhD Dissertation, University of California, Berkeley.

Leemans, R. and Prentice, I.C. (1987). Description and simulation of tree-layer composition and size distributions in a primeval *Picea-Pinus* forest. *Vegetatio*, **69**, 147–56.

Lindgren, A. and Axelsson, B. (1980). *STAND—A simulation model of the long-term development of a pine stand*. Project Technical Report 28. Stockholm: Swedish Coniferous Forest Project.

Mielke, D.L., Shugart, H.H. and West, D.C. (1978). *A Stand Model for Upland Forests of Southern Arkansas*. ORNL/TM–6225. Oak Ridge, Tennessee: Oak Ridge National Laboratory.

Monserud, R.A. (1976). Simulation of forest tree mortality. *Forest Science*, **22**, 438–44.

Monserud, R.A. (1984). Problems with site index: An opinioned review. In *In Forest Land Classification: Experiences, Problems, Perspectives*, ed. J. Bockheim, pp. 167–80. Symposium Proceedings. Madison, Wisconsin: University of Wisconsin Cooperative Extension.

Moore, A.D. and Noble, I.R. (1990) An individualistic model of vegetation stand dynamics. *Journal of Environmental Mangement* **31**:61–81.

Munro, D.D. (1974). Forest growth models. A prognosis. In *Growth Models for Tree and Stand Simulation*, ed. J. Fries, pp. 7–21. Research Notes 30. Stockholm: Department of Forest Yield Research, Royal College of Forestry.

Newnham, R.M. (1964). *The Development of a Stand Model for Douglas-Fir*. PhD Thesis, University of British Columbia, Vancouver.

Noble, I.R. and Slatyer, R.O. (1978). The effect of disturbance on plant succession. *Proceedings of the Ecological Society of Australia*, **10**, 135–45.

Olson, J.S., Watts, J.A. and Allison, L.J. (1983). *Carbon in Live Vegetation of Major World Ecosystems*. ORNL–5862. Oak Ridge, Tennessee: Oak Ridge National Laboratory.

O'Neill, R.V., DeAngelis, D.L., Waide, J.B. and Allen, T.F.H. (1986). *A Hierarchical Concept of Ecosystems*. Princeton, New Jersey: Princeton University Press.

Pastor, J. and Post, W.M. (1986). Influence of climate, soil moisture, and succession on forest carbon and nitrogen cycles. *Biogeochemistry*, **2**, 3–27.

Pastor, J. and Post, W.M. (1988). Response of northern forests to CO_2-induced climate change. *Nature*, **344**, 55–8.

Prentice, I.C. (1986a). Vegetation response to past climatic variation. *Vegetatio*, **67**, 131–41.

Prentice, I.C. (1986b). Some concepts and objectives of forest dynamics research. In *Forest Dynamics Research in Western and Central Europe*, ed. J. Fanta, pp. 32–41. Wageningen, The Netherlands: PUDOC.

Prentice, I.C., Bartlein, P.J. and Webb, T. III. (1991). Vegetation changes in eastern North America since the last glacial maximum: A response to continuous climate forcing. *Ecology*, **72**:2038–2056.

Prentice, I.C., Cramer, W., Harrison, S.P., Leemans, R., Monserud, R.A. and Solomon, A. M. (1992). A global biome model based on plant physiology and dominance, soil properties, and climate. *Journal of Biogeography,* 19:117–134.

Prentice, I.C. and Leemans, R. (1990). Pattern and process and the dynamics of forest structure: a simulation approach. *Journal of Ecology*, **78**, 340–55.

Prentice, I.C. and Solomon, A.M. (1991) Vegetation models and global change. In: Bradley, R.S. (ed.). *Global Changes of the Past*. pp. 365–383. Boulder, Colorado: UCAR/OIES.

Prentice, I.C., Sykes, M.T., and Cramer, W. (1991). The possible dynamic responses of northern forests to greenhouse warming. *Global Ecology and Biogeography Letters* **1,** 129–135.

Prentice, I.C., Sykes, M.T. and Cramer, W. (1992). A simulation model for the transient effects of climate change on forest landscapes. *Ecolgical Modelling*, (in press).

Prentice, I.C., van Tongeren, O. and de Smidt, J.T. (1987). Simulation of heathland vegetation dynamics. *Journal of Ecology*, **75**, 203–19.

Prentice, I.C., Webb, R.S., Ter-Mikhaelian, M.T., Solomon, A.M., Smith, T.M., Pitovranov, S.E., Nikolov, N.T., Minin, A.A., Leemans, R., Lavorel, S., Korzukhin, M.D., Hrabovsky, J.P., Helmisaari, H.O., Harrison, S.P., Emanuel, W.R. and Bonan, G.B. (1989). *Developing a Global Vegetation Dynamics Model: Results of a Summer Workshop*. IIASA Research Reports 89(7). International Institute for Applied Systems Analysis, Laxenburg, Austria.

Renka, R. (1982). *Interpolation of Data on the Surface of a Sphere*. ORNL/CSD–108. Oak Ridge, Tennessee: Oak Ridge National Laboratory.

Richards, F.J. (1959). A flexible growth function for empirical use. *Journal of Experimental Botany*, **10**, 290–300.

Rudis, V.A. and Ek, A.R. (1981). Optimization of forest spatial patterns: Methodology for analysis of landscape pattern. In *Forest Island Dynamics in Man-Dominated Landscapes*, ed. R.L. Burgess and D.M. Sharpe, pp. 241–56. New York: Springer-Verlag.

Sellers, P.J., Mintz, Y., Sud, Y.C. and Dalcher, A. (1986). A simple biosphere model (SiB) for use within general circulation models. *Journal of the Atmospheric Sciences*, **43**, 505–31.

Shugart, H.H. (1984). *A Theory of Forest Dynamics*. New York: Springer-Verlag.

Shugart, H.H., Antonovsky, M.Ya., Jarvis, P.G. and Sandford, A.P. (1986). CO_2 climatic change and forest ecosystems. In *The Greenhouse Effect, Climatic Change, and Ecosystems*, ed. B. Bolin, B.R. Döös, J. Jäger and T.A. Warrick, pp. 475–521. SCOPE 21. Chichester, U.K.: John Wiley.

Shugart, H.H. and Noble, I.R. (1981). A computer model of succession and fire response of the high-altitude *Eucalyptus* forest of the Brindabella Range, Australian Capital Territory. *Australian Journal of Ecology*, **6**, 149–64.

Shugart, H.H. and West, D.C. (1977). Development of an Appalachian deciduous forest succession model and its application to assessment of the impact of the chestnut blight. *Journal of Environmental Management*, **5**, 161–79.

Shugart, H.H. and West, D.C. (1979). Size and pattern of simulated forest stands. *Forest Science*, **25**, 120–2.

Smith, A.G. (1965). Problems of inertia and threshold related to post-glacial habitat changes. *Proceedings of the Royal Society*, **B161**, 331–42.

Smith, T.M. and Huston, M. (1989). A theory of the spatial and temporal dynamics of plant communities. *Vegetatio*, **83**, 49–69.

Smith, T.M. and Urban, D.L. (1988). Scale and resolution of forest structural pattern. *Vegetatio*, **74**, 143–50.

Solomon, A.M. (1986). Transient response of forests to CO_2-induced climate change: Simulation modeling experiments in eastern North America. *Oecologia*, **68**, 567–79.

Solomon, A.M. and West, D.C. (1986). Atmospheric carbon dioxide change: Agent of future forest growth or decline? In *Effects of Changes in Stratospheric Ozone and Global Climate*, ed. J. Titus, pp. 23–38. Volume 3. Washington, D.C.: U.S. Environmental Protection Agency.

Spurr, S.H. (1952). *Forest Inventory*. New York: Ronald Press.

Spurr, S.H. and Barnes, B.V. (1980). *Forest Ecology*. 3rd edition. New York: John Wiley.

Swift, L.W., Jr. (1976). Algorithm for solar radiation on mountain slopes. *Water Resources Research*, **12**, 108–12.

Turner, S.J. and Walker, B.H. ed. (1990). *The Land-Atmosphere Interface*. Global Change Report 10. International Geosphere-Biosphere Program, Stockholm.

Waring, R.H. (1983). Estimating forest growth and efficiency in relation to canopy leaf area. *Advances in Ecological Research*, **13**, 327–54.

Waring, R.H. and Schlesinger, W.H. (1985). *Forest Ecosystems: Concepts and Management*. New York: Academic Press.

Watt, A.S. (1947). Pattern and process in the plant community. *Journal of Ecology*, **35**, 1–22.

Webb, T., III. (1986). Is vegetation in equilibrium with climate? How to interpret late-Quaternary pollen data. *Vegetatio*, **67**, 75–92.

Webb, T., III. (1987). The appearance and disappearance of major vegetation assemblages: Long-term vegetational dynamics in eastern North America. *Vegetatio*, **69**, 177–87.

Weller, D.E. (1987a). A reevaluation of the −3/2 power rule of plant self-thinning. *Ecological Monographs*, **57**, 23–43.

Weller, D.E. (1987b). Self-thinning exponent correlated with allometric measures of plant geometry. *Ecology*, **68**, 813–21.

White, P.S. (1979). Pattern, process, and natural disturbance in vegetation. *Botanical Review*, **45**, 229–99.

Whitmore, T.C. (1982). On pattern and process in forests. In *The Plant Community as a Working Mechanism*, ed. E.I. Newman, pp. 45–60. Oxford, U.K.: Blackwell.

Whittaker, R.H. and Levin, S.A. (1977). The role of mosaic phenomena in natural communities. *Theoretical Population Biology*, **12**, 117–39.

Woods, K.D. and Davis, M.B. (1989). Paleoecology of range limits: Beech in the upper peninsula of Michigan. *Ecology*, **70**, 681–96.

Woodward, F.I. (1987). *Climate and Plant Distribution*. Cambridge: Cambridge University Press.

13

Rapid Simulations of Vegetation Stand Dynamics with Mixed Life-Forms

Mark R. Fulton

Introduction

The goal formulated by the global vegetation modeling project at The International Institute for Applied Systems Analysis, IIASA, (Prentice *et al.* 1989) was to develop a mechanistic vegetation model suitable for exploring processes of global vegetation change in a changing environment over time-scales of decades to centuries, taking into account the processes of plant growth, regeneration, mortality, disturbance regimes, and dispersal that mediate the response to environmental change. Gap models, a class of vegetation dynamics simulation models, have desirable features for use in such a project. They operate at the relevant time-scale and have proven useful in clarifying processes of forest succession and response to disturbance (Shugart 1984) and climatic change (Solomon 1986) on local and regional scales in a variety of environments. So far, however, gap models have mainly been applied to forests and have not included the mixture of life-forms characteristic of most plant communities. (However, see Burton and Urban 1989; and Coffin and Lauenroth 1990.) Another limitation of gap models for global simulation purposes is that they appear to include an excessive amount of detail; for example, vegetation patches are described in terms of explicit characteristics of all individual plants rather than in terms of canopy and population structure, which are the effective variables governing competition (Fulton 1991).

A desirable tool for global vegetation modeling would be a stand simulator that (1) is based on gap model principles (Shugart 1984) of local competition for light and nutrients; (2) realistically incorporates growth-forms smaller than trees; and (3) has better computational efficiency than conventional gap models. Prentice et al. (1989) concluded that a patch model based on interacting size-structured plant populations, along the lines of Fulton's (1991) efficient forest dynamics model, would be a promising solution. This paper summarizes the development of a stand simulator meeting these requirements, including conceptual issues considered during model development, a description of the model developed (hereafter called GVM: Generic Vegetation Model), and examples of simulation exercises carried out with the model.

Conceptual Issues

Patch Models

In patch or gap models (Shugart 1984; Prentice 1986), the interaction of individual trees is modeled on a small (0.01–0.1 ha) patch of ground representing the approximate range of influence of a tree. The trees modify their environment by shading and use of water and nutrients, and the environment in turn modifies the establishment, growth, and mortality of trees growing within the patch. This has proven to be a robust concept for the explanation of forest dynamics over time-scales from decades to centuries. Because of the wide applicability of models of this type and because of the match between the useful time-scales of the model and the time-scales of interest in the prediction of vegetation change in response to human-induced climate change, extension of this concept toward a more general-purpose model of vegetation dynamics has been a goal in global vegetation model development (Prentice et al. 1989). Two (sometimes conflicting) criteria have guided this extension: increased computational efficiency and applicability to growth-forms other than trees.

Layer-Based Models

Gap models (Shugart 1984) include each tree as a unique individual in the state description of the patch. In the patch models UVM (Universal Vegetation Model; Prentice *et al.* 1989) and FLAM (Forest LAyer Model; Fulton 1991), the state description is simplified by taking advantage of the relation between competitive asymmetry and height. These models simulate the interaction of height-class–structured populations of each species on a patch. This state description can incorporate mixed growth-forms more easily than an individual-based model because of the impracticality of including descriptions of each individual in, for example, a field of grass. Individuals of a given species within a height class (layer) are assumed to experience the same light environment and the same probability of mortality. Individuals are promoted from one layer to another by a stochastic function of a calculated average height growth. (See below.) UVM keeps track of the number of individuals of each species in each height class; stem volume/individual and leaf area/individual are taken as constants for each height class and species. In FLAM the leaf area and stem volume can vary independently of the number of individuals in a layer. FLAM has been shown to effectively imitate the dynamics of an individual-based gap model with a considerable savings in computation time (Fulton 1991).

The UVM had stability problems, with stems getting "trapped" within layers or being promoted too rapidly between layers (I.C. Prentice, pers. comm.). The formulation used in FLAM avoids these problems, in part because trees grow (and therefore increase their leaf area) continuously within a height class. M. Korzukhin (pers. comm.) observed that this problem and its resolution are common to many systems in which a continuous process is approximated by a discrete process. The stability problems of the UVM may be solvable by an intermediate level of detail in the formulation of growth: a formulation that has leaf area as a fixed function of diameter but allows stem growth within a height class.

As a modeled tree grows, certain functional relationships between height, diameter, and leaf area are implicitly or explicitly applied to control the relative growth of these variables. A layer model speeds up evaluation of these relationships by a discrete size approximation. For each layer and species (or functional type), a "representative tree" is defined which has height halfway between the top and the bottom of the layer. This representative tree, which has fixed size, also has fixed relationships between height, diameter, and (in some models) leaf area. These relationships are calculated at program initialization and are applied to all trees of that species within the layer. Experimentation showed that this procedure can accommodate relationships of almost arbitrary complexity. For the purposes of the discussion to follow, a representative tree should be distinguished from an average tree, which is a tree of mean stem volume and leaf area for a given species and layer at any given time step in a model run.

Growth and Promotion of Trees Between Height Classes

In the UVM and FLAM the function for "promotion" from one layer to the next assumes a rectangular distribution of heights within the layer to calculate promotion probabilities. Promotion is simulated as a binomially distributed random variable; the probability of promotion is simply the notional height growth of an average tree in the layer divided by the range of the height class (assuming that this ratio is less than 1.0). FLAM actually retains slightly more information about height distributions than is implied by the promotion routine; for example, if the height of the average tree in the layer is below that of the representative tree (halfway between the top and the bottom of the layer), then the distribution of heights should in principle be skewed toward the lower end of the layer and the probability of promotion should be lower. An improved promotion routine could use information about the average stem volume to correct the assumed height distribution within the layer. It may be possible to apply this distribution deterministically instead of stochastically. A deterministic procedure would avoid the calculation of random numbers used for promotion in the UVM and FLAM and would save computation time.

The UVM and FLAM include provisions for "skipping" layers during promotion; it is possible for trees to be promoted in jumps of two or more layers in a single time step. This feature allows wide latitude in the choice of time steps and layer structure, but may not be necessary for most applications. If a time step of 1 year is used, a choice of fairly thick layers (2–5 m) compatible with computational efficiency would make provision for layer skipping unnecessary. I obtained reasonable simulations of central Swedish forest dynamics with layers 5 m thick and time steps of 1–5 years (Fulton, 1991).

Dynamics of Herbaceous Species and the Scaling of Recruitment

The nature of the dynamics of different plant species in a model is partly a function of the units taken to define the population. If the population of herbaceous plants is considered to be the number of ramets (sensu Harper 1977), then a model with an annual time step is not modeling growth of herbaceous plants but rather their population dynamics. "Reproduction" would then be tied to the net assimilation of plants on the patch as well as the usual "bath" propagule rain (sensu Whittaker and Levin 1977) implicitly assumed in conventional gap models. This assimilation-dependent reproduction could simulate either clonal growth or localized seed dispersal of small growth-forms.

Mortality probabilities are partly a function of size. Other things being equal, small plants are more subject to mortality from a variety of causes than are large plants. The imposition of an additional mortality probability to all plants in the lowest layer(s) could approximate this effect.

Competition among Small Growth-Forms

The usual assumption in a gap model is that any individual affects conditions (light and soil) over the entire patch. This assumption breaks down severely when small plants (e.g., 0.01–1.0 m tall) grow in a simulated gap of a size appropriate to mature trees; a 0.01–0.1-ha patch is an inappropriate size to resolve competition among smaller plants. Spatial resolution is increased in the model ZELIG (Smith and Urban 1988) with a grid of smaller patches with weighted averaging of conditions in adjacent plots. This solution considerably increases computational requirements. A model by van Tongeren and Prentice (1986) used competition for cover as a basis for heathland vegetation dynamics. Ground area covered on a spatially explicit grid was taken as a proxy for acquired resources. It may be possible to approximate this mechanism in a layer model by doing away with the spatial explicitness; the small plants in the lowest layer could compete for free horizontal space within the layer. Competition would occur during establishment and among plants increasing their cover within a timestep. Space occupied could be scaled to leaf area and the leaf area index (LAI) of the individual canopy; a plant with 1.0 m^2 leaf area and a canopy LAI of 2.0 would occupy 0.5 m^2 of space.

The "Small" Tree Problem

Standard tree-growth equations in gap models return (either implicitly or explicitly) a stem volume increment/leaf area. The cost of increased leaf area is ignored. This approximation works as long as the usual practice of introducing trees at breast height or greater is followed, because wood biomass is much greater than leaf biomass for full-sized trees. When such a growth equation is applied to seedlings 1.0–3.0 cm high, unreasonably large growth increments are returned. This is because in small trees a large proportion of the biomass growth is in leaves; in large trees this proportion is small enough (and relatively constant enough) to be ignored. A growth equation that returned an increment of aboveground biomass/leaf area would avoid this problem. The biomass increment would be distributed to stem and leaves according to a functional relationship. (See below.)

H.H. Shugart (pers. comm.) pointed out another possible problem: the leaf area may increase significantly within a growing season for some tree seedlings. This could be solved by either shortening the model timestep for small growth-forms or by explicitly integrating the growth equations with respect to time.

Gap and patch models use various functional relationships between plant features such as height, diameter, and leaf area to model plant growth. The standard allometric (power functions) equations are not ideal for such relationships, particularly in a model that includes both seedlings and adult trees. For example, in the equation L = kWc (where L = leaf area, W = stem biomass, and k and c are constants), dL/dW approaches infinity as W approaches 0.0 when c is less than 1. This becomes important when modeled seedlings start at 1–3 cm tall; unreasonably large height growth increments result.

In some cases, there are relationships between plant parts that are more robust than power functions. According to the pipe-model of Shinozaki *et al.* (1964), the leaf area of a tree is proportional to the sapwood cross-sectional area of the supporting stem. Two additional premises could be used to relate leaf area to basal area: (1) the proportion of heartwood increases with age from 0.0 and (2) the maximum leaf area is constrained by self-shading within the crown. These premises lead to a function relating leaf area to diameter squared with an initial slope proportional to the ratio of leaf area to sapwood cross-sectional area and an asymptotic approach to the maximum leaf area. An asymptotic function of this type was used by Leemans and Prentice (1989) to simulate height as a function of diameter.

Design Features of a GVM

The GVM is a modified version of the forest simulator FLAM. The preceding considerations guided the design of the GVM in the incorporation of new features and the modification of existing features.

The GVM recognizes three different growth types: trees, perennial herbs, and annual herbs. The way each type is handled is dictated by the model timestep of 1 year. Trees grow continuously and survive from year to year, as in standard patch models. Perennial herbs are of fixed stem volume and leaf area but can survive from one year to the next. Annuals are of fixed stem volume and leaf area and die at the end of a growing season. The growth equations for perennials and annuals do not actually simulate the growth of the plants but are used to calculate a growth dependent reproduction term, as discussed under "Dynamics of herbaceous species and the scaling of recruitment."

An optional feature of GVM is a *cover competition layer* that includes a mechanism of competition for area in the reproduction of all growth-forms. The area occupied within the lower layer is calculated as discussed under "competition among small growth forms," above and the expected reproduction of each species or functional type is reduced by the factor $1.0-P$ where P is the proportion of the patch area occupied. The factor $1.0-P$ also reduces the growth of trees within this layer. Because small plants have higher mortality, an additional mortality factor can be imposed on all plants in this layer.

The growth equations used in GVM are formulated to return an above-ground biomass increment/leaf area. In perennials and annuals this only influences the growth-dependent reproduction rate because these types have fixed biomass/individual in the model. For trees, the biomass is distributed to stem volume and leaf area according to two functional relationships. The first is a relationship between the height and diameter of a tree:

$$H = H_{max}(1.0-\exp(-sD/H_{max})) \qquad (13.1)$$

where H is height (m), D basal stem diameter (m), H_{max} is maximum height, and s is a constant. This function has a fixed initial slope (s) and an asymptotic approach to a maximum height. The second functional relationship is between the leaf area and diameter of a tree:

$$L = L_{max}D^2/(D^2+L_{max}/m) \qquad (13.2)$$

where L is leaf area, L_{max} is maximum leaf area, and m is a constant. This function also has an asymptotic approach to a maximum, but the reduction in increment per unit time from the initial slope is more rapid than in Equation 1.

A deterministic procedure is used for promotion of trees from one layer to the next. A trapezoidal or triangular distribution of tree heights is calculated that depends on the relative stem volumes of the average and representative trees (see "Layer-Based Models," above) of that layer and species. The number of trees promoted is determined by notionally moving this distribution upward by a height increment calculated from the biomass increment and the functional relationships between height, diameter, and leaf area (see "promotion," below). A comparison with an earlier version of the GVM (using the FLAM promotion subroutine which simulates a random binomial variate to determine promotion) showed that the current version runs approximately 35% more quickly when initialization time is discounted. There was no apparent loss of realism in the simulations when height distributions and leaf area density profiles were compared.

Model Description

The GVM models the leaf area (LA), stem volume (V), and number of stems (N) for each species/functional type in a small number of height classes (layers). The timestep is 1 year. The area of the modeled patch and the number and spacing of height classes are set at execution time through the input file. The model is coded in FORTRAN 77. Several features of the state description (e.g., the vertical distribution of leaf area) and growth equations (e.g., the height/diameter allometry, the expression of maintenance costs, the photosynthesis/light response function) are taken directly from the model FORSKA (Leemans and Prentice 1989).

State Description

The state of each patch is expressed as

$N_{l,j,t}$ = the number of stems of species j in layer l at time t,

$V_{l,j,t}$ = the product of basal diameter squared times height for all the stems of species j in layer l at time t (m^3), and

$L_{l,j,t}$ = total leaf area for species j in layer l at time t (m^2)

for all layers and species (functional types). The above-ground biomass (B) is calculated for some purposes:

$$B_{l,j,t} = \mu_j V_{l,j,t} + \lambda_j L_{l,j,t} \quad (13.3)$$

where μ_j is the stem biomass (kg)/product of basal diameter squared times height (which includes both stem density [dry weight basis] and a form factor) for species j, and λ_j is the biomass (kg)/leaf area for species j (the inverse of specific leaf area). Leaf area for each species in a given layer is assumed to be distributed evenly from the top of the layer to the top of the top bole layer. The top bole layer is calculated for each species and layer at program initialization according to a species-specific constant which gives the proportion of stem height usually occupied by leaves. Leaf area density is thus taken to be uniform within a layer, and light extinction follows the usual Beer's law approximation:

$$I_z = I_o \exp(-kLAI_z) \tag{13.4}$$

where Iz is available light at depth z, Io is the light at the top of the canopy, k is the extinction coefficient and LAIz is the cumulative leaf area index from the top of the canopy to depth z.

The proportion of plant cover in the bottom layer is calculated at each time step and is used to control recruitment of all growth types and growth of trees:

$$P_t = CL_j L_{1,j,t}/A \tag{13.5}$$

where CL_j is the area of ground/leaf area occupied by species j and A is the area of the plot. This cover competition mechanism can be turned off in the GVM program by setting the input parameter NCL to 0; in which case the model runs by using only the whole patch-scaled environmental modifications to control growth and recruitment.

Growth

For all growth types, an increment of aboveground biomass is calculated according to:

$$\Delta B_{l,j} = \text{MAX}\,(L_{l,j}(G_{l,j}M_j - \delta_j Z_{l,j}), 0.0) \tag{13.6}$$

where $B_{l,j}$ is the above-ground biomass growth, $G_{l,j}$ is the growth-modifying factor for species j in layer l, M_j is the maximum growth/leaf area for species j, δ_j is the respiration required to maintain a unit length of active stem tissue, $Z_{l,j}$ is the mean leaf height for plants of species j in layer l (the mean of the height of the top of the layer and the height of the top bole layer), and $L_{l,j}$ is the total leaf area of plants of species j in layer l. For annuals, this quantity is only used to calculate growth efficiency to determine reproductive rates. For perennials, the growth efficiency is also used to calculate the probability of mortality. In trees, growth is expressed as an increase in stem and leaf biomass and as promotion from one height class to the next. Above-ground biomass of trees is partitioned into leaves and stem according to

$$B_{l,j,t+1} = B_{l,j,t} + \Delta B_{l,j} \tag{13.7}$$

$$V_{l,j,t+1} = CVB_{l,j} B_{l,j,t+1}/\mu_j \tag{13.8}$$

$$L_{l,j,t+1} = CLB_{l,j} B_{l,j,t+1}/\lambda_j \tag{13.9}$$

where $CVB_{l,j}$ is the ratio of stem biomass to above-ground biomass for a representative tree with height equal to the height of the middle of the layer (calculated at program initialization from the equations in "the 'small tree' problem," above.)

$CLB_{l,j}$ is the ratio of leaf biomass to aboveground biomass for a representative tree (as above).

The growth-modifying factors (G) are calculated on the basis of a light-response function and a maximum biomass/area function. (The soil competition model of Botkin *et al.* 1972 is retained). The light-response function is

$$GI_j = (kI_z - c_j)/(kI_z + \alpha_j - c_j) \tag{13.10}$$

where GI_j is the light growth-multiplier for species j, k is the light extinction coefficient, I_z is the light falling on the leaves at level z, and α_j and c_j are the half-saturation point and compensation points for species j. For plants in the cover competition layer, this relationship is applied directly by using I = light at the top of the layer. For other layers, the actual GI is the integral:

$$GI_{l,j} = \int_{T_{l-l}}^{t_1} (kI_{z-cj})/(kI_z + \alpha_j - c_j)\, dz \tag{13.11}$$

This integral accounts for shading within the layer and has an analytical solution due to the assumption of constant leaf area density within the layer. The soil competition factor (GB) is calculated as

$$GB_{l,j} = 1.0 - W_{tot}/W_{max} \tag{13.12}$$

where W_{max} is the maximum biomass/area (a model parameter). W_{tot} is either (1) the total above-ground biomass/area (for plants in layers above the cover competition layer) or (2) the total above-ground biomass of plants above the cover competition layer/area (for plants in the cover competition layer). The purpose of this separation is to have plants in the cover competition layer only competing for space in the regeneration phase. The overall growth-modifying factor is the product of GI and GB. For "trees" in the cover competition layer, an additional term of $1.0 - P_t$ is multiplied into the growth-modifying factor to account for the increased occupation of space associated with growth in this layer.

The relative growth efficiency (RGE) is calculated to determine probability of mortality (Leemans and Prentice 1989), and, for species confined to the cover competition layer to determine growth-dependent establishment. It is defined as the ratio of actual biomass increment ($B_{l,j}$) to the maximum biomass increment attainable by the leaf area. The maximum biomass increment (ΔB_{max}) is

$$\Delta B_{max} = M_j L_{l,j}(kI_0 - c_j)/(kI_0 + \alpha_j - c_j) \tag{13.13}$$

where I_0 is the light at the top of the canopy.

Promotions

Annuals and perennials are taken to have a constant height (due to the model timestep of 1 year) and are not promoted from one layer to another. Trees can be promoted according to the following procedure. The stem volume of species j at the top ($VT_{l,j}$) and bottom ($VT_{l-1,j}$) of each layer l are calculated at program initialization. The average stem volume of an individual ($\overline{V}$) in the layer is $V_{l,j}/N_{l,j}$. The height of the average individual ($\overline{H}$) is calculated by the linear approximation

$$H = (T_l - T_{l-1})(\overline{V} - VT_{l-1,j})/(V\,T_{l,j} - VT_{l-1,j}) + T_{l-1} \quad (13.14)$$

where T_l and T_{l-1} are the heights of the top and the bottom of layer l, respectively. The average height and the number of individuals uniquely specify a distribution of heights within the layer if the shape of the distribution is assumed to be bounded by a straight line and the area is the number of individuals in the layer. For example, if $\overline{H} >= (2/3)*(T_l - T_{l-1}) + T_{l-1}$, the distribution is triangular at the top of the layer (see Fig. 13.1) with the lower tip of the triangle at $T_l - 3.0*(T_l - H)$. When $H = (1/2)*(T_l - T_{l-1}) + T_{l-1}$ the distribution is the rectangular one assumed in the UVM and FLAM. The promotion subroutine is called after the mortality subroutine, so the individuals available for promotion are the survivors. If the distribution is moved upward by the quantity ΔH (the height growth of the average tree in the layer), the number promoted is the area (rounded off to the nearest integer) that crosses the height T_l. ΔH is calculated by

$$\Delta H = CHB_{l,j} \Delta B_{l,j} \quad (13.15)$$

where $CHB_{l,j}$ is a constant calculated at program initialization for each species and layer from the allometric relationships between height, diameter, and leaf area. Instead of promoting stem volume and leaf area equal to that of an average tree in the layer for each promoted tree, GVM promotes stem volume and leaf area calculated for trees with height = T_l. This device assures that $\overline{H}$ will always be between T_l and T_{l-1} for any layer l.

Establishment

Establishment occurs when the light available at the top of the cover competition layer is greater than the light compensation point for the species (c_j). Expected establishment (EXR) is the sum of two terms: a "bath" propagule rain (constant/area for a given species) and, for species confined to the cover competition layer, a growth-dependent term (maximum reproduction/individual of a given species). Actual reproduction is drawn from a Poisson distribution with mean EXR. The bath term is $EXBR_j*A$, where $EXBR_j$ is the expected establishment/m^2 for species j and A is the area of the patch in m^2. The growth-dependent term is

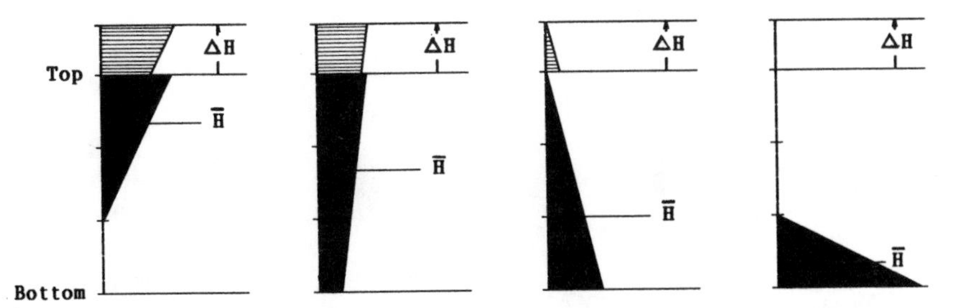

Fig. 13.1. Height distributions assumed given different average heights, and calculation of number promoted given the height increment. If the area of the figure representing the distribution is taken to be the total number of individuals of a given species in the layer, then the number promoted is the area crossing the top of the layer if the distribution is raised by the distance ΔH.

$EXGR_j*N_{l,j}*RGE_{l,j}$, where $EXGR_j$ is the expected maximum reproduction/individual for species j, $N_{l,j}$ is the number of ramets in the cover competition layer, and $RGE_{l,j}$ is the relative growth efficiency for species j in the cover competition layer.

Mortality

All annuals die at the end of the growing season. Perennials and trees are killed as a random function of their relative growth efficiency. If $RGE_{l,j}$ is greater than a species-specific threshold (θ_j) then the probability of mortality PM $= U\theta_j$. If $RGE_{l,j}$ is less than or equal to θ_j then PM $= U_{l,j}$. The actual mortality is simulated as a binomially distributed random variable with an expected value $N_{l,j}$*PM. Additional mortality (K) can be imposed on all plants in the cover competition layer, in which case PM is reset to PM + K−PM*K.

Model Exercises

Three model exercises, caricatures of forest dynamics in central Sweden, are reported here. Parameter lists for each of the exercises (including sources) are in the appendix.

Forest Dynamics with Recruitment of Saplings

The first exercise modeled a spruce (*Picea abies*) and pine (*Pinus sylvestris*) forest starting from an initial state with no trees, and introducing trees with a height of 1.3 m. It is designed to compare with the simulations made by Prentice and Leemans (1989) with FORSKA and Fulton (1991) with FLAM. The GVM captures the essential features of the forest structure dynamics modeled by FORSKA in spite of the layer approximation and the simplified growth equations (Figs. 13.2 and 13.3). Plots of leaf area dynamics (Fig. 13.4) show that spruce

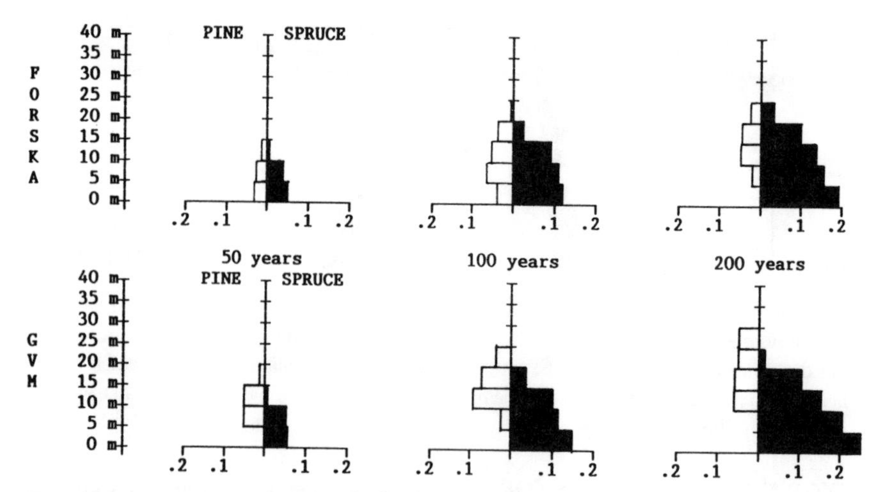

Fig. 13.2. Leaf area density (m^2/m^3) profiles from the first model exercise of GVM compared with a similar run of FORSKA.

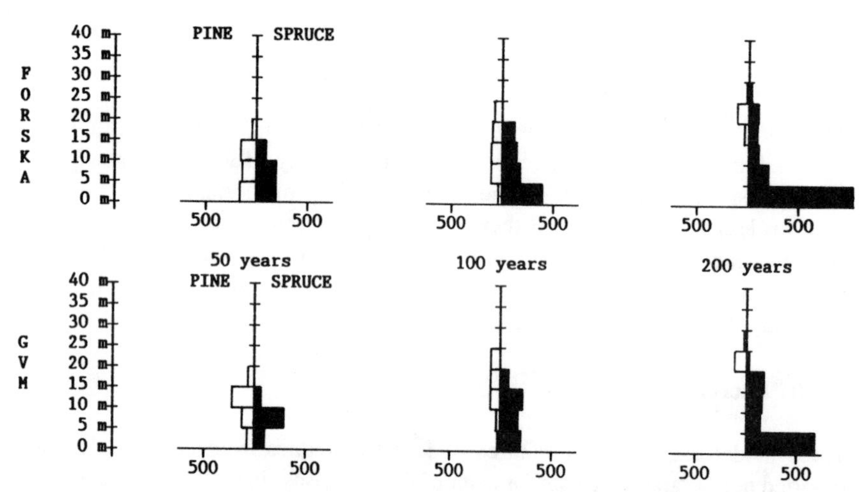

Fig. 13.3. Height distributions (stems/ha/height class) from the first model exercise of GVM, compared with a similar run of FORSKA.

exceeds pine in total leaf area within 40 years. The simulation on the GVM was about 25 times faster than the comparable simulation on FORSKA.

Forest Dynamics with Recruitment of Seedlings

The second model exercise simulated the same pine/spruce forest as the first, but trees were started at 2-cm notional height in a 0.5-m-thick bottom layer. Leaf area density and height distributions are shown in Figure 13.5; years 60, 110,

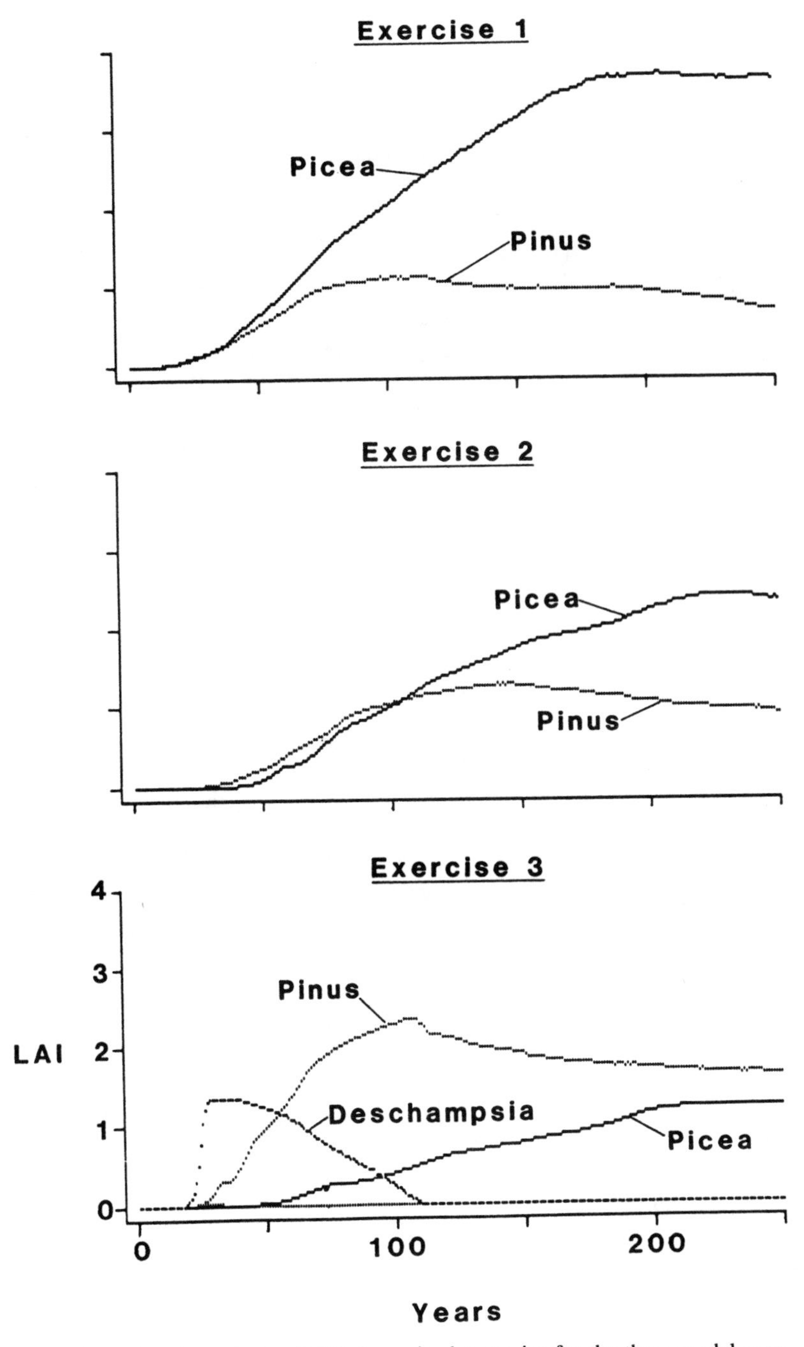

Fig. 13.4. Leaf area index (LAI) dynamics by species for the three model exercises.

and 210 are shown to make the results more nearly comparable with the first exercise. The canopy development of spruce is relatively slow in this exercise, but the leaf area density profiles are qualitatively similar. Pine, with a larger growth constant and a steeper initial slope for the height/diameter function, escapes from the understory layer much more rapidly to dominate the canopy. (See also Fig. 13.6). This is seen in the simulations of Prentice and Leemans (1989), but the effect is considerably amplified here by starting the trees at a much smaller size; spruce did not achieve canopy dominance even after 250 years—a longer delay than is explained by a linear extrapolation of height growth rates alone. In the postwindstorm dynamics simulated by Prentice and Leemans (1989), starting trees at 1.3 m is probably reasonable due to the advance regeneration characteristic of many of these forests. The spruce saplings to a large extent escape being killed by the storm. The different pattern shown in the present simulation may be a more accurate reflection of postfire dynamics. The height distribution of spruce shows waves of recruitment, which may be an artifact of the layer structure of the model, but the overall trends are correct.

Forest Dynamics with Competition from Smaller Growth Forms

The third model exercise included some understory plants in a cover competition layer. The species included *Deschampsia flexuosa, Vaccinium myrtillus,* and *Vaccinium vitus-idaea*. These are common components of the understory of central Swedish forests. Parameters were estimated by deductions from literature and field observations. (See appendix.) The results should be taken as an indication of the behavior of the model with a variety of understory species. Recruitment rates of pine and spruce were increased to compensate for competition with the understory plants, and the trees were started at 2-cm notional height as in the second exercise. A small disturbance rate (0.1) was imposed on the cover competition layer; this prevented any one species (such as *Deschampsia;* see below) from suppressing the regeneration of the other species.

The tree canopy development (Fig. 13.6) is qualitatively similar to that seen in the previous exercise, but the early dominance of pine and suppression of spruce are even more pronounced due to the more rapid escape of pine from the competition in the lowest layer. The LAI dynamics (Fig. 13.4) show a rapid initial dominance of *Deschampsia,* once enough stems are established for local (*growth dependent*) reproduction to be significant. The decline is driven by shading as pine begins to dominate the overstory. Neither of the *Vaccinium* species ever achieves significant leaf area. These species may be more shade tolerant than the parameters imply, or possibly the assumed reproductive rates are too low. The comparison of leaf area dynamics among the three model exercises in Figure 13.6 underscores the sensitivity of these models to the way trees are introduced and the possible sensitivity of forest dynamics to interactions in the understory.

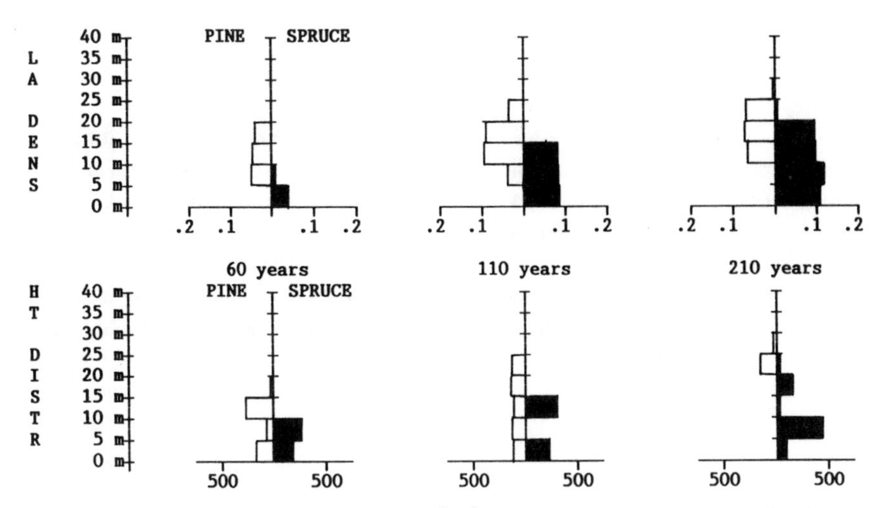

Fig. 13.5. Leaf area density profiles (m^2/m^3) and height distributions (stems/height class) of the upper eight layers of the second model exercise of GVM.

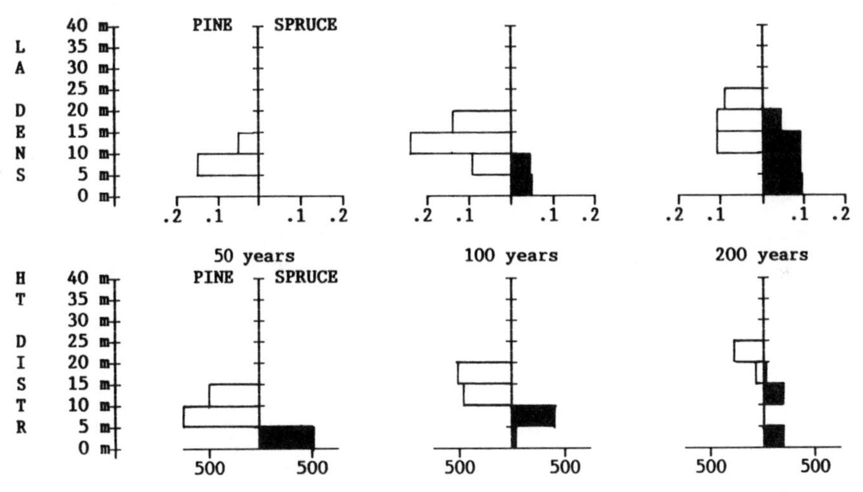

Fig. 13.6. Leaf area density (m^2/m^3) profiles and height distribution (stems/hectare/height class) of the upper eight layers of the third model exercise of GVM (tree species only).

In an early version of this simulation, the growth-dependent reproduction rate of *Deschampsia* was set higher (>4 per stem). This led to oscillatory dynamics of *Deschampsia* and complete suppression of all other species. This should not have been surprising, because when the growth rate of *Deschampsia* is set high enough, its dynamics will dominate and the model behaves as a difference equation of population growth:

$$N < N_{max}: N_{t+1} = N_t(1 + r[1 - N_t/N_{max}])(1 - m)$$
$$N >= N_{max}: N_{t+1} = N_t(1 - m)$$

where m is the mortality rate and r is the growth-dependent reproduction rate. Many simple difference models of population dynamics oscillate when the growth rate is set high enough (May 1981). If this behavior is not realistic for the system being simulated, then the growth-dependent reproduction rate should be kept low. When m is close to 0.0, the model begins to oscillate at $r>2.0$, the same threshold as the logistic difference equation (May 1981). A model that incorporated the spatial contagion characteristic of clonal growth would probably not exhibit this type of instability.

Conclusions

A vegetation stand simulator using height-class structured populations (layers) in the state description offers several advantages for modeling processes of vegetation change. The mechanisms of growth and local competition for resources characteristic of conventional gap models are retained without including every individual in the state description, while computation time can be speeded up by one or two orders of magnitude. The layer structure simplifies the inclusion of a field layer and competition between small trees and field-layer plants.

The scaling of competition among plants of different sizes is an important technical problem in gap models; the area of influence of an individual increases with height. An approximate solution is to use a patch size appropriate to the largest individuals and have the small plants competing for free horizontal space within the field layer. Including the cost of increased leaf area in the growth equations (and using appropriate functional relationships between height, diameter, and leaf area) makes a single set of equations appropriate for large trees and small tree seedlings.

The GVM, a model based on these ideas, succeeded in (1) mimicking the output of the gap model FORSKA when set up with similar parameters and initial conditions (in about 1/30th of the computation time); (2) giving a reasonable simulation with trees introduced as small seedlings instead of saplings; and (3) giving a reasonable simulation with some common species included in a field layer. The GVM is therefore a successful prototype for the generic model of the dynamics of local vegetation composition and structure within the framework for modeling global vegetation change outlined in Prentice *et al.* (1989).

Acknowledgments

Comments from and conversations with a number of people improved this work, among them: I.C. Prentice, A.M. Solomon, R. Leemans, H.H. Shugart, T.M.

Smith, M. Korzukhin, and C. Skarpe. I.C. Prentice read several versions of the manuscript.

Appendix: Parameters for Model Exercises

Parameter names and explanations:

Model Structure

NPATS- Number of replicate patches
NLAYS- Number of layers (height classes)
T_1- Height of layer 1
T_{NLAYS}- Height of top layer
NCL- Cover competition layer on (1) or off (0)

Physical Parameters

I_0- Average illumination at top of canopy ($mol/m^2/s$)
k- Light extinction coefficient
A- Area of a modeled patch (m^2)
W_{max}- Maximum stem volume (D^2H) per unit area (m)
K- Additional mortality in cover competition layer

Species Parameters

GRTYPE- Growth type (1="tree", 2="perennial", 3="annual")
M- Maximum assimilation/leaf area ($kg/m^2/year$)
α- Half-saturation point in light-response equation
c- Light compensation point
δ- Stemwood maintenance respiration (/yr)
H_{max}- Maximum height (m)
s- Initial slope of height diameter function
m- Initial slope of leaf area D^2 function
L_{max}- Maximum leaf area (m^2)
λ- Dry weight/projected leaf area (kg/m^2)
μ- Dry weight/D^2H (kg/m^3)
CL- Area occupied/leaf area in cover competition layer
LFSTEM- Proportion of stem height with leaves

V_0- Initial D^2H (m^3)
L_0- Initial leaf area (m^2)
EXBR- Expected bath reproduction (/m^2/yr)
EXGR- Growth-dependent reproduction(/individual/yr)
θ- Threshold relative growth efficiency (mortality)
U_0- Growth-independent mortality rate
U_1- Mortality rate when RGE<θ

Parameters for First Model Exercise:

Model Structure	
NPATS	20
NLAYS	8
T1	5.0
T2	10.0
.	.
.	.
T8	40.0
NCL	0

Physical Parameters	
I_0	400.0[a]
k	0.4[a]
A	1000.0[a]
W_{max}	0.2[b]
K	0.0

Species Parameters

	Pinus sylvestris	*Picea abies*
GRTYPE	1	1
M	2.005[c]	0.25[c]
α	330.0[a]	100.0[a]
c	59.0[a]	22.0[a]
δ	0.005[b]	0.0011[b]
H_{max}	35.1[a]	42.9[a]
s	111.0[b]	96.0[b]
m	800.0[b]	1600.0[b]
L_{max}	78.6[b]	140.4[b]
λ	0.29[d]	0.26[d]
μ	250.0[e]	230.0[e]
CL	1.0[f]	1.0[f]
LFSTEM	0.5[f,g]	0.9[f,g]
V_0	1.8E-4[h]	2.4E-4[h]
L_0	0.11[h]	0.29[h]
EXBR	0.001[b]	0.001[b]
EXGR	0.0	0.0
θ	0.025[a]	0.025[a]
U_0	0.0046[a]	0.0046[a]
U_1	0.4646[a]	0.4646[a]

Parameters for Second Model Exercise
(same as for first exercise except as noted)

Model Structure			Species Parameters	
			Pinus	*Picea*
NLAYS	9			
T1	0.5			
T2	5.5	Vo	6.5E-10[i]	8.6E-10[i]
T3	10.5	Lo	2.6E-5[i]	6.9E-5[i]
.	.			
.	.			
T9	40.5			

Parameters for Third Model Exercise
(same as for second exercise except as noted)

Model Structure		Physical Parameters	
NCL	1	K	0.1

Species Parameters

Parameters for pine and spruce same as for second exercise except that EXBR was raised to 0.01.

	Vaccinium myrtillus	*Vaccinium vitus-idaea*	*Deschampsia flexuosa*
GRTYPE	1	1	2
M	0.25[j]	0.5[j]	2.0[j,l]
α	100.0[j]	150.0[j]	100.0[j]
c	22.0[j]	33.0[j,f]	22.0[j]
δ	0.005[j]	0.005[j]	0.0
H_{max}	0.4[f]	0.3[f,k]	0.4[f]
s	80.0[d]	80.0[d]	0.0[l]
m	1600.0[d]	1600.0[d]	0.0[l]
L_{max}	0.4[d]	0.004[d]	0.0[l]
λ	0.09[d]	0.26[d]	0.15[d]
μ	210.0[d]	210.0[d]	140.0[d]
CL	1.0[f]	1.0[f]	0.5[f]
LFSTEM	1.0[f]	1.0[f]	1.0[f]
V_0	1.0E−9[j]	1.0E−9[j]	6.5E−10[j]
L_0	1.0E−4[j]	1.0E−4[j]	2.5E−5[j]
EXBR	0.005[j]	0.005[j]	0.005[j]
EXGR	1.0[m]	0.3[j]	2.0[n]
θ	0.04[j]	0.04[j]	0.3[j]

U_0	0.259[f,o]	0.259[f,o]	0.3[j]
U_1	0.52[j]	0.52[j]	0.8[j]

Appendix Notes

a. Leemans and Prentice (1987).

b. Leemans and Prentice (1987), units changed.

c. Calculated assuming the growth constant (which applies to stem volume) is correct for a tree with height $H=H_{max}/2.0$, and applying the leaf area–diameter relationship to get biomass produced/leaf area.

d. Measurements on local plants.

e. Wood density values from Busgen *et al.* (1929), volume of trunk+branches approximated by a prolate spheroid.

f. Field observations.

g. Leemans (1986).

h. Calculated from functional relationships using initial height of 1.3 m.

i. Calculated from function relationships using initial height of 0.02 m.

j. Guess.

k. Ritchie (1955).

l. Value arbitrary; important values are those controlling population.

m. Field observations, and inference from Ritchie (1956) on rate of clonal spread in heathlands.

n. See text.

o. Internode counts of locally growing plants give ages of 5–9 years for larger ramets; value calculated assuming 5% survival after 10 years.

References

Burton, P. J. and Urban, D. L. (1989). Enhanced simulation of early secondary forest succession by incorporation multiple life form interaction and dispersal. In *Forests of the World, Diversity and Dynamics (Abstracts),* ed. E. Sjögren, Studies in Plant Ecology 18. Uppsala, Sweden: Svenska Växtgeografiska Sällskapet.

Busgen, M. and E. Münch. (1929). *The Structure and Life of Forest Trees.* London: Chapman and Hall.

Coffin, D. P. and Lauenroth, W. K. (1990). A gap dynamics simulation model of succession in a semiarid grassland. *Ecological Modelling,* **49,** 229–36.

Fulton, M. R. (1991). A computationally efficient forest succession model: Design and initial tests. *Forest Ecology and Management,* 42, 23–34.

Harper, J. L. (1977). *Population Biology of Plants.* London: Academic Press.

Leemans, R. (1986). Structure of the primaeval coniferous forest of Fiby. In *Forest Dynamics Research in Western and Central Europe,* (ed.), J. Fanta pp. 221–30. Wageningen: PUDOC, Wageningen.

Leemans, R. and Prentice, I. C. (1987). Description and simulation of tree-layer composition and size distributions in a primaeval *Picea-Pinus* forest. *Vegetatio,* **69,** 147–56.

Leemans, R. and Prentice, I. C. (1989). FORSKA, a general forest succession model. Meddelanden frσn Växtbiologiska Institutionen, Uppsala, Sweden.

May, R. M. (1981). Models for single populations. In *Theoretical Ecology,* ed. R. M. Max, Oxford: Blackwell.

Prentice, I. C. (1986). The design of a forest succession model. In *Forest Dynamics Research in Western and Central Europe,* ed. J. Fanta, pp. 253–56. Wageningen: PUDOC, Wageningen.

Prentice, I. C., and Leemans, R. (1990). Pattern and process and the dynamics of forest structure: A simulation approach. *Journal of Ecology,* **78,** 340–55.

Prentice, I. C., Webb, R. S., Ter-Mikaelian, M. T., Solomon, A. M., Smith, T. M., Pitovranov, S. E., Nikolov, N. T., Minin, A. A., Leemans, R., Lavorel, S., Korzukhin, M. D., Hrabovszky, J. P., Helmisaari, H. O., Harrison, S. P., Emanuel, W. R., and Bonan, G. B. (1989). Developing a global vegetation dynamics model: Results of an IIASA summer workshop. RR-89-7, International Institute for Applied Systems Analysis, Laxenburg, Austria.

Ritchie, J. C. (1955). Biological Flora of the British Isles. *Vaccinium vitis-idaea* L. *Journal of Ecology,* **43,** 701–08.

Ritchie, J. C. (1956). Biological Flora of the British Isles. *Vaccinium myrtillus* L. *Journal of Ecology,* **44,** 291–99.

Shinozaki, K., Yoda, K., Hozumi, K., and Kira, T. (1964). A quantitative analysis of plant form—the pipe model theory. I—Basic analyses. *Japanese Journal of Ecology,* **14**(3), 97–104.

Shugart, H. H. (1984). *A Theory of Forest Dynamics.* New York: Springer-Verlag.

Smith, T. M. and Urban, D. L. (1988). Scale and resolution of forest structural pattern. *Vegetatio,* **74,** 143–150.

Solomon, A. M. (1986). Transient response of forests to CO_2-induced climate change: Simulation modeling experiments in eastern North America. *Oecologia,* **68,** 567–79.

van Tongeren, O. and Prentice, I. C. (1986). A spatial simulation model for vegetation dynamics. *Vegetatio,* **65,** 163–73.

Whittaker, R. H. and Levin, S. A. (1977). The role of mosaic phenomena in natural communities. *Theoretical Population Biology,* **12,** 117–39.

14

Plant Functional Types

T.M. Smith, H.H. Shugart, F.I. Woodward, and P.J. Burton

Introduction

Biomes or ecosystem types have long been recognized as occupying domains in an environmental space defined by temperature and moisture gradients. Within particular biomes, the factors governing structure and dynamics may vary considerably. For example, species composition and age structure in plant communities are largely controlled by asymmetric competition for an above-ground resource (light) and symmetric competition for below-ground resources (moisture and nutrients). In mesic forests, the principal constraint seems to be the availability of light. As a forest environment tends from mesic to xeric conditions, the effective constraint shifts from above-ground to below-ground (Tilman 1988; Smith and Huston 1989). In still drier environs, the forest yields to grassland, where the principal constraint is below-ground. This suggests that there are general trends or patterns in the relative influence of environmental constraints in structuring ecosystems.

There is an increasing interest in understanding the factors responsible for ecosystem scaling, in terms of both the dominant temporal responses and the spatial patterns. Current discussions on biosphere dynamics and global ecology (Bolin *et al.* 1986; Rosswall *et al.* 1987) have sharpened the research interest in time- and space-scales in ecological systems, as have discussions of hierarchy theory (Allen and Starr 1982; Allen and Hoekstra 1984; O'Neill *et al.* 1986; Urban *et al.* 1987). An appreciation of scale seems a clear prerequisite to unifying the dynamics of atmospheric processes and the dynamics of terrestrial ecosystems. The categorization of controlling factors at different space- and time-scales in particular ecosystems has been the topic of several reviews (Delcourt *et al.* 1983; Pickett and White 1985; Shugart, this volume).

There is also a clear need for computer models capable of simulating the responses of major ecosystems to novel environmental stress at global scales. At present there exists considerable cooperation at the international level among ecologists, and the initial development of computer models to predict both forests (Shugart *et al.* 1991) and grasslands (Coffin and Lauenroth 1990) is underway. These models can be changed to synthesize new information and insights as they develop. For example, since 1984 there has been an international effort to develop a *unified boreal forest model* (called FORBORS) capable of predicting the dynamic change in response to environmental stress (including climate change) for the northern forests of North America and Eurasia (Shugart *et al.* 1991). Significant parts of the FORBORS model were developed in conjunction with the doctoral dissertations of Dr. G.B. Bonan at the University of Virginia and Dr. R. Leemans at Uppsala University. The North American model (LOKI, Bonan 1990) and European forest model (FORSKA, Leemans and Prentice, 1989) were merged and the resultant *unified boreal zone model* has been published as part of a major review of the patterns and processes in the boreal forests of the world (Shugart *et al.* 1991). Computer models of the same genre as those developed in the boreal model global network have already been used in North America to project continental-scale responses of forests to climatic change (Solomon and Webb 1985; Solomon 1986). A major comparative effort using individual-based models (see Prentice *et al.*, this volume) to investigate the dynamics of several different forests (using the ZELIG model, Smith and Urban 1988) and grasslands (using the STEPPE model, Coffin and Lauenroth 1990) is currently underway.

While these cooperative models have enjoyed some success in simulating relatively large-scale patterns of responses (particularly those of forests) to climatic change and other stresses, they have been limited to applications in well-studied sites and regions with relatively low diversity of dominant plants because they are dependent on model parameterizations at the species level. The problems of using the species as the basic unit of vegetation for models functioning at the global scale are obvious. A species-by-species treatment of vegetation at the continental scale has restrictions imposed by data requirements. It is not clear whether or not it is appropriate to consider the species as the focal unit when addressing questions of how features such as physiognomic structure, leaf area, and net primary productivity will respond to climate change.

However, if one chooses not to consider the species in global studies of vegetation, one must address the question of how to define vegetation units which are not species yet are descriptive in terms of the physiological responses to environmental factors and the associated life-history characteristics required to predict vegetation response to environmental change. If it were possible to develop such models, then, presumably, a set of more detailed models with greater parameter estimation requirements (including in some cases species-level parameters for individual-based simulation models) could be used to inspect the robustness of larger-scale models at well-studied sites.

Prediction of the global distribution of leaf area index (LAI) in Woodward's chapter (this volume) was achieved by combining a simple plant-canopy-based general model and climatic information. The approach proved to be generally successful in delimiting zones of vegetation in terms of LAI. However, poor predictions emerged in some areas (such as tundra) and in other areas with markedly seasonal climates. An improved set of global predictions of vegetation type from climate was achieved by incorporating more detailed biological information about the vegetation. Such information included the length of the growing season (for tundra) and the lowest annual temperature (for temperate deciduous forest).

These additional, apparently important, vegetation properties are a conceptual basis for the development of descriptors or dimensions of *functional types* of plant groups. The reason for developing functional types, rather than including information at the species level, is that the latter information is far too sparse on a global or even a regional scale. Clearly the number of functional types should be much smaller than the number of plant species (indeed, smaller than even the number of species of dominant plants) to have significant applicability and utility in generic or regionally specific models.

Using a Dynamic Model as a Basis for Functional Types

A general photosynthesis and growth model was developed in chapter 4 (Woodward). For global predictions, the coefficients incorporated in the equations in chapter 4 need to be quantified. These coefficients (for example, those which define the influence of the environment on stomatal conductance) can be considered to constitute a dimension or characteristic of a classification of functional types. Such coefficients are *intensive* in that they incorporate detailed and precise information about a single process. For the model described in chapter 4, the intensive characteristics are:

1. Leaf longevity
2. Threshold temperature for low-temperature mortality
3. Threshold temperature for growth
4. Daylength for growth
5. Seed longevity
6. Photosynthate allocation to leaves, stems and roots
7. Environmental coefficients for stomatal conductance
8. Environmental coefficients for mesophyll conductance
9. Environmental coefficients for leaf growth
10. Environmental coefficients for respiration

These intensive functional-type dimensions are contrasted by *extensive* functional-type characteristics in which the characteristic is not a model parameter but a descriptor which implies extensive properties. Such a descriptor is the term *arboreal*. Plant types which are arboreal have woody stems, are long-lived, tall, and, when mature, they can control the environmental responses of subordinates. The arboreal descriptor may be divided into trees or shrubs and then further subdivided into angiosperm or gymnosperm. A gymnosperm tree further implies a tracheidal xylem, narrow or needle leaves (in most cases), low stomatal and mesophyll conductance, and a low wood density. In some cases the extensive descriptors imply estimates of intensive coefficients (e.g., photosynthetic maxima).

Important descriptors for the development of an extensive descriptor-based classification of functional types are:

1. Physiognomy
2. Desiccation features
3. Life span
4. Pollination
5. Seed dispersal
6. Photosynthetic pathway
7. Shade tolerance
8. Fire tolerance
9. Nutrient stress tolerance

Extensive dimensions are primarily, but not entirely, related to either life cycle or structural and morphological properties. In this sense, these dimensions are like those at the basis of several of the classic classifications of vegetation shown in Table 14.1.

The descriptors for extensive dimensions can be nested hierarchically. By contrast, the intensive descriptors are model parameters and cannot be subdivided in the same manner. However, the model on which the intensive descriptors are based may be limited to a particular time- and space-scale by the underlying assumptions in the model. Further, one can easily imagine a hierarchy of mechanistic models each of which could be used as the basis of a system of intensive descriptors for developing a functional classification.

A Review of Theoretical Treatments of Functional Types

The approach of relating characteristics of plants to adaptations for specific environments has formed the basis of a number of plant classification systems (Tab. 14.1). Approaches include r, K, and adversity strategies (MacArthur and Wilson 1967; Southwood 1977; Greenslade 1983); early and late successional species (Budowski 1965, 1970; Whittaker 1975; Bazzaz 1979; Finegan 1984);

Table 14.1. The use of different plant growth-form criteria in major systems of growth-form classification.

	Warming (1909) p. 136	Warming (1909) p. 137–8	Du Rietz (1931)	Raunkiaer (1934)	Schmid (1956–57)	Ellenberg and Mueller-Dombois (1967)	Golubev (1968)[b]	Whittaker (1975)	Halle at al. (1978)	Box (1981)
System organization:	Hierarch.	Hierarch.	?Hierarch.	Hierarch.	Hierarch.	Hierarch.	Parallel	Hierarch.	Hierarch.	Hierarch.
System purpose:	Veg.func.	Veg.desc.	General	Veg.dist.	Veg.desc.	General	General	General	Architect	Veg.pred.
Attributes used:										
Taxonomic affinity	2					3		2		2
Climatic/regional affinity										2
Life-history attributes:										
trophic level						1				
life span		1		1		1	1			
growth seasonality										
deciduousness		1	1			2,3,4	1	2	4,6,9	2
iteroparity						1				
reproductive age							1			
vegetative reproduction							1			
dispersal mode										
pollination mechanism		2					1			
Above-ground structural attributes:										
woodiness/lignification		1	3	1	1	1,2,4	1	1	1	
thorns						4		2		2
bark features						4				
stature/height[c]		1,2	1	1,2	2	1,2,3,4,5	1	1,2		1,2
foliage profile/growing point		1				3,4		2		1
no. stems and their equivalence[c]		1				2,3		2	2,3	1
branching locations/patterns			1		2	?2,3	1	2	3,4,5	
modularity							1		3,7	
support		1				1,3		1	1,10	
leaf size/shape	2		1		3	2,3,5	1	2		1,2

(continued)

Table 14.1. (continued)

	Warming (1909) p. 136	Warming (1909) p. 137–8	Du Rietz (1931)	Raunkiaer (1934)	Schmid (1956–57)	Ellenberg and Mueller-Dombois (1967)	Golubev (1968)[b]	Whittaker (1975)	Halle at al. (1978)	Box (1981)
System organization:	Hierarch.	Hierarch.	?Hierarch.	Hierarch.	Hierarch.	Hierarch.	Parallel	Hierarch.	Hierarch.	Hierarch.
System purpose:	Veg.func.	Veg.desc.	General	Veg.dist.	Veg.desc.	General	General	General	Architect	Veg.pred.
Attributes used:										
Above-ground structural attributes:										
succulence						2,3,4				1
stem dieback				2	1,2	1,4		2	7	
bud location/protection			1	1		1,3	1			
position of inflorescence						4			3,8	
Rooting or below-ground attributes:										
substrate	2	1								
root architecture, especially visible					3	2,4	1			
storage organs			2			2	1			
rhizomatous growth			2			2				
Environmental tolerances/habitat:										
salt	2									
acidity	2									
drought	1,2					2				2
flooding/anoxia	1,2									
cold	2									2
shade						5				
infertility										

[a]1–8=criterion used, designating the level in hierarchical systems at which used.

[b]Plus assorted aspects of each criterion, and others not listed.

[c]stature was considered the distinguishing feature behind undefined use of the term "tree"; the multistemmed attribute was considered the distinguishing feature behind undefined use of the term "shrub."

exploitative and conservative (Bormann and Likens 1981); ruderal, stress tolerant, and competitive (Grime 1977, 1979); gap and nongap species (Hartshorn 1978; Brokaw 1985a,b); structural classification (Raunkiaer 1934; Webb *et al.* 1970; Halle' 1974; Halle' and Oldemann 1975; Walker *et al.* 1981); and vital attributes (Noble and Slatyer 1980). Most of these schemes are based on the constraints imposed on plants by different environmental conditions, such as resource availability or disturbance regime. After classifying environmental conditions, most schemes categorize plant strategies based on responses to those environmental conditions and the associated life-history characteristics, either explicitly or implicitly considering the consequences of resource allocation.

A major problem with many of these attempts to relate plant (adaptive) strategies to environmental conditions has been the circularity of defining the environmental conditions in terms of the plant response. The problem is inherent in the concept of r- and K-selection (MacArthur and Wilson 1967; Whittaker and Goodman 1979; Greenslade 1983) in which K, the carrying capacity of the environment, is defined in terms of the number or biomass of a particular species.

Another problem with many schemes of vegetation classification by strategy, and perhaps the most important in terms of predicting vegetation response to environmental change, is that the characterization of the species (i.e., their strategies) is dependent on their role within the present vegetation (e.g., community) or environment. For example, the categories of early vs. late successional are actually a description of the species' role in the temporal dynamics of a community. They provide no insight as to why the species respond in time the way they do and therefore are ineffective in predicting how the same species may respond in a different context, either under different environmental conditions or in the presence of other species. The role of a species in the community or ecosystem is a function of the prevailing environmental conditions (e.g., climate, geology) and the suite of species present; it is context sensitive. (For discussions see Smith and Huston 1989.) What is needed is an approach for defining key characteristics and processes of plants which determine their response to environmental conditions, competitive interactions, and ultimately their distribution in time and space.

We will discuss three such approaches which have examined patterns in the primary processes of dispersal, establishment, growth, and reproduction as they relate to environmental conditions for the purposes of predicting vegetation patterns in time and space. The three approaches differ in the plant attributes and processes examined, but all share the general philosophy that patterns at higher levels of organization (i.e., communities, ecosystems) can be predicted as a function of patterns of attributes at the level of the species (which are defined at the level of the individual plant).

Vital Attributes

Vital attributes (Noble and Slatyer 1980) are those attributes of a species which are vital to its role in a vegetation replacement sequence. Noble and Slatyer defined three categories relating to the following: (1) method of arrival and persistence of the species at the site during and after a disturbance; (2) ability to establish and grow to maturity in the developing community; and (3) time taken for the species to reach critical life-stages. The characteristics include such features as longevity, persistence of seed bank, and the ability to establish following disturbance or in the presence of established vegetation. Species are then classified into *types* based on the combination of attributes exhibited. Although there may be differing mechanisms responsible for a particular vital attribute the classification relates only to the outcomes of those mechanisms.

Qualitative predictions of species or type (strategy—defined as a combination of characters) as a function of time following disturbance (i.e., replacement sequence) can then be made and the consequences of different disturbance regimes can be explored (Fig. 14.1).

This approach does not explicitly consider the response of types to environmental factors but relates establishment and growth under conditions associated with disturbance and the presence of vegetation. The scheme is intended to deal primarily with successional dynamics on a particular site with a stable physiography in the absence of changes in the underlying environmental conditions (e.g., climate). However, these restrictions are not inherent in the approach, and the vital-attributes scheme has been modified to simulate vegetation dynamics in both time and space (Noble *et al.* 1988).

Roles: Life History and Regeneration

Competition for occupancy of canopy gaps has been shown to be an important construct in understanding the dynamics of natural forests, starting with the classic work of Watt (1925, 1947). Trees attain sufficient size to alter their own microenvironment and that of subordinate trees. The species, shapes, and sizes of trees in a forest can have a direct influence on the local forest environment. The environment, in turn, has a profound influence on the performance of different species, shapes, and sizes of trees. Thus, there can be a feedback from the canopy tree to the local microenvironment and subsequently to the seedling and sapling regeneration that may result in the next canopy.

This environmental alteration due to a canopy tree is perhaps most easily observed in the case of the forest light environment. The nature of the leaf area profile and canopy geometry are dominant factors determining the amount and nature of the light at the forest floor (Anderson 1964; Cowan 1968, 1971). The apparency of the tree/light interaction should not diminish the importance of other important tree/environment interactions. Trees can also alter the local environment with respect to the nature of throughfall (Zinke 1962; Helvey and Patric 1965), soil moisture (Shear and Stewart 1934; Swift *et al.* 1979), soil nutrient availability (Zinke 1962; Challinor 1968), and a myriad of other factors.

Shugart (1984, 1987) treats the question of tree/environment interactions by

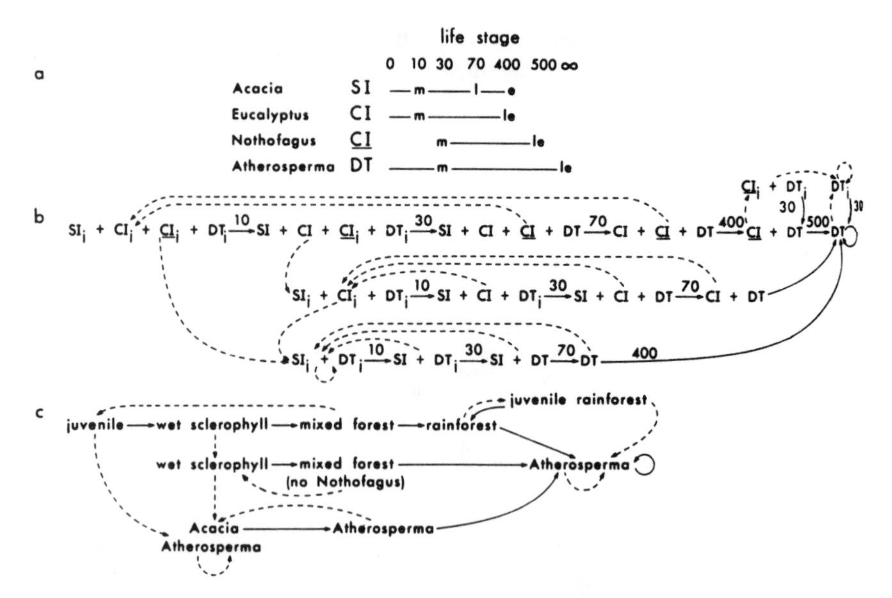

Fig. 14.1. Replacement series for Tasmanian forest, using vital attribute model.

considering the minimal categories of gap competition in trees. This categorization amounts to characterizing trees according to two dichotomies:

1. Does the species require a canopy gap for successful regeneration—Yes or No?
2. Does the species typically generate a gap with the death of a mature individual of typical size—Yes or No?

In this pair of dichotomies, there are implied two times two or four resultant tree strategies which Shugart categorized as species roles and numbered from one to four. Of these roles, two are self-reinforcing: role 1 species that both require gaps and generate them and role 4 species that neither require nor generate gaps.

It is appropriate to discuss the associated life-history traits that are expected from this relatively simple dichotomy as a preamble to illustrating the relatively rich dynamics that attend monospecies and mixed species forests. The four species roles are:

Role 1 species: species that require a canopy gap for successful regeneration and can generate such gaps. The generation of gaps is dependent on the height of the canopy (as well as on sun angle and crown diameter), and to succeed at the role 1 mode of survival one would expect selection in the direction of being larger than the other trees in the forest. Successful role 1 species would have reproductive strategies that would increase the likelihood of regeneration in gaps. Expected attributes would include seed with light-level-related germination triggers, seed storage in the vicinity of a gap-generating parent tree, and reproduction or germination timed to match seasonal gap-generating disturbance factors. One would expect efficient dispersal mechanisms to position propagules in locations that could potentially be beneath canopy gaps. Rapid growth in full light would allow the species to dominate gaps when they become available, as would strong apical dominance or a tendency to gain height rapidly. Mortality patterns that would abruptly create a gap (e.g., tendency to windthrow) would also reinforce the success of role 1 species.

Role 2 species: species that do not require a canopy gap for successful regeneration but attain sufficient size to generate gaps. Species would be expected to be tolerant of shading and to regenerate successfully under shaded conditions. Successful species would gain dominance by being relatively long-lived (thus increasing site occupancy) and by producing seedlings and saplings that could survive relatively shady conditions and block other species from occupying gaps by being present in such sites as advanced regeneration. One would expect mortality patterns that featured a relatively slow release of the gap (e.g., the production of standing dead 'snags' even after the canopy tree is dead) to favor self-replacement. Growth forms that spread to occupy more area as a canopy tree matures (as opposed to an emphasis on height growth) should be favored.

Role 3 species: species that require a canopy gap for successful regeneration but do not attain a sufficient size to generate such gaps. This role includes what Van Steenis (1958) referred to as 'nomad' species. Since the mature seed-producing individuals do not generate gaps when they die, role 3 species would be expected to have extremely efficient dispersal mechanisms since their survival is tied to having propagules in position to occupy gaps as they occur in locations that are removed from the seed source. Rapid growth in full light would allow the species to dominate gaps when they become available, as would strong apical dominance or a tendency to gain height rapidly. The growth rates of the species should be relatively high to allow rapid occupancy of colonized gaps, and the time to reach seed-bearing size should be relatively short. Dispersal by animals that frequent gaps would be expected to occur often in species of this role. As for role 1 species, one would expect role 3 species to have attributes including seed dormancy with light-level-related germination triggers, and reproduction or germination timed to match seasonal gap-generating disturbance factors.

Role 4 species: species that neither require nor generate gaps. Clearly species of this role should be shade tolerant to a considerable degree and should have adaptations to allow regeneration under low-light conditions. The high degree of shade tolerance implies that individuals of this role would be able to survive on relatively thin margins of productivity minus respiration. Thus one would expect slow growth rates, and masting in the production of flowers and seeds. Regenerating seeds might be expected to be large if the low-light establishment of the seedling is subsidized from seed energy reserves.

The different roles of trees with respect to gap colonization produce essentially different biomass and numbers when monospecies plots are simulated at small spatial scales (ca. 0.1 ha) using an individual-based forest model (Shugart 1984). The long-term behavior of numbers and biomass for typical example species is shown in Figure 14.2.

Growth, Tolerance, and Competitive Ability

Another approach to plant strategies or plant types has focussed on growth processes. The approach examines patterns of plant functional response to environmental factors (and associated life-history characteristics) at the level of the individual plant and is based on the simple premise that characteristics which enable a plant to maximize photosynthesis and growth under a given set of environmental conditions limit its ability to maximize these same responses under different environmental conditions. No plant can be optimally adapted to all environmental conditions.

This approach to physiological constraints in plants has provided the basis for a number of contributions that have addressed the costs and benefits associated with adaptation to a given set of environmental conditions and the consequences for leaf size and shape (Parkhurst and Loucks 1972; Givnish 1978, 1979), leaf type in arid environments (Orians and Solbrig 1977), plant height (Givnish 1982; Chazdon 1986), photosynthesis (Mooney and Gulmon 1979; Cowan 1986), the use of multiple resources (Chapin *et al.* 1987), herbivore defense and nutrient use (Mooney and Gulmon 1982; Bryant *et al.* 1983), and the effect of light response and life history on plant succession (Huston and Smith 1987).

One of the more general constraints surrounding plant response to environmental conditions is the trade-off between maximum growth rate under high-resource conditions and the ability to continue photosynthesis and growth (i.e., survive) under conditions of low-resource availability (Parsons 1968; Grime 1977; Chapin 1980) (Fig. 14.3). This pattern is a function of an array of constraints at various scales ranging from biochemical processes to carbon allocation at the level of the whole plant.

This trade-off between growth rate and tolerance forms the basis of Grime's dichotomy between competitive (high growth rate—low tolerance) and stress-tolerant (low growth rate—high tolerance) plants along an axis of productivity. However, if we partition this axis into its components of resource availability (light, water, and nutrients) and temperature we can begin to explore the consequences of this simple premise for vegetation patterns along environmental gradients.

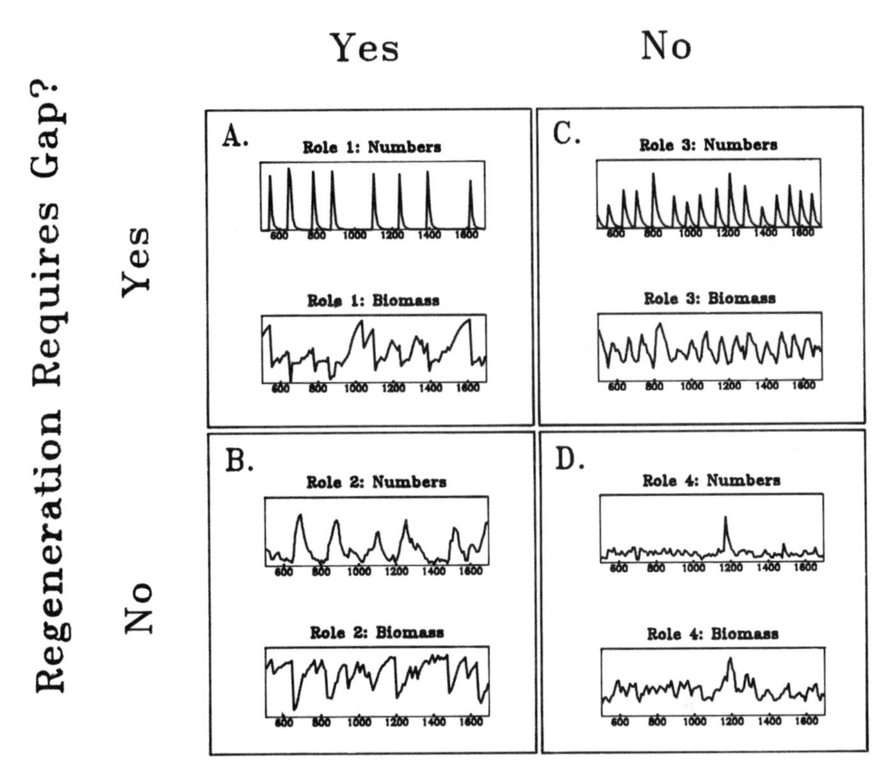

Fig. 14.2. Successional dynamics of diameter distributions simulated by the ZELIG model for a 1-hectare forest. Successional sequence was initiated in year 0 with a nonforested plot but with an adequate seed supply for all species.

The most widely used form of the *trade-off model* is the concept of shade tolerance and shade intolerance. In general, shade-intolerant plants have a high light compensation and light saturation point, a fast growth rate, and are small and rather short-lived in comparison with shade-tolerant plants. (See Bazzaz 1979; Huston and Smith 1987.) These characteristics are defined at the level of the individual and we can examine their consequences at higher levels of organization (i.e., populations, communities, ecosystems) by using individual-based models (Shugart 1984).

The population dynamics of a hypothetical monoculture of a shade-intolerant (Fig. 14.4a) and a shade-tolerant (Fig. 14.4b) species for a single simulated plot is shown in Figure 14.4. A number of contrasts in dynamics can be seen as a direct consequence of the individual-level characteristics. The shade-intolerant

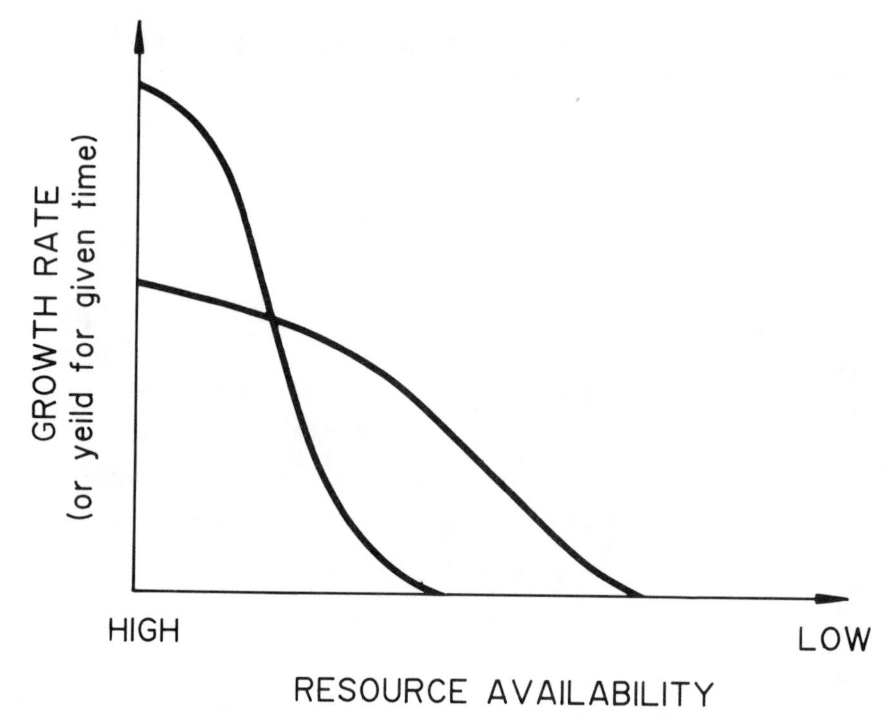

Fig. 14.3. Trade-off between maximum growth rate under high resource conditions and ability to tolerate resource limitation.

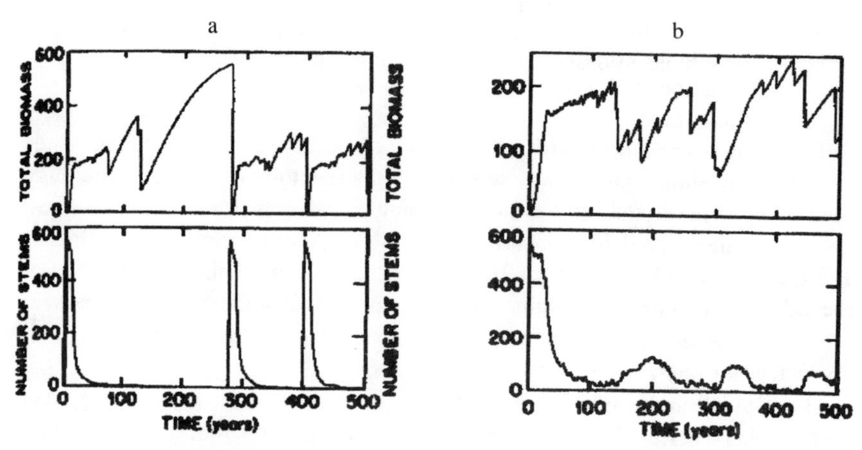

Fig. 14.4. Population dynamics of a hypothetical monoculture of (a) a shade-intolerant and (b) a shade-tolerant species for a single simulated plot.

species exhibits a faster dynamic with an initially higher rate of biomass accumulation (a result of a higher growth rate). However, the maximum biomass of the plot is lower than the value reached for the shade-tolerant species (a result of lower maximum size and a higher light compensation point). A more dramatic difference in the temporal dynamics is the cyclic pattern of biomass dynamics for the shade-intolerant species in contrast to the shade-tolerant one. This cyclic behavior is a result of the inability of the shade-intolerant species to regenerate under canopy cover because of its high light compensation point. In contrast, the lower light compensation point of the shade-tolerant species allows recruitment under canopy cover. This results in a more continuous size-class distribution and dampens the decline in total biomass following the death of a canopy dominant. When the two species are grown together, the competitive effects of both species can be seen. The combined (community) dynamic is a function of the two differing functional responses to light, and the addition of further species with different light response/growth strategies will change the observed dynamics. (For further discussion see Huston and Smith 1987.)

Although the different growth rates and life-history characteristics resulting in the above patterns are associated with light response, the physiological/morphological characteristics exhibited by a plant are a consequence of adaptations to the array of environmental factors influencing growth processes. Thus the response of plants to these various factors cannot be approached independently.

Smith and Huston (1989) have expanded this approach of plant strategies based on trade-offs (growth and tolerance) to include patterns of plant response to multiple resources. Figure 14.5 shows a hypothesized trade-off between maximum relative growth rate and strategies of light and water response. This surface encompasses the two important patterns discussed in the trade-off model: (1) maximum relative growth rate decreases with increasing tolerance, and (2) there is a limit to the combined tolerance for low availability of both water (belowground resources) and light. The surface was used to define a number of plant types and their associated resource-response functions and life-history characteristics. These functional plant types were then used to examine the consequences of these patterns of plant response at the level of the individual to population and community dynamics just as was done with the earlier simulations.

Figure 14.6 shows the temporal dynamics of 15 plant types along a simulated moisture gradient. The results can be expressed as either the successional dynamics at any one point along the moisture gradient or the distribution of late succession dominants across the moisture gradient. The combined temporal and spatial dynamics of these late successional dominants are shown (Fig. 14.6). One very important observation from these results is that the role a species plays in the temporal dynamics of the community is dependent on the set of prevailing environmental conditions. A plant species or type which is an early successional transient under high moisture availability can be a late successional dominant under more xeric conditions. The ability to capture this behavior of changing

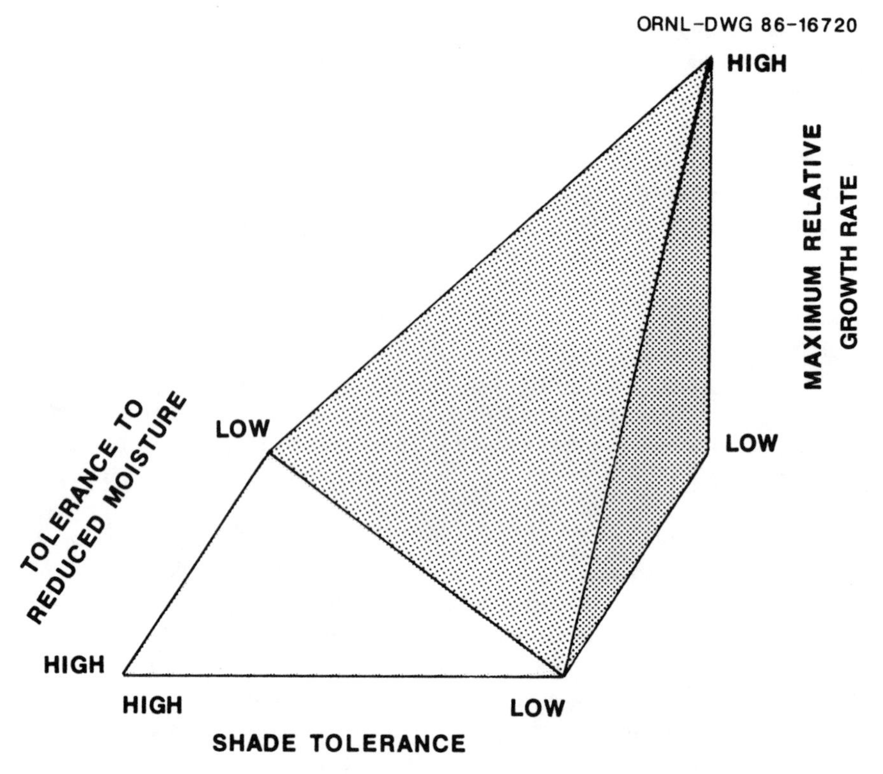

Fig. 14.5. Trade-off between maximum relative growth rate and tolerance to low availability of light and water.

roles in the system as a function of differing environmental conditions is a result of considering fundamental patterns of plant response to environmental conditions rather than defining plant types based on their role in a given community or set of environmental conditions.

Discussion

The species "strategies" that interact to produce vegetation pattern can be thought of as a nested hierarchy of system rules that blend in a continuous fashion from one hierarchical level to another. Plant ecologists have categorized these rules to various degrees, usually as theories about how plant processes in interaction with environmental factors produce vegetation pattern. These abstractions are usually constrained (either explicitly or implicitly in the domain of their successful applications) to particular space- and time-scales. One theory can "work" in producing successful predictions at one scale, while another, with different funda-

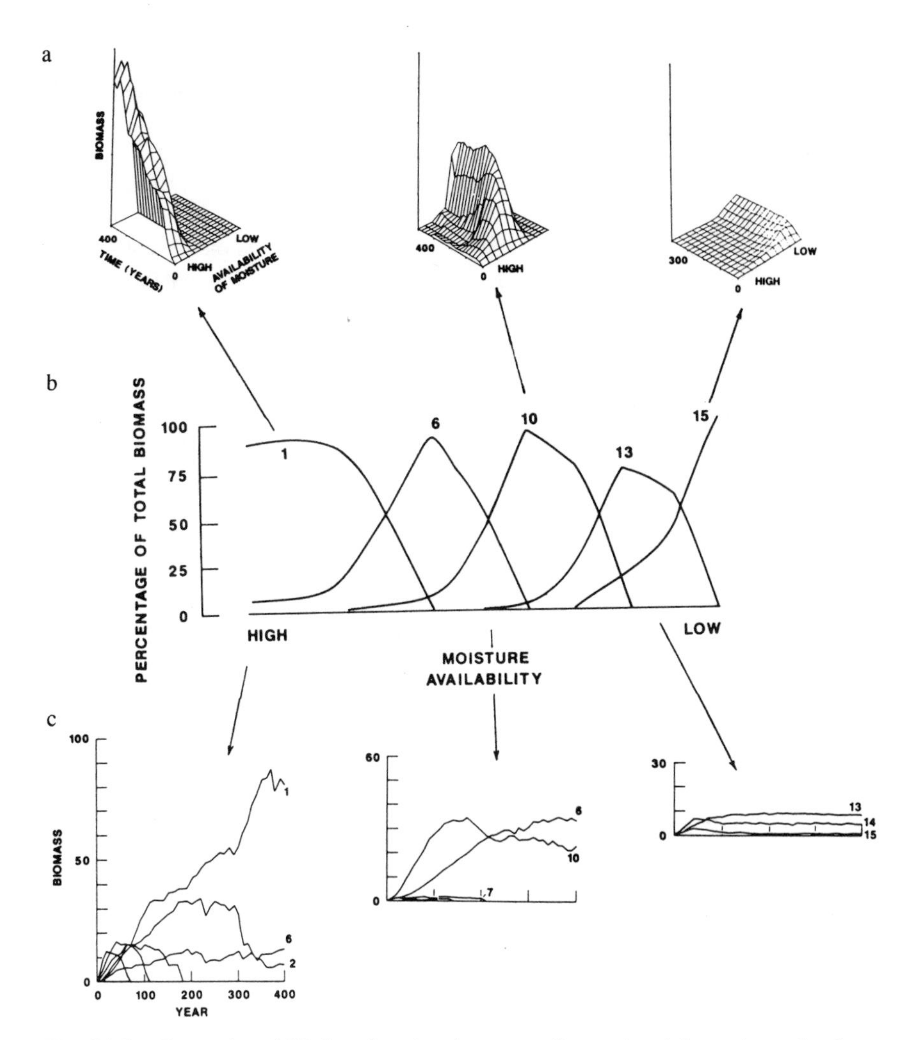

Fig. 14.6. Dynamics of 15 plant functional types: a. Successional dynamics under three different water availabilities. b. Distribution of late successional types along a simulated water gradient. c. Distribution of type 10 as a function of time and water availability.

mental assumptions, can also work, albeit at a different scale. This latter feature is consistent with a hierarchical concept of vegetation pattern.

Both Holdridge (1967) and Box (1981) have developed global systems of vegetation-climate classification. The two systems differ in the units of vegetation description (Holdridge—biome/vegetation type; Box—physiognomic/morphologic types), but they are both correlative approaches relating present vegetation patterns to simple climate data (mean precipitation and temperature). These approaches are able to relate vegetation and climate patterns under present conditions, but they do not examine the processes responsible for the observed correlations and as such they may encounter difficulties in predicting vegetation response to changing climate. This is especially true if changes in the climate alter the present correlation among environmental features, therefore resulting in environmental conditions not presently encountered within a particular land unit (e.g., continental).

A second problem relates to defining vegetation units based on present patterns of species distribution (e.g., Holdridge). This approach carries with it the assumption that the present species assemblages will move as units under changing environmental conditions. Both current vegetation theory and paleobotanical studies show this assumption to be false. In contrast, the approaches we have reviewed begin by defining key plant processes and how those processes vary as a function of environmental conditions. By defining the patterns of response independent of any given environmental vector, the approach can predict responses to environmental conditions beyond the present patterns. None of the above approaches directly addresses the question of plant response to gradients of temperature, the obvious next step in the development of any functional approach to plant classification for the purposes of modeling vegetation response to global climate change. The link between patterns of plant response to environmental factors which have a direct physiological effect yet are not resources (e.g., temperature) and the major resource gradient of water, nutrients, and light is the next logic step in this development.

References

Allen, T.F.H. and Starr, T.B. (1982). *Hierarchy: Perspectives for Ecological Complexity*. Chicago: University of Chicago Press.

Allen, T.F.H. and Hoekstra, T.W. (1984). Nested and non-nested hierarchies: a significant distinction for ecological systems. In *Proceedings of the Society for General Systems Research. I. Systems Methodologies and Isomorphies*, ed. A.W. Smith, pp. 175–80. Lewiston, New York: Intersystems Publications, Coutts Library Service.

Anderson, M.C. (1964). Light relations of terrestrial plant communities and their measurement. *Biological Reviews*, **39**, 425–86.

Bazzaz, F.A. (1979). The physiological ecology of plant succession. *Annual Review of Ecology and Systematics*, **10**, 351–71.

Bolin, B., Döös, B.R., Warrick, R.A. and Jaeger, J. (1986). *The Greenhouse Effect, Climatic Change, and Ecosystems*. SCOPE 29. Chichester: John Wiley.

Bonan, G.B. (1990). Carbon and nitrogen cycling in North American boreal forests. II. Biogeographic patterns. *Canadian Journal of Forest Research*, **20**, 1077–88.

Bormann, F.H. and Likens, G.E. (1981). *Pattern and Process in a Forested Ecosystem*. New York: Springer-Verlag.

Box, E.O. (1981). *Macroclimate and Plant Forms: An Introduction to Predictive Modeling in Phytogeography*. The Hague: Junk.

Brokaw, N.V.L. (1985a). Gap-phase regeneration in a tropical forest. *Ecology*, **66**, 682–7.

Brokaw, N.V.L. (1985b). Treefalls, regrowth, and community structure in tropical forests. In *The Ecology of Natural Disturbance and Patch Dynamics*, ed. S.T.A. Pickett and P.S. White, pp. 101–8. New York: Academic Press.

Bryant, J.P., Chapin, F.S. and Klein, D.R. (1983). Carbon/nutrient balance of boreal plants in relation to vertebrate herbivory. *Oikos*, **40**, 357–68.

Budowski, G. (1965). Distribution of tropical American trees in the light of the successional process. *Turrialba*, **15**, 40–2.

Budowski, G. (1970). The distinction between old secondary and climax species in tropical Central American lowland forests. *Tropical Ecology*, **11**, 44–8.

Challinor, D. (1968). Alteration of soil surface characteristics by four tree species. *Ecology*, **49**, 286–90.

Chapin, F.S. (1980). The mineral nutrition of wild plants. *Annual Review of Ecology and Systematics*, **11**, 233–60.

Chapin, F.S., Bloom, A.J., Field, C.B. and Waring, R.H. (1987). Plant responses to multiple environmental factors. *Bioscience*, **37**, 49–57.

Chazdon, R.L. (1986). The costs of leaf support in understory palms: Economy versus safety. *American Naturalist*, **127**, 9–30.

Coffin, D.P. and Lauenroth, W.K. (1990). A gap dynamics simulation model of succession in a semi-arid grassland. *Ecological Modelling*, **49**, 229–36.

Cowan, I.R. (1968). The interception and absorption of radiation in plants. *Journal of Applied Ecology*, **5**, 367–79.

Cowan, I.R. (1971). Light in plant stands with horizontal foliage. *Journal of Applied Ecology*, **8**, 579–80.

Cowan, I.R. (1986). Economics of carbon fixation in higher plants. In *On the Economy of Plant Form and Function*, ed. T.J. Givnish, pp. 133–70. Cambridge: Cambridge University Press.

Delcourt, H.R., Delcourt, P.A. and Webb, T. III. (1983). Dynamic plant ecology: The spectrum of vegetation change in space and time. *Quaternary Science Review*, **1**, 153–75.

Finegan, B. (1984). Forest succession. *Nature*, **312**, 109–14.

Givnish, T.J. (1978). On the adaptive significance of compound leaves, with particular reference to tropical trees. In *Tropical Trees as Living Systems*, ed. P.B. Tomlinson and M.H. Zimmermann, pp. 351–80. Cambridge: Cambridge University Press.

Givnish, T.J. (1979). On the adaptive significance of leaf form. In *Topics in Plant Population Biology*, ed. O.T. Solbrig, S. Jain, G.B. Johnson and P.H. Raven, pp. 375–407. New York: Columbia University Press.

Givnish, T.J. (1982). On the adaptive significance of leaf height in forest herbs. *American Naturalist*, **120**, 353–81.

Greenslade, P.J.M. (1983). Adversity selection and the habitat templet. *American Naturalist*, **122**, 352–65.

Grime, J.P. (1977). Evidence for the existence of three primary strategies in plants and its relevance to ecological and evolutionary theory. *American Naturalist*, **111**, 1169–94.

Grime, J.P. (1979). *Plant Strategies and Vegetation Processes*. New York: John Wiley.

Halle', F. (1974). Architecture of trees in the rain forest of Morobe District, New Guinea. *Biotropica*, **6**, 43–50.

Halle', F. and Oldemann, R.A.A. (1975). *Essay on the Architecture and Dynamics of Growth of Tropical Trees*. Penerbit University, Kuala Lumpur, Malaysia.

Hartshorn, G.S. (1978). Tree falls and tropical forest dynamics. In *Tropical Trees as Living Systems*, ed. P.B. Tomlinson and M.H. Zimmermann, pp. 617–38. Cambridge: Cambridge University Press.

Helvey, J.D. and Patric, J.H. (1965). Canopy and litter interception of rainfall by hardwoods of the eastern United States. *Water Resources Research*, **1**, 193–290.

Holdridge, L.R. (1967). *Life Zone Ecology*. Tropical Science Center, San Jose, Costa Rica.

Huston, M.A. and Smith, T.M. (1987). Plant succession: Life history and competition. *American Naturalist*, **130**, 168–98.

Leemans, R. and Prentice, I.C. (1989). A General Forest Succession Model. *Meddelanden*, **2**, 1–45. Växtbiologiska institutionen, Uppsala, Sweden.

MacArthur, R.H. and Wilson, E.O. (1967). *The Theory of Island Biogeography*. New Jersey: Princeton University Press.

Mooney, H.A. and Gulmon, S.L. (1979). Environmental and evolutionary constraints on the photosynthetic characteristics of higher plants. In *Topics in Plant Population Biology*, ed. O.T. Solbrig, S. Jain, G.B. Johnson and P.H. Raven, pp. 316–37. New York: Columbia University Press.

Mooney, H.A. and Gulmon, S.L. (1982). Constraints on leaf structure and function in reference to herbivory. *Bioscience*, **332**, 198–206.

Noble, I.R., Moore, A.D. and Strasser, M.J. (1988). *Predicting Vegetation Dynamics Based on Structural and Functional Attributes*. Proceedings, International Symposium on Vegetation Structure, Utrecht.

Noble, I.R. and Slatyer, R.O. (1980). The use of vital attributes to predict successional changes in plant communities subject to recurrent disturbances. *Vegetatio*, **43**, 5–21.

O'Neill, R.V., DeAngelis, D.L., Waide, J.B. and Allen, T.F.H. (1986). *A Hierarchical Concept of the Ecosystem*. Princeton, New Jersey: Princeton University Press.

Orians, G.H. and Solbrig, O.T. 1977. A cost-income model of leaves and roots with special reference to arid and semiarid areas. *American Naturalist*, **111**, 677–90.

Parkhurst, D.G. and Loucks, O.L. (1972). Optimal leaf size in relation to environment. *Journal of Ecology*, **60**, 505–37.

Parsons, R.F. (1968). The significance of growth rate comparisons for plant ecology. *American Naturalist*, **102**, 295–7.

Pickett, S.T.A. and White, P.S. ed. (1985). *The Ecology of Natural Disturbance and Patch Dynamics*. New York: Academic Press.

Raunkiaer, C. (1934). *The Life-forms of Plants and Statistical Plant Geography*. Oxford: Oxford University Press.

Rosswall T., R.G. Woodmansee and P.G. Risser, Eds. (1987). *Scales and Global Change*. Scope 35. NY: John Wiley & Sons.

Shear, G.M. and Stewart, W.D. (1934). Moisture and pH studies of the soil under forest trees. *Ecology*, **15**, 350–8.

Shugart, H.H. (1984). *A Theory of Forest Dynamics*. New York: Springer-Verlag.

Shugart, H.H. (1987). Dynamic ecosystem consequences of tree birth and death patterns. *Bioscience*, **37**, 596–602.

Shugart, H.H., Leemans, R. and Bonan, G.B. (1991). *A Systems Analysis of the Global Boreal Forest*. Cambridge: Cambridge University Press.

Smith, T.M. and Huston, M. (1989). A theory of the spatial and temporal dynamics of plant communities. *Vegetatio*, **83**, 49–69.

Smith, T.M. and Urban, D.L. (1988). Scale and resolution of forest structural pattern. *Vegetatio*, **74**, 143–50.

Solomon, A.M. (1986). Transient response of forests to CO_2-induced climate change: simulation experiments in eastern North America. *Oecologia*, **68**, 567–79.

Solomon, A.M. and Webb, T. III. (1985). Computer-aided reconstruction of late-Quaternary landscape dynamics. *Annual Review of Ecology and Systematics*, **16**, 63–84.

Southwood, T.R.E. (1977). Habitat, the templet for ecological strategies. *Journal of Animal Ecology*, **46**, 337–65.

Swift, L.W., Swank, W.T., Mankin, J.B., Luxmore, R.J. and Goldstein, R.A. (1979). Simulation of evapotranspiration and drainage from mature and clear-cut deciduous forests and young pine plantation. *Water Resources Research*, **11**, 667–73.

Tilman, D. (1988). *Resource Competition and Community Structure*. Princeton, New Jersey: Princeton University Press.

Urban, D.L., O'Neill, R.V. and Shugart, H.H. (1987). Landscape ecology. *Bioscience*, **37**, 119–27.

Van Stennis, C.G.G.J. (1958). Rejuvination as a factor for judging the status of vegetation types: the biological nomad theory. In *Study of tropical vegetation: Proceedings of the Kandy Symposium*, pp. 212–15. UNESCO, Paris.

Walker, B.H., Ludwig, D., Holling, C.S. and Peterman, R.M. (1981). Stability of semi-arid savanna grazing systems. *Journal of Ecology*, **69**, 473–98.

Watt, A.S. (1925). On the ecology of British beechwoods with special reference to their regeneration. *Journal of Ecology*, **13**, 27–73.

Watt, A.S. (1947). Pattern and process in the plant community. *Journal of Ecology*, **35**, 1–22.

Webb, L.J., Tracey, J.G., Williams, W.T and Lance, G.N. (1970). Studies in the numerical analysis of complex rain-forest communities. V. A comparison of the properties of floristic and physiognomic-structural data. *Journal of Ecology*, **58**, 203–32.

Whittaker, R.H. (1975). *Communities and Ecosystems*. New York: MacMillan.

Whittaker, R.H. and Goodman, D. (1979). Classifying species according to their demographic strategy. I. Population fluctuations and environmental heterogeneity. *American Naturalist*, **113**, 185–200.

Zinke, P.J. (1962). The pattern of influence of individual trees on soil properties. *Ecology*, **43**, 130–3.

15

Vegetation Functional Classification Systems as Approaches to Predicting and Quantifying Global Vegetation Change

J.P. Grime

Introduction

Correlations between plant form and climate are evident even to the most casual observer of vegetation. There have been several attempts to formalize these relationships by developing a system in which morphological attributes of the dominant plants in various parts of the world are predicted from selected climatic variables and then tested against field reality (Holdridge 1947; Box 1981). Further refinements of this correlative approach are being made, and with the application of modern analytical methods (e.g., Givnish 1986), it is probable that eventually the predictive models will be allied to a thorough mechanistic understanding of the way in which climate determines plant morphology and vegetation structure.

It might be supposed from the optimistic tone of the preceding paragraph that plant ecologists are well placed to devise predictions of the likely consequences of climate change for vegetation. However, the step from climate and vegetation classification to forecasts of change is extremely hazardous, for two main reasons:

1. Classification systems such as those of Holdridge (1947) and Box (1981) refer to the vegetation occurring in circumstances where, over long periods, natural selection has established an equilibrium between climate and the genetic resources of each region. However, over a large and rapidly increasing proportion of the land surface, vegetation is prevented from attaining such equilibrium by natural disturbances and by human intervention. In many areas of intensive exploitation, the natural vegetation dominants are now confined to local refugia or have been eliminated completely; in such cases, as documented for northern England by Hodgson (1986), the impact of land use now strongly outweighs that of climate as a vegetation determinant.

2. Extreme caution is required in reviewing the results of computer predictions of vegetation response to changed climate (e.g., Box 1981:123–131). In particular, it is vital to recognize that these predictions assume that there will be no major impediments to species dispersal and vegetation substitutions. There is a considerable fund of information (e.g., Ridley 1930; Salisbury 1942; van der Pijl 1972) indicating large differences in the capacities of plant populations to disperse and to expand their geographical limits. Perhaps most important of all is the need to consider the strong possibility (explained later in this chapter) that vegetation changes in response to climate forcing will occur at much faster rates in those parts of the landscape where *equilibrium vegetation* no longer survives and the plant cover is a recent and ephemeral product of intensive forms of land use.

In view of these problems, it seems necessary to devise additional functional classifications which do not focus exclusively upon the vegetation of pristine, mature ecosystems but instead include *all* major floristic components, including those of recent origin or subject to disruptive influences, such as agriculture, urbanization, commercial forestry, and dereliction. It seems imperative also that such classifications should extend beyond recognition of climate-related morphometric traits to include those facets of life history, physiology, and reproductive biology which allow predictions of both the direction and rate of population responses to climate change.

In this chapter reference will be made to three different classification systems with the potential to assist prediction. These systems, henceforward described as strategy theories (sensu Macleod 1894; Ramenskii 1938; Hutchinson 1959), refer to (1) primary strategies of established plants; (2) strategies of growth attunement to climate; and (3) regenerative (juvenile) strategies. Each of these systems has been described in detail elsewhere (Grime 1979, 1983, 1988; Grime *et al.* 1988). Here attention will be confined to basic features and to their broad implications for prediction of effects of climate change. At the end of this chapter, an attempt is made to summarize the present opportunity for a complementary interaction between strategy theories and more conventional morphological systems of vegetation classification.

Primary Strategies of Established Plants

For both plants and animals (Grime 1974; Southwood 1977), major recurring axes of adaptive specialization in life history and in the physiology of the adult (established) organism appear to be associated with variation in the duration and quality of the opportunities which habitats provide for resource capture, growth, and reproduction. In plants, three very contrasting conditions (Tab. 15.1) fa-

Table 15.1. Conditions of resource supply associated with three primary strategies.

1. Ruderal (ephemeral)	Temporarily abundant
2. Competitor	Continuously abundant but subject to local and/or progressive depletion as resources are exploited
3. Stress tolerator	Continuously scarce

voring, respectively, either stress tolerators, competitors, or ruderals have been recognized; however, these are only the extremities of the range of environments and strategies occurring in nature. The full spectrum of habitat conditions and their associated strategies can be described as an equilateral triangle (Grime 1974) in which the relative importance of competition, stress, and disturbance is represented by three sets of contours. This model allows recognition of not only the three extremes of plant specialization described in Table 15.1 but also a range of intermediate strategies associated with less extreme equilibria between stress, disturbance, and competition.

A review of the full implications of the model is beyond the scope of this chapter. Essential features of the primary strategies are summarized in Table 15.2, which also draws attention to the attributes which are most relevant to predictions of plant-population responses to climate change.

Life span is a particularly important criterion in attempts to predict the *rate* of vegetation change in response to climate variation. The potential for the most rapid changes in genotypic and species composition clearly resides in vegetation of frequently and severely disturbed habitats where the prevailing strategy is that of the ephemeral. Under a changing climate the continuous flux of individuals in ruderal vegetation may be expected to provide abundant opportunity for incursions by species better suited to the prevailing conditions.

Slower rates of change are likely in circumstances of high productivity but low vegetation disturbance; here the dominant strategy, that of the competitor (sensu Grime 1977), is represented by fast-growing but often relatively long-lived clonal herbs, shrubs, and trees.

The turnover of individuals and the scope for rapid response to climate change will reach a minimum in relatively undisturbed vegetation of low productivity (e.g., stress-tolerant lichens, mosses, and small shrubs of skeletal habitats or slow-growing forest trees on shallow, nutrient-poor soils). In temperate and cold regions, inertia in populations of evergreen stress-tolerant species may be further encouraged by the presence of mechanisms of tissue acclimation to seasonal climatic variation (Tab. 15.2). In theory, acclimation could permit survival of long-lived individuals through major climatic perturbations.

Strategies of Growth Attunement to Climate

Plant species and populations differ considerably in their abilities to exploit particular seasonal patterns of temperature and moisture supply. It is desirable

Table 15.2. Some characteristics of plant strategies relevant to prediction of responses to climate change.

	Competitive	Stress tolerant	Ruderal
1. Life-forms	Herbs, shrubs, and trees	Lichens, bryophytes, herbs, shrubs, and trees	Herbs, bryophytes
2. Morphology	High dense canopy of leaves; extensive lateral spread above and below ground	Extremely wide range of growth forms	Small stature, limited lateral spread
3. Life span[a]	Long or relatively short	Long–very long	Very short
4. Longevity of leaves and roots	Relatively short	Long	Short
5. Leaf phenology	Well-defined peaks of leaf production coinciding with periods of maximum potential productivity	Evergreens, with various patterns of leaf production	Short phase of leaf production in period of high potential productivity
6. Reproduction[a]	Established plants usually reproduce each year	Intermittent reproduction over a long history	Prolific reproduction early in life history
7. Proportion of annual production devoted to seeds[a]	Relatively small	Small	Large
8. Perannation	Dormant buds and seeds	Stress-tolerant leaves and roots	Dormant seeds
9. Maximum potential relative growth rate[a]	Rapid	Slow	Rapid
10. Photosynthesis and uptake of mineral nutrients	Strongly seasonal, coinciding with long continuous period of vegetative growth	Opportunities, often uncoupled from vegetative growth	Opportunistic, coinciding with vegetative growth

continued

Table 15.2. (continued)

	Competitive	Stress tolerant	Ruderal
11. Acclimation of photosynthesis, mineral nutrition, and tissue hardiness to seasonal change in temperature, light, and moisture supply[a]	Weakly developed	Strongly developed	Weakly developed
12. Storage of photosynthate and nutrients	Most photosynthate and mineral nutrients are rapidly incorporated into vegetative structure but a proportion is stored and forms the capital for expansion of growth in the following growing season	Storage systems in leaves, stems and/or roots	Confined to seeds
13. Defense against herbivory	Often ineffective	Usually effective	Often ineffective
14. Litter decomposition	Rapid	Slow	Rapid
15. Associated regenerative strategies[b]	V, S, W, B_s	V, B_j, W	S, W, B_s
16. Role in secondary successions in productive habitats	Relatively early	Late	Early

[a]Particularly relevant to predictions of vegetation response to climate change.

[b]Key to regenerative strategies (see Table 15.3): V, vegatative expansion; S, seasonal regeneration in vegetation gaps; W, numerous small, widely dispersed seeds or spores; B_s, persistent seed or spore bank; B_j, persistent juveniles.

therefore to build into the predictive framework aspects of our current understanding of interspecific differences in performance and tolerance with respect to these two major climate variables. A great deal of research has been conducted (1) on the relative cold-hardiness of species of different geographical distribution and origins and (2) on the efficiency of C3, C4 and CAM (Crassalacian Acid Metabolism) photosynthetic systems under various temperature regimes. Although these types of studies provide information relevant to vegetation change, further insights are required. In order to devise adequate predictions it is necessary to recognize the extent to which growth itself (i.e., the construction of plant tissue through division and expansion of cells) is capable of radically different strategies of attunement to particular seasonal patterns of temperature and moisture supply.

Recent developments in our understanding of how the mechanism of plant growth varies with climate have occurred through a curiously indirect route. The initial stimulus for this research was the need to explain geographical patterns of variation in plants with respect to nuclear DNA amount.

From a number of investigations (Stebbins 1956; Bennett and Smith 1976; Levin and Funderburg 1979) it is clear that the amount of DNA in the nucleus of tropical species is consistently small when compared with the wide range observed in temperate floras. In crop plants Bennett (1976) has drawn attention to a latitudinal trend in DNA amounts, while in native geophytes and grasses of the British flora a tendency has been reported (Grime and Mowforth 1982) for species with enlarged nuclear DNA amounts to be concentrated in southern England and to have European distributions centered on the Mediterranean.

The phenomenon of interspecific variation in DNA amount has been placed in an ecological context by evidence that, in a range of species common in northern England, nuclear DNA amounts are correlated with shoot phenology (Grime and Mowforth 1982). Evidence of the predictive value of nuclear DNA amount with respect to phenology is provided in Figure 15.1, which is based upon measurements of seasonal change in the shoot biomass of naturally established species in various plant communities in the Sheffield region. These data show that from early spring to midsummer, changes in the identity of the most actively expanding species range from *Hyacinthoides non-scripta* (DNA per nucleus = 42.4 pg) in March to *Chamaerion angustifolium* (DNA = 0.7 pg) in June, with a continuous reduction in nuclear DNA amount between the two.

In an attempt to explain the correlation between nuclear DNA amount and the timing and rate of shoot growth in the spring, it has been suggested by Grime and Mowforth (1982) that climatic selection has operated upon nuclear DNA amount through the differential sensitivity of cell division and cell expansion to low temperature. According to this hypothesis, species from continuously warm climates, or species which exploit the summer season in cool temperate climates such as that of the British Isles, have been subject to natural selection for low nuclear DNA amount and small cell size. These are features which coincide with a potentially short cell cycle, and which are conducive to rapid production of

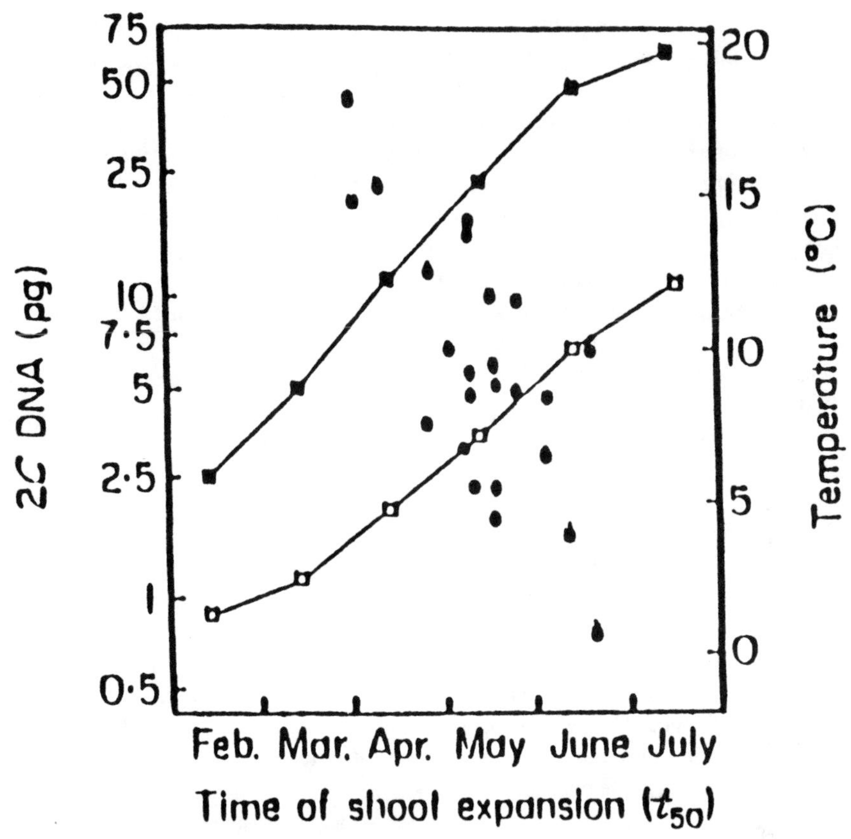

Fig. 15.1. The relationship between DNA amount and the time of shoot expansion (t_{50}) in 24 plant species commonly found in the Sheffield region (redrawn from Grime and Mowforth 1982). t_{50} is based upon measurements conducted throughout the year and describes the time at which 50% of the annual maximum in shoot dry mass is attained. Temperature at Sheffield is expressed as the long-term average for each month of daily minima (□) and maxima (■) in air temperature 1.5 m above the ground.

plant tissue in conditions which allow continuous division and expansion of cells. This strategy of growth is attuned to summer conditions and is clearly vulnerable to the potentially limiting effect of cold spring temperatures upon mitosis. It seems reasonable to suggest that this could explain the delayed shoot growth of the species of low DNA amount shown in Figure 15.1.

A quite different strategy of growth is proposed to explain the capacity of many geophytes and grasses of high DNA amount to expand shoots in the spring. Here it is suggested that rapid early growth is achieved by expansion of large cells (with large nuclei) which are formed, but not expanded, during preceding warmer conditions. This growth strategy circumvents the limitations imposed upon mitosis by low spring temperatures, but it may be expected to restrict the performance of the plant toward the latter part of the growing season. This is because, in species with large cells and large nuclear DNA amounts, relatively slow rates of summer growth may be expected to arise from the extended cell cycle (Bennett 1971). Moreover, in the geophytes with massive nuclei (e.g., *Allium ursinum* and *Hyacinthoides nonscripta*) and in some species of moderately high DNA amount (e.g., *Bromus erectus*, *Lolium perenne,* and *Ranunculus bulbosus*), the summer is marked by a quiescent phase in which development appears to be mainly restricted to meristematic activity preparatory to future episodes of cell expansion. This phenomenon has been particularly well documented in geophytes (Hartsema 1961), but there is also evidence of its importance in grasses and cereals, where it is commonly described by agriculturalists as *stored growth* (Salter and Goode 1967).

Further research into the predictive value of measurements of nuclear DNA amount is required. It is already evident, however, that some predictions can be attempted. A recent report forecasting the impact of climate warming on the British flora (Grime and Callaghan 1988) concludes as follows: "In general we expect grasses and geophytes with high nuclear DNA amounts to remain important in drier habitats but where moisture supply is sufficient to sustain summer growth a progressive trend towards species with low DNA amounts seems likely." The conditional nature of this conclusion emphasizes the need for predictions of future climates to be defined with respect to both temperature *and* moisture supply.

Recent evidence (Grime *et al.* 1985) suggests that differences in DNA amount between species in a plant community may provide an index of the extent to which differences in the phenology of shoot growth contribute to niche differentiation and species coexistence. Figure 15.2 shows a clear association between DNA amount and leaf extension rate in the early spring in a species-rich grassland community in northern England. There appears to be a strong possibility that differences in DNA amount provide a clue to the role of phenological differences in the *combining ability* of some familiar grassland codominants. The complementary seasonal peaks of growth in ryegrass–white clover swards coincide with a considerable difference between these species (*Lolium perenne* DNA = 9.9 pg, *Trifolium repens* DNA = 3.0 pg). A similar pattern is evident (Al-Mashhadani 1980) within the red fescue–brown bent grasslands of upland Britain (*Festuca rubra* DNA = 13.9 pg, *Agrostis capillaris* DNA = 5.6 pg). In lowland limestone or chalk grasslands in Western Europe, phenological differences consistent with those predictable from nuclear DNA amounts have been observed between the codominants *Bromus erectus* (DNA = 22.5 pg) and *Brachypodium pinnatum* (DNA = 2.3 pg) (Law 1974).

In cool temperate conditions, therefore, it would appear that phenological differences, coinciding with differences in DNA amount, play a major role in the structure of plant communities and are the result of close attunement of cell division and cell expansion to seasonal patterns in temperature and moisture

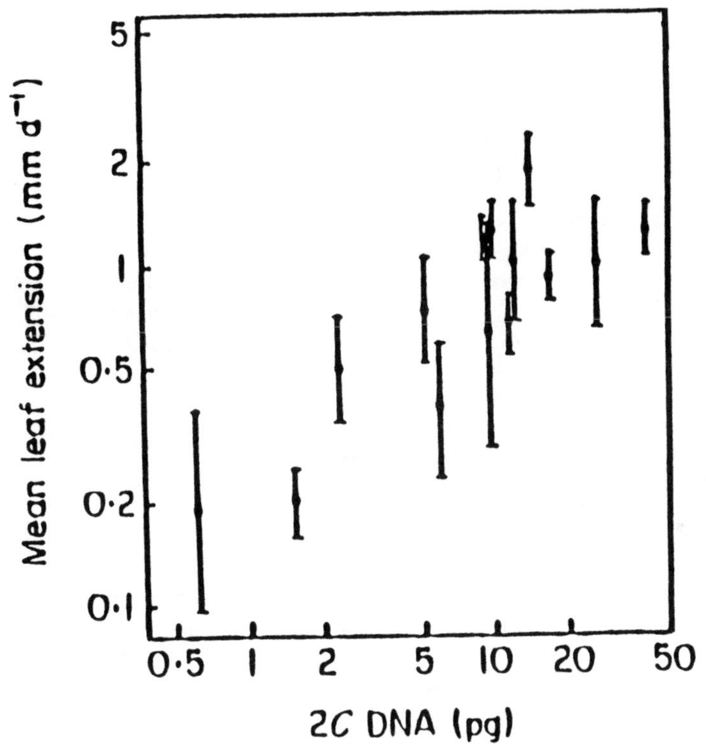

Fig. 15.2. The relationship between DNA amount and the mean rate of leaf expansion over the period March 25 to April 5, 1983, in 14 grassland species coexisting in the same turf (redrawn from Grime *et al.* 1985); 95% confidence limits are indicated by vertical lines.

supply. On this basis, we may expect changes in the timing of growth and in the relative abundance of species of high and low DNA amounts to provide a sensitive indicator of the first effects of climate change on existing plant communities.

Regenerative Strategies

It is now widely recognized that functional classifications of plants necessitate separate consideration of the strategies exhibited in the established and regenerative phases. The need for this distinction has emerged from the work of numerous biologists including Stebbins (1951, 1971, 1974), Grubb (1977), Gill (1978), and Grime (1979), who have recognized the particular selection processes and design constraints which determine the characteristics of juvenile plants. This has led to the hypothesis (Tab. 15.3) that there are five major types of regenerative

Table 15.3. Five regenerative strategies of widespread occurrence in terrestrial vegetation.

Strategy	Habitat conditions to which strategy apppears to be adapted
1. Vegetative expansion (V)	Productive or unproductive habitats subject to low intensities of disturbance
2. Seasonal regeneration in vegetation gaps (S) (seed banks transient)	Habitats subjected to seasonally predictable disturbance by climate or biotic factors
3. Regeneration involving persistent seed or spore bank (B_s)	Habitats subjected to spatially predictable but temporally unpredictable disturbance
4. Regeneration involving numerous widely dispersed seeds or spores (W) (seed banks usually transient)	Habitats relatively inaccessible (cliffs, tree trunks, etc.) or subjected to spatially unpredictable disturbance
5. Regeneration involving persistent juveniles (B_j) (seed banks transient or nonexistent)	Unproductive habitats subjected to low intensities of disturbance

strategy, which differ in features such as propagule size, mobility, and dormancy, and confer different and predictable capacities upon populations and species.

Earlier in this chapter it has been suggested that vegetation responses to climate change may be strongly affected by the relative abilities of plant species to disperse to new habitats. It is therefore pertinent to consider whether the classification system outlined in Table 15.3 provides a basis for prediction of species mobility.

The common regenerative strategy which involves the release of large numbers of small, buoyant, wind-dispersed propagules (W in Table 15.3) is obviously related to the capacity for long-distance dispersal in many mosses, pteridophytes, and particular families of flowering plants, e.g., Onagraceae, Orchidaceae, Salicaceae, and Betulaceae. It cannot be assumed, however, that small wind-dispersed spores or fruits will inevitably confer an advantage under changing climatic conditions. The scant food reserves of small propagules make them exceedingly vulnerable to establishment-failure due to dominance by established vegetation. In consequence successful colonization through W usually depends upon the occurrence of open microhabitats. In landscapes with a closed cover of vegetation this may result in situations where W is ineffective and the most successful colonizing species are large-seeded trees such as *Quercus*, *Castanea,* and *Carya* in which the capacity of the seedling to withstand competition compensates for the low output of propagules and for dependence upon birds as dispersal vectors.

On first inspection, the regenerative strategy which involves the accumulation of dormant seeds or spores in the soil (B_s in Table 15.3) seems to provide low potential for a mobile response to a changing climate mosaic. However, studies of common British plants (Salisbury 1953; Grime 1986) reveal that many of the species which exhibit the strongest colonizing potential under currently prevailing conditions (e.g., *Holcus lanatus*, *Matricaria matricarioides*) have no obvious mechanisms of dispersal and depend upon transport in soil attached to animals, humans, or vehicles. As pointed out by Salisbury (1953) in his *infection pressure* theory of colonizing ability, a critical feature of such passive dispersal is the dependence upon (1) an exceedingly high output of propagules and (2) a high frequency of earth-moving activities.

We may conclude, therefore, that no simple generalizations can be made with regard to the role of regenerative strategies in vegetation responses to climate. However, the relative mobility of populations and species *in particular areas* will be predictable from knowledge of the regenerative strategies represented in the local flora and the prevailing forms of landscape management.

Conclusion

From the computer models relating macroclimate to world vegetation types (e.g., that of Box 1981), predictions of the *direction* of vegetation change in response to a change in climate can be made with respect to mature "equilibrium" ecosystems. Plant Strategy Theory (sensu Grime 1979) broadens the perspective to include nonequilibrium conditions and provides inferences concerning *rates* of vegetation change. These additional predictions draw upon knowledge of rates of population turnover and depend critically upon relationships between vegetation dynamics, habitat productivity, and intensity of vegetation disturbance.

In temperate regions a further dimension to the predictions can be made by reference to phenological data relating shoot growth to seasonal patterns in temperature and moisture supply. Nuclear DNA amounts appear to provide a valuable marker for different phenological types and provide a basis for prediction of changes in the relative abundance of populations within communities.

Finally, but not least, classification of plants with respect to their regenerative strategies can allow further refinement of predictions of vegetation response to climate change. However, the identity of the regenerative strategies capable of facilitating vegetation response to climate change will vary regionally and will be strongly affected by the dominant patterns of land use.

References

Al-Mashhadani, Y.D. (1980). *Experimental investigations of competition and allelopathy in herbaceous plants*. PhD thesis, University of Sheffield.

Bennett, M.D. (1971). The duration of meiosis. *Proceedings of the Royal Society of London*, **178B**, 277–99.

Bennett, M.D. (1976). DNA amount, latitude and crop plant distribution. *Environmental and Experimental Botany*, **16**, 93–108.

Bennett, M.D. and Smith, J.P. (1976). Nuclear DNA amounts in angiosperms. *Philosophical Transactions of the Royal Society*, **B274**, 227–74.

Box, E.O. (1981). *Macroclimate and Plant Forms: An Introduction to Predictive Modeling in Phytogeography*. The Hague: Junk Publishers.

Gill, D.E. (1978). On selection at high population density. *Ecology*, **59**, 1289–91.

Givnish, T.J. ed. (1986). *On the Economy of Plant Form and Function*. Cambridge: Cambridge University Press.

Grime, J.P. (1974). Vegetation classification by reference to strategies. *Nature*, **250**, 26–31.

Grime, J.P. (1977). Evidence for the existence of three primary strategies in plants and its relevance to ecological and evolutionary theory. *American Naturalist*, **111**, 1169–94.

Grime, J.P. (1979). *Plant Strategies and Vegetation Processes*. Chichester: Wiley.

Grime, J.P. (1983). Prediction of weed and crop response to climate based upon measurements of nuclear DNA content. *Aspects of Applied Biology*, **4**, 87–98.

Grime, J.P. (1986). The circumstances and characteristics of spoil colonization within a local flora. *Philosophical Transactions of the Royal Society*, **B314**, 637–54.

Grime, J.P. (1988). The C-S-R model of primary plant strategies—Origins, implications and tests. In *Evolutionary Plant Biology*, ed. L.D. Gottlieb and S. Jain, pp. 371–93. London: Chapman and Hall.

Grime, J.P. and Callaghan, T.V. (1988). *Direct and indirect effects of climate change on plant species, ecosystems and processes of conservation and amenity interest*. Contract Report to the Department of the Environment. Ref. T07020b1. London: Department of the Environment.

Grime, J.P., Hodgson, J.G. and Hunt, R. (1988). *Comparative Plant Ecology. A Functional Approach to Common British Species*. London: Unwin Hyman.

Grime, J.P. and Mowforth, M.A.G. (1982). Variation in genome size—An ecological interpretation. *Nature*, **299**, 151–3.

Grime, J.P., Shacklock, J.M.L. and Band, S.R. (1985). Nuclear DNA contents, shoot phenology and species coexistence in a limestone grassland community. *New Phytologist*, **100**, 435–45.

Grubb, P.J. (1977). The maintenance of species-richness in plant communities: The importance of the regeneration niche. *Biological Reviews*, **52**, 107–45.

Hartsema, A.M. (1961). Influence of temperature on flower formation and flowering of bulbous and tuberous plants. In *Handbuch der Pflanzenphysiologie. 16: Ansenfaktoren in Wachstum und Entwicklung*, ed. W. Ruhland, pp. 123–67. Berlin: Springer.

Hodgson, J.G. (1986). Commonness and rarity in plants with special reference to the Sheffield flora. Part II. The relative importance of climate, soils and land use. *Biological Conservation*, **36**, 253–74.

Holdridge, L.R. (1947). Determination of world plant formations from simple climatic data. *Science*, **105**, 367–8.

Hutchinson, G.E. (1959). Homage to Santa Rosalia or why are there so many kinds of animals? *American Naturalist*, **93**, 145–59.

Law, R. (1974). *Features of the biology and ecology of* Bromus erectus *and* Brachypodium pinnatum *in the Sheffield region*. PhD thesis, University of Sheffield.

Levin, D.A. and Funderburg, S.W. (1979). Genome size in angiosperms; temperate versus tropical species. *American Naturalist*, **114**, 784–95.

Macleod, J. (1894). Over de beruchting der bloemen in het Kempisch gedeelte van Vlaanderen. Deel 11. *Botanische Jaarboek*, **6**, 119–511.

Ramenskii, G.L. (1938). *Introduction to the Geobotanical Study of Complex Vegetations*. Moscow: Selkhozgiz.

Ridley, H.N. (1930). *The Dispersal of Plants Throughout the World*. Ashford: Reeve.

Salisbury, E.J. (1942). *The Reproductive Capacity of Plants*. London: Bell.

Salisbury, E.J. (1953). A changing flora as shown in the study of weeds or arable land and waste places. In *The Changing Flora of Britain*, ed. J.E. Lousley, pp. 130-9. Arbroath: Buncle.

Salter, P.J. and Goode, J.E. (1967). *Crop responses to water at different stages of growth*. Research Review 2. East Malling, Kent: Commonwealth Bureau of Horticulture and Plantation Crops.

Southwood, T.R.E. (1977). Habitat, the templet for ecological strategies? *Journal of Animal Ecology*, **46**, 337–65.

Stebbins, G.L. (1951). Natural selection and the differentiation of angiosperm families. *Evolution*, **5**, 299–324.

Stebbins, G.L. (1956). Cytogenetics and evolution of the grass family. *American Journal of Botany*, **43**, 890–905.

Stebbins, G.L. (1971). Adaptive radiation of reproductive characters of angiosperms. II. Seeds and seedlings. *Annual Review of Ecology and Systematics*, **2**, 237–60.

Stebbins, G.L. (1974). *Flowering Plants: Evolution Above the Species Level*. London: Arnold.

van der Pijl, L. (1972). *Principles of Dispersal in Higher Plants*. Berlin: Springer-Verlag.

16

Modeling Crop Responses to Environmental Change

Cynthia Rosenzweig

This chapter explores potential biophysical responses of crops to climate change. Environmental variables affect crops at two levels: they physically delimit the geographical extent of the crop and they determine in large part the level of crop yield. Since both of these aspects are important in predicting the impacts of environmental change on agriculture, both will be addressed throughout the chapter. First, the primary environmental variables which define the limits to agricultural crop growth and production are surveyed. Second, the principal methods for predicting climate change impacts on crop geography and production are described, along with their limitations. Examples are then given of several case studies which use environmental variables and prediction techniques to demarcate present and future crop geography and to estimate potential crop production changes. Economic and social factors also contribute to the geographic distribution of agricultural production to a large extent, but these are not considered here.

Environmental Variables

The primary environmental variables which define the limits to growth and production of agricultural commodities (food and fiber) and govern the changes in the geographic distribution of crop production are solar radiation, temperature, water, soil, and nutrients. Atmospheric carbon dioxide concentration is an additional variable critical to crop growth; its direct effects may influence crop productivity and hence crop distribution. Responses to these variables are often interactive, e.g., nutrient uptake is temperature dependent and soil moisture often governs crop responses to temperature.

Solar Radiation

Crops require solar radiation for growth and production of usable biomass, since solar radiation provides the primary energy for photosynthesis. At high latitudes, the angle of incidence of the radiation received at the Earth's surface is low and thus the intensity of the sunlight is less than at many agriculturally important areas at more southern latitudes. However, the amount of light received in high latitudes is adequate for production of some crops in summer, if other growth factors are not limiting. The interrelationships between light, air temperature, and soil temperature are particularly important in these regions. Agricultural crops are grown as far north as Fairbanks, Alaska (Dinkel 1984), and the Southern Uplands of Scotland (Parry and Carter 1985). At low latitudes, poor light conditions are often present during the rainy season.

Light has another effect on crop development, independent of photosynthesis, which strongly influences the geographic distribution of many crop species and varieties, namely, photoperiodism. Photoperiodism is the process by which the relative duration of light and dark periods controls flowering.

Photoperiodism was first demonstrated by Garner and Allard in 1920. Oats and winter barley are long-day plants; i.e., they flower in response to days longer than some critical length. Soybean plants, on the other hand, are short-day plants that bloom only when daylengths are shorter than some maximum. Soybean cultivars differ in their photoperiod sensitivity; this sensitivity is the major determinant of the range of latitudes for which each cultivar is adapted. Daylength has been shown to affect the duration of soybean reproductive stages (Thomas and Raper 1976), pod and seed addition rates (Fisher 1963), and carbon partitioning to fruit (Cure *et al.* 1982). Temperature also influences the time required for soybeans to reach floral differentiation, floral expression, and subsequent reproductive stages under inductive photoperiods (Shibles *et al.* 1975).

Temperature

While temperature minima may be the controlling factor of the geographic distribution of unmanaged species (Woodward 1987), temperature limits crop production at both high and low extremes. This is probably due to the need for agricultural crops to produce usable (usually reproductive) biomass rather than merely to survive. High temperatures directly affect agricultural production through heat stress at critical phenological stages. Severe high-temperature stress at the silking and tasseling stages in maize, when the number of kernels on the ear is determined, results in significant yield decreases (Shaw 1983). Johnson and Kanemasu (1983) have shown that excessive high temperatures (over 30–32°C) are particularly damaging to wheat during grain filling.

Regression analysis has shown that July and August daily maximum temperatures greater than 33.3°C are negatively correlated with yield in the U.S. Corn

Belt (Thompson 1975) and that temperatures exceeding 37.7°C caused severe damage to corn in 1980 (McQuigg 1981). Analysis of wheat yields has shown similar responses to high temperatures (Ramirez *et al.* 1975). In soybeans, temperature stress above 40°C at flowering results in flower and pod abortion (Mederski 1983).

Low temperatures also cause damage to crops. Extreme cold is the primary cause of winterkill in wheat, particularly in the absence of snow. Other factors which affect the susceptibility of wheat to winterkill include drought, wind erosion, flooding, hail, disease, and insect pests (Caprio 1984). Winter temperature minima define the boundaries of citrus fruit production, because low minimum temperatures cause severe damage to citrus trees. Boundary shifts of citrus production toward higher latitudes may be an early signal of agricultural response to climate warming.

An isolated freezing event can be even more damaging to crops if there has been a lack of previous gradually decreasing temperatures for cold hardening. Therefore, if winter temperatures rise as predicted for the greenhouse warming (Hansen *et al.* 1986), unexpected cold damage may result due to lower levels of cold hardening.

Vernalization is another temperature effect which is important for the distribution of crops. This is a requirement for a low-temperature period before floral induction. Thus, for a crop such as winter wheat, winter temperatures must be low enough for vernalization but high enough to remain above the crop threshold survival temperature in its dormant vegetative state.

Many cereal grasses require vernalization. Winter wheat responds to and even requires vernalization, whereas spring wheat shows little response. Vernalization of winter wheat occurs between 0°C and 11°C with an optimum at 3°C (Evans *et al.* 1975) for a duration of 6–8 weeks (Ahrens and Loomis 1963). In regions such as Mexico and the Mediterranean where winter temperatures rarely satisfy these requirements, spring wheat is sown in the fall and harvested in early summer.

Studies with general circulation models (GCMs) show that temperatures should warm more in winter than in summer with increased CO_2 (Hansen *et al.* 1986). This is partially because reduced snow cover allows for more absorption of incident solar radiation. As winter temperatures increase, low temperatures needed for vernalization of some crops may not occur in some locations. Thus, winter wheat cultivars may be replaced by spring types in these places.

Precipitation

Precipitation, and the related variable soil moisture, are probably the most important factors determining the location, timing, and productivity of crop growth. Both too little and too much moisture are critical. Interannual variability of precipitation is a major cause of variation in crop yields. During the 1930s, severe droughts reduced U.S. Great Plains yields of wheat and corn by as much as 50%.

Failure of the monsoon in 1987 caused yield shortfalls in Pakistan, Bangladesh, and India. Drought also reduces vegetative cover which in turn exacerbates both wind and water erosion, thus affecting future crop productivity.

Excessive precipitation may also prohibit growth of crops such as wheat, which does not possess the structural integrity to stand, and is susceptible to insects and diseases under rainy conditions. Little wheat is grown where precipitation is greater than 1200 mm yr^{-1}. Soft wheat types, which are used for cakes and cookies, are grown in areas with moderate precipitation (760–1200 mm yr^{-1}). Hard wheat types for bread flour are grown in drier areas. If hydrologic regimes alter with climate change, distribution of wheat types may change because of these relationships.

Soils

The basic soil requirements of crop plants relate to temperature, moisture, aeration, fertility, depth, texture and stoniness, acidity, salinity and toxicity, tilth, slope, and flooding (FAO 1978). For a given crop, ranges of these characteristics may be defined for (1) optimal production; (2) marginal growth where yield will be diminished; and (3) unsuitability, when the crop cannot be grown under present technology. These have been estimated for 12 major crops under rain-fed cultivation (FAO 1978). This type of tabulation is useful for evaluating soil resource suitability for projected shifts in geographical crop regions.

Nutrients

Crops require both micronutrients (molybdenum, copper, zinc, manganese, boron, iron, and chlorine) and macronutrients (sulfur, phosphorus, magnesium, calcium, potassium, nitrogen, oxygen, carbon, and hydrogen) to produce usable biomass. These are provided by the soil, water, and air where crops are grown. However, many nutrients can be added to the crop-growing medium as fertilizer, thus freeing crops from some natural geographical constraints. The inherent fertility of a soil, while composed of many individual characteristics, is basically related to its cation-exchange capacity and the relative degree of base saturation. This information is available for most soils and may be used to classify the suitability of soils for agricultural production.

The processes by which many nutrients become available to plants are dependent on climate variables (Russell 1973). For example, nitrification, the process by which bacteria oxidize ammonium salts to nitrates, is slowed by soils which are too cold or too wet. The rate of phosphate uptake is low at low soil temperatures. On the other hand, high soil temperatures have a depressing effect on legume nodulation of nitrogen-fixing bacteria. These types of relationships will also affect crop spatial distribution.

Atmospheric Concentration of CO_2

Most plants growing in atmospheric CO_2 concentrations higher than ambient exhibit increased rates of net photosynthesis (i.e., total photosynthesis minus respiration). CO_2 enrichment also narrows the stomata of some crop plants, reducing transpiration per unit leaf area while enhancing photosynthesis. Thus it often improves water-use efficiency, which is defined as the ratio between crop biomass accumulation or yield and the amount of water lost through evapotranspiration. Experimental effects of CO_2 on crops have been reviewed by Acock and Allen (1985) and Cure (1985); see also Körner, this volume. In a compilation of greenhouse and other experimental studies, Kimball (1983) estimated a mean crop-yield increase of 33 $\pm$ 6% for a doubling of CO_2 concentration from 300 to 600 ppm.

Plant species differ in their photosynthetic mechanisms and thus in their responses to increased CO_2. In species with the C3 pathway characteristic of nontropical plants, a considerable fraction of the CO_2 initially fixed into carbohydrate is reoxidized back to CO_2 in a process known as photorespiration (Tolbert and Zelitch 1983). Photosynthesis rates in C3 species respond positively to increased CO_2 levels because photorespiration is suppressed in these conditions. In C4 plants which are particularly characteristic of tropical and warm arid regions, CO_2 is first trapped inside the leaf and then concentrated in the cells in which photosynthesis occurs (Black 1973). These plants are more efficient photosynthetically than C3 plants and show less response to increasing CO_2.

Important crop plants with the C3 photosynthetic pathway are wheat, rice, and soybean. With increased levels of CO_2 in the atmosphere, these may be substituted for C4 crops because of their relatively greater photosynthetic enhancement. C4 crops of economic importance are corn, sorghum, sugarcane, and millet. Such crops may suffer from increased infestation by C3 weeds.

The direct effects of CO_2 occur at less-than-optimal levels of other environmental variables, such as light, water, temperature, nitrogen, and salinity. The interactions of CO_2 and temperature effects are particularly important because of the predicted direct relationship between these variables, but the interactions have not been studied extensively. Experiments have shown differing results in this regard. In some studies, plants have responded to increasing air temperatures along with CO_2 enrichment with rising leaf temperatures (Chaudhuri *et al.* 1987) and higher yields (Kimball *et al.* 1986). In another study, spring wheat plants grew best at high CO_2 levels and low temperatures (Wall and Baker 1987).

Yields are relatively higher with high CO_2 for water-stressed plants than for nonstressed plants (Gifford 1987). Nitrogen-limited plants also respond positively to high CO_2 (Wong 1979). Thus some marginal agricultural areas, e.g., semiarid regions, may improve in stability in food production with increased atmospheric CO_2 (Gifford 1987).

In order to project the impact of increasing CO_2 on agricultural production, these beneficial direct effects must be taken into consideration along with the climatic effects of the radiatively active trace gases. GCMs project warming in virtually all regions, while hydrologic regimes may become either wetter or drier. The climatic effects on agriculture may thus be either negative or positive depending on location. The assessment of the relative contributions of the direct effects of CO_2 and the predicted climate changes to agricultural crop responses remains a crucial research question.

Methods for Projecting the Effects of Climate Change on Crop Geography and Production

Spatial Approach

One way to project climate-change effects on the spatial distribution of agricultural plants is to identify and quantify the critical environmental limits of specific crops, apply climate-change scenarios, and calculate the resulting spatial shifts in crop regions. In this method, the spatial patterns of crop attributes are compared with the spatial patterns of climate variables or soils and then recompared with climate-change scenarios. This approach provides an approximation of possible changes in crop areas from a biological perspective but does not address potential changes in yield and production.

The environmental requirements of crop growth may be expressed as ranges of measured climatic variables, e.g., surface air temperature, precipitation, and solar radiation, or as derived variables known as agroclimatic indices. The agroclimatic indices are more closely related to crop growth and distribution than simple climatic variables alone. Examples of agroclimatic indices are length of potential growing period (either frost-free days or wet/dry seasons); growing degree units (accumulated temperatures above a threshold) (Newman 1980); potential evapotranspiration (temperature and solar radiation) (Thornthwaite 1948); and the Palmer drought index (a water balance based on precipitation and temperature) (Palmer 1965).

An example of the spatial approach is found in Rosenzweig (1985). Environmental requirements for classification of North American wheat-growing regions were defined (Tab. 16.1) and used to project regional changes under a GCM doubled-CO_2 scenario. Results showed that areas of wheat production increase, particularly in Canada, due to increased growing degree units. (See Fig. 16.1.) While global warming may alleviate some climatic constraints to growth of some crops, actual expansion of agricultural production may be limited by the absence of soils suitable for agriculture. For example, the rocky nature of the Canadian Shield would preclude to a large extent agricultural expansion in northern Saskatchewan (Williams *et al.* 1988), and the peaty, poorly drained soils of many north boreal regions can hardly support agriculture.

Table 16.1. Environmental requirements for wheat used in classification of wheat-growing regions of North America (from Rosenzweig 1985).

Length of growing season (days)	90
Growing degree units per growing season	1200
Minimum and base temperature	4°C
Maximum temperature	32°C
Mean minimum temperature in January	
Spring wheat	<−12°C
Winter wheat	≥−12°C
Vernalization requirement	
Winter wheat—at least one mean monthly surface temperature	≤5°C
Fall-sown spring wheat—mean monthly temperature for all months	>5°C
Annual precipitation (mm yr^{-1})	
No wheat grown	≥1200
Soft wheat	760–1200
Hard wheat	0–760
Dry moisture conditions	0–380
Adequate moisture conditions	380–760

Potential Production

Methods to estimate maximum yield or total potential agricultural productivity have also been developed, and may be used in climate-change impact analysis. However, these approaches tend to smooth the high degree of variability in production attributable to other factors that affect crop growth such as nutrients and management. Also, high-temperature effects on crop biomass and crop yields are often opposite. While annual net primary productivity of perennial vegetation should increase with increased temperatures, many agricultural crops are annuals which respond to higher temperatures by hastening the rate of development and by decreasing yields.

Turc and Lecerf (1972) derived a climatic index of agricultural potential which combines precipitation, evapotranspiration, solar radiation, and temperature. This index was used by Williams (1985) to estimate the sensitivity of agricultural productivity in Alberta, Canada, to climatic change. Warming induced by CO_2 doubling gave a net increase in bioresource productivity of about 16%.

Another technique for determining production potential is the method developed for the FAO Agro-Ecological Zone Project (FAO 1978). The three factors that contribute to maximum yield potential are temperature, length of growing season, and incident solar radiation (which is in turn a function of cloud cover, season, and latitude). Doorenbos and Kassam (1979) expanded this method to quantify the relationships between water use and crop yields on a regional basis.

Statistical Regression Analysis

Statistical techniques, such as regression analysis, have been used to link climatic variables and crop yields for given regions quantitatively (Ramirez *et al.* 1975;

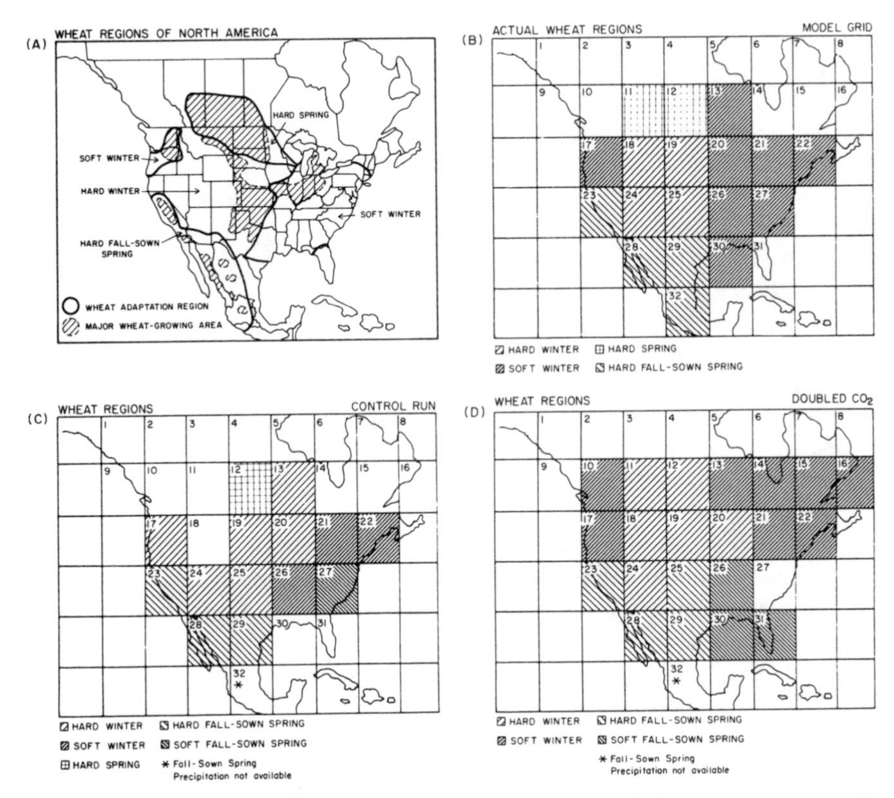

Fig. 16.1. A. Major wheat-growing areas of North America. Source: U.S. Wheat Associates and Foreign Agricultural Service, USDA. B. Actual North American wheat-growing regions on the Goddard Institute for Space Studies (GISS) GCM grid. C. Simulated North American wheat regions using the GISS GCM control run. D. Simulated wheat regions using the GISS GCM doubled-CO_2 run (from Rosenzweig 1985).

Thompson 1975). This approach has been most successful where one or two environmental factors dominate in effect on crop yields. Climatic indices, described above, are often used in regression equations.

Limitations of this technique include the lack of attention to processes by which climate affects plants and the specificity of location, past climate, and technology inputs. Problems include the assumption of linearity of crop response to climate variables, the interdependence of climatic variables, the inference of causation from cross-correlation, and, most basically, the complex interactions of location, season, cultivar, and management that occur wherever crops are grown. Thus, the use of statistical regression techniques may not be particularly appropriate for analysis of climate-change agricultural impacts.

Dynamic Process Crop-growth Models

Crop-growth models that simulate the response of agricultural plants to climate, soil, and management factors may be used with climate-change scenarios to explore consequences for yields and phenology. Results from such studies may then be used in economic models of agricultural sectors to estimate changes in the structure of production (see, for example, Pitovranov *et al.* 1988) or comparative advantage among regions (Environment Canada 1987).

Dynamic-process crop-growth models exist for many crops. These models use a physiological approach by which crop growth and yields respond to the major climatic, edaphic, and management factors governing production (Wisiol and Hesketh 1987). There exists a continuum of empirical and functional relationships in the structure of these models.

Crop models were developed from relationships among current climate variables and crop responses. These relationships may or may not hold under differing climatic conditions and this adds uncertainty to the predictions of crop responses to climate change. Many of the models employ simple, empirically based relationships. Other uncertainties reside in the lack of simulation of weeds, diseases, insect pests, and catastrophic weather events.

A single crop model may be used over a range of sites to study potential crop responses to climate change scenarios. This allows for investigation of the interannual variability in crop response to climate-change scenarios and definition of changing levels of risk of crop failure. This is often the critical factor determining presence or absence of crop production in a region.

In addition, crop-growth models provide a means with which to study the relative contributions of the direct and climatic effects of increasing CO_2 on crop growth. Changes in photosynthesis and transpiration can be embedded functionally in the models, which can then be run with climate-change scenarios. Rosenzweig (1987) studied the relative magnitudes of these effects on wheat yields in the Southern Great Plains with the CERES-Wheat model (Ritchie and Otter 1985) and found that water regime defined the sign and magnitude of the combined effects. Under moist conditions, the positive physiological effects compensated for the less-favorable climate, while under more arid conditions, the physiological effects were not able to compensate for effects of moisture stress. (See Fig. 16.2.) Results such as these are preliminary, given the uncertainties in GCM-generated climate-change scenarios and the need for better understanding of crop responses to CO_2 under field conditions.

Adjoint Method and Risk Assessment

Parry *et al.* (1988) advocate an approach to climate-change impact analysis they term *the adjoint method*. Using this method, one specifies the sensitivity of a crop attribute to climate variables, determines which climate variations are the most critical, and then compares the climatic variations projected for the future with the critical levels. An example of the use of the adjoint approach is found in Williams *et al.* (1988); this study identified the frequency of days of blowing dust as critical to cereal production in a Canadian wheat-growing region. An

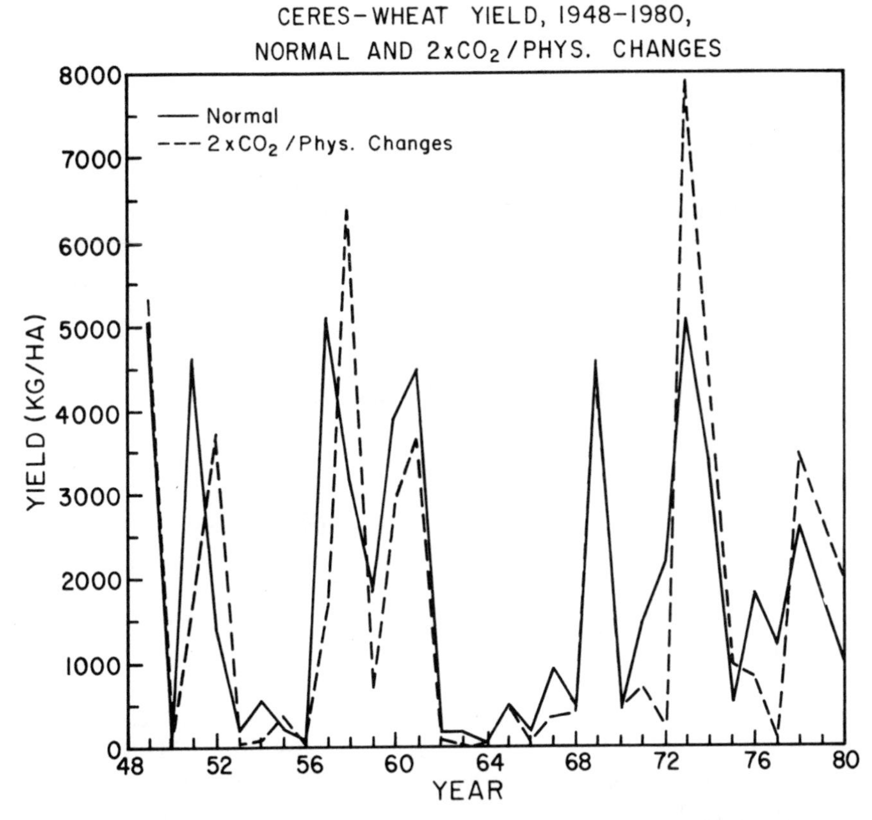

Fig. 16.2. Simulated CERES-Wheat yield for central Kansas, 1948–80, for observed climate and GISS GCM doubled-CO_2 climate change scenario with direct effects of CO_2 on photosynthesis and water use.

advantage of the adjoint approach is that it helps to identify the thresholds of climate change which are critically important. Since there are large uncertainties in the prediction of future climate change, an additional advantage to the adjoint approach is that it is independent of climate scenarios generated by GCMs for the initial analysis. A limitation of this approach (and many of the other approaches described above) is the lack of consideration of explicit technological improvements and adjustment mechanisms. Adjustments by farmers such as adoption of improved crop varieties and establishment of irrigation systems are likely to have significant effects on the delineation of the critical thresholds of climate sensitivity.

Another aspect of the adjoint approach is that it views the impacts of climate change in terms of change in the level of risk. These risks are often related to changes in the frequency of extreme meteorological events which can have a large detrimental effect on crop yields. It is important to study both the changes in climate variability and the related changes in risks of crop failures.

Mearns *et al.* (1984) found that the relationships between changes in mean temperature and the corresponding changes in the probabilities of extreme temperature events are nonlinear and that relatively small changes in mean temperature can result in relatively large changes in event probabilities. Parry and Carter (1985) assessed the risk of crop failure resulting from low levels of accumulated temperature for oat farming in southern Scotland and concluded that minor climatic variations in the United Kingdom have induced substantial changes in magnitudes of agricultural risks.

Case Studies of Crop Geography and Production

Winter Wheat in China

The north-south extent of wheat production in China is largely determined by temperature, while the east-west extent is mainly the result of annual and seasonal precipitation (Zheng and Newman 1986). Zheng and Newman (1986) suggest that the northern boundary of winter wheat production in China can be determined by the annual mean extreme minimum temperature plus the negative accumulation of winter daily mean temperature less than 0°C. They set a range of minimum temperatures of less than or equal to −20 to −22°C and a range of accumulated negative cold temperatures of −400 to −700°C. Winter wheat will not grow in China when temperatures are lower than the lower limit of these ranges; it will survive consistently when temperatures are higher than the upper limits of these ranges. In between, survival of winter wheat depends on other factors, such as drought and snow cover. These relationships can be used to project Chinese wheat regions for predicted climate-change scenarios.

Rice Production in Japan

Yoshino *et al.* (1988) evaluated the effects of specified climate scenarios on the location of rice production and on rice yields, among other aspects of Japanese agriculture. With the warmer conditions predicted in a GCM doubled-CO_2 scenario, a great expansion of cultivable rice area was predicted by using agroclimatic indices. A process-based model was used to simulate rice yields as a function of solar radiation incident on the crop and temperature conditions during its growth. The study concluded that a doubled CO_2 climate, as predicted by a GCM, would result in a small rise in rice yields, caused by a decreased frequency of cool summer events. This could lead to a rice surplus that would require government measures for its disposal. If farmers were to adapt to the warmer temperatures with more heat-tolerant rice varieties even higher surpluses are to be expected.

Maize Production in the U.S. Corn Belt

Blasing and Solomon (1983) identified growing degree units and annual precipitation characteristics of the current Corn Belt and then mapped the projected Corn Belt corresponding to climate change predicted for a doubling of CO_2 based on GCM results published in Manabe and Stouffer (1980). In this scenario, the Corn Belt shifts in a northeasterly direction, but movement is limited by the Great Lakes. Warmer and drier conditions in the southwestern portions of the Corn Belt favor the replacement of corn with winter wheat.

Conclusions

Climate has a significant impact on the geographic distribution and productivity of agricultural crops. Numerous methods for predicting how climate change may affect agriculture have been devised. Many of these methods have been utilized in preliminary studies of agricultural response to global warming.

The principal uncertainties in the prediction of the impacts of climate change on agriculture reside in the contribution of the direct effects of increasing CO_2, in potential changes in climate variability, and in the effects of adjustment mechanisms that farmers may adopt in light of climatic changes. Research has often addressed the climatic and vegetational effects of increasing CO_2 separately, even though biological and atmospheric processes are linked in many ways and have synergistic effects. The potential beneficial effects of increasing CO_2 on photosynthesis and water use must be weighed in relation to the magnitudes of the climatic changes predicted by the GCMs in order to more completely understand the nature and direction of physical crop responses.

Although potential changes in climate variability have rarely been studied in relation to the increasing radiatively active trace gases, they are particularly crucial for analysis of agricultural impact analysis. Also uncertain are the effects of adjustment mechanisms such as the breeding of new crop types and improvements in technology by which farmers may alleviate the effects of climate change and thus affect crop productivity and distribution.

Methods of studying climate change, increasing CO_2, and agriculture are improving as experimental and modeling techniques continue to be developed. Iterations of agricultural assessment studies should continue to bring clearer understanding to this important field of research.

References

Acock, B. and Allen, L.H., Jr. (1985). Crop responses to elevated carbon dioxide concentrations. In *Direct Effects of Increasing Carbon Dioxide on Vegetation*, ed. B.R.

Strain and J.D. Cure, pp. 33–97. DOE/ER-0238. Washington, D.C.: U.S. Department of Energy.

Ahrens, J.F. and Loomis, W.E. (1963). Floral induction and development in winter wheat. *Crop Science*, **3**, 463–6.

Black, C. (1973). Photosynthetic carbon fixation in relation to net CO_2 uptake. *Annual Review of Plant Physiology*, **24**, 253–86.

Blasing, T.J. and A. Solomon, A.M. (1983). *Response of the North American Corn Belt to Climate Warming*. DOE/NBB-0040. Washington, D.C.: U.S. Department of Energy.

Caprio, J.M. (1984). *Study to Improve Winterkill Parameters for a Winter Wheat Model*. NASA Contract NAS 9-16007. Final Project Report. Bozeman, Montana: Montana State University.

Chaudhuri, U.N., Burnett, R.B., Kanemasu, E.T. and Kirkham, M.B. (1987). Effect of elevated levels of CO_2 on winter wheat under two moisture regimes. 040 in Series: *Response of Vegetation to Carbon Dioxide*. Washington, D.C.: U.S. Departments of Energy and Agriculture.

Cure, J.D. (1985). Carbon dioxide doubling responses: A crop survey. In *Direct Effects of Increasing Carbon Dioxide on Vegetation*, ed. B.R. Strain and J.D. Cure, pp. 99–116. DOE/ER-0238. Washington, D.C.: U.S. Department of Energy.

Cure, J.D., Patterson, R.P., Raper, C.D., Jr. and Jackson, W.A. (1982). Assimilate distribution in soybeans as affected by photoperiod during seed development. *Crop Science*, **22**, 1245–50.

Dinkel, D.H. (1984). Improved crop potential for northern latitudes to occur with small increases of air and soil temperatures. In *The Potential Effects of Carbon Dioxide-Induced Climatic Changes in Alaska*, ed. J.H. McBeath, pp. 175–7. Miscellaneous Publication 83-1. University of Alaska–Fairbanks.

Doorenbos, J. and Kassam, A.H. (1979). *Yield Response to Water*. FAO Irrigation and Drainage Paper No. 33. Rome: Food and Agriculture Organization of the United Nations.

Environment Canada. (1987). *Implications of Climatic Warming for Canada's Comparative Position in Agricultural Production and Trade*. LEG-27. Land Evaluation Group. Guelph, Ontario: Atmospheric Environment Service.

Evans, L.T., Wardlaw, I.F. and Fischer, R.A. (1975). Wheat. In *Crop Physiology*, ed. L.T. Evans, pp. 101–49. Cambridge: Cambridge University Press.

FAO. (1978). *Report on the Agro-Ecological Zones Project. Volume 1. Methodology and Results for Africa*. Rome: Food and Agriculture Organization of the United Nations.

Fisher, J.E. (1963). The effects of short days on fruit set as distinct from flower formation in soybeans. *Journal of Botany*, **41**, 871–3.

Garner, W.W. and Allard, H.A. (1920). Effect of the relative length of day and night and other factors of the environment on growth and reproduction in plants. *Journal of Agricultural Research*, **18**, 553–606.

Gifford, R.M. (1987). Exploiting the "fertilizing" effect of higher levels of atmospheric carbon dioxide. Presented at *Climate and Food Security—An International Symposium*, ISA/IRRI/AAAS, New Delhi, February 1987.

Hansen, J., Lacis, A., Rind, D., Russell, G., Fung, I., Ashcraft, P., Lebedeff, S., Ruedy, R. and Stone, P. (1986). The greenhouse effect: Projections of global climate change. In *Effects of Changes in Stratospheric Ozone and Global Climate. Volume 1: Overview*, pp. 199–218. Washington, D.C.: U.S. Environmental Protection Agency.

Johnson, R.C. and Kanemasu, E.T. (1983). Yield and development of winter wheat at elevated temperatures. *Agronomy Journal*, **75**, 561–5.

Kimball, B.A. (1983). Carbon dioxide and agricultural yield: An assemblage and analysis of 430 prior observations. *Agronomy Journal*, **75**, 779–88.

Kimball, B.A., Mauney, J.R., Radin, J.W., Nakayama, F.S., Idso, S.B., Hendrix, D.L., Akey, D.H., Allen, S.G., Anderson, M.B. and Hartung, W. (1986). Effects of increasing atmospheric CO_2 on the yield and water use of crops. 039 in Series: *Response of Vegetation to Carbon Dioxide*. U.S. Departments of Energy and Agriculture.

Manabe, S. and Stouffer, R.J. (1980). Sensitivity of a global climate model to an increase in CO_2 concentration in the atmosphere. *Journal of Geophysical Research*, **85**, 5529–54.

McQuigg, J.D. (1981). Climate variability and crop yield in high and low temperature regions. In *Food-Climate Interactions*, ed. W. Bach, J. Pankrath and S.H. Schneider, pp. 121–38. Dordrecht: Reidel.

Mearns, L.O., Katz, R.W. and Schneider, S.H. (1984). Extreme high-temperature events: Changes in their probabilities with changes in mean temperature. *Journal of Climate and Applied Meteorology*, **23**, 1601–13.

Mederski, H.J. (1983). Effects of water and temperature stress on soybean plant growth and yield in humid temperate climates. In *Crop Reactions to Water and Temperature Stress in Humid, Temperate Climates*, ed. C.D. Raper and P.J. Kramer, pp. 35–48. Boulder: Westview Press.

Newman, J.E. (1980). Climate change impacts on the growing season of the North American "Corn Belt." *Biometeorology*, **7**, Part 2: 128–42.

Palmer, W.C. (1965). *Meteorological Drought*. Research Paper No. 45. Washington, D.C.: U.S. Department of Commerce.

Parry, M.L. and Carter, T.R. (1985). The effect of climatic variations on agricultural risk. *Climatic Change*, **7**, 95–110.

Parry, M.L., Carter, T.R. and Konijn, N.T. ed. (1988). *The Impact of Climatic Variations on Agriculture. Volume I: Assessments in Cool Temperate and Cold Regions*. Dordrecht: Kluwer Academic Publishers.

Pitovranov, S.E., Iakimets, V., Kiselev, V.I. and Sirotenko, O.D. (1988). The effects of climatic variations of agriculture in the subarctic zone of the USSR. In *The Impact of Climatic Variations on Agriculture. Volume 1: Assessments in Cool Temperate and Cold Regions*, ed. M.L. Parry, T.R. Carter and N.T. Konijn, pp. 615–722. Dordrecht: Kluwer Academic Publishers.

Ramirez, J.M., Sakamoto, C.M. and Jensen, R.E. (1975). Wheat. In *Impacts of Climatic Change on the Biosphere*, pp. 4-37 to 4-90. CIAP Monograph 5, Part 2, Climatic Effects.

Ritchie, J.T. and Otter, S. (1985). Description and performance of CERES-Wheat: a user-oriented wheat yield model. In *ARS Wheat Yield Project*, ed. W.O. Willis, pp. 159–75. ARS–38. USDA-ARS.

Rosenzweig, C. (1985). Potential CO_2-induced climate effects on North American wheat-producing regions. *Climatic Change*, **7**, 367–89.

Rosenzweig, C. (1987). Climate change impact on wheat: The case of the High Plains. In *Proceedings of the Symposium on Climate Change in the Southern United States: Future Impacts and Present Policy Issues*, ed. M. Meo, pp. 135–60. Washington, D.C.: U.S. Environmental Protection Agency.

Shaw, R.H. (1983). Estimates of yield reductions in corn caused by water and temperature stress. In *Crop Reactions to Water and Temperature Stresses in Humid, Temperate Climates*, ed. C.D. Raper and P.J. Kramer, pp. 49–66. Boulder: Westview Press.

Shibles, R., Anderson, I.C. and Gibson, A.H. (1975). Soybean. In *Crop Physiology*, ed. L.T. Evans, pp. 151–89. Cambridge: Cambridge University Press.

Thomas, J.F. and Raper, C.D. (1976). Photoperiodic control of seed filling for soybeans. *Crop Science*, **16**, 667–72.

Thompson, L.M. (1975). Weather variability, climatic change, and grain production. *Science*, **188**(4188), 535–41.

Thornthwaite, C.W. (1948). An approach toward a rational classification of climate. *Geographical Review*, **38**, 55–89.

Tolbert, N.E. and Zelitch, I. (1983). Carbon metabolism. In *CO_2 and Plants, The Response of Plants to Rising Levels of Atmospheric Carbon Dioxide*, ed. E.R. Lemon, pp. 21–64. AAAS Selected Symposium 84. Boulder: Westview Press.

Turc, L. and Lecerf, H. (1972). Indice climatique de potentialité agricole. *Science du Sol*, **2**, 81–102.

Wall, G.W. and Baker, D.N. (1987). Effects of temperature and elevated carbon dioxide concentrations on the phenology of spring wheat. *Agronomy Abstracts 1987*, p. 18, American Society of Agronomy, Madison.

Williams, G.D.V. (1985). Estimated bioresource sensitivity to climatic change in Alberta, Canada. *Climatic Change*, **7**, 55–79.

Williams, G.D.V., Fautley, R.A., Jones, K.H., Stewart, R.B. and Wheaton, E.E. (1988). Estimating effects of climatic change on agriculture in Saskatchewan, Canada. In *The Impact of Climatic Variations on Agriculture. Volume 1: Assessments in Cool Temperate and Cold Regions*, ed. M.L. Parry, T.R. Carter and N.T. Konijn, pp. 219–379. Dordrecht: Kluwer Academic Publishers.

Wisiol, K. and Hesketh, J.D. (1987). *Plant Growth Modeling for Resource Management. Volume I: Current Models and Methods. Volume II: Quantifying Plant Processes*. Boca Raton, Florida: CRC Press, Inc.

Wong, S.C. (1979). Elevated atmospheric partial pressure of CO_2 and plant growth. I. Interactions of nitrogen nutrition and photosynthetic capacity in C_3 and C_4 species. *Oecologia* (Berl.), **44**, 68–74.

Woodward, F.I. 1987. *Climate and Plant Distribution*. Cambridge: Cambridge University Press.

Yoshino, M., Horie, T., Seino, H., Tsujii, H., Uchijima, I. and Uchijima, Z. (1988). The effect of climate variations on agriculture in Japan. In *The Impacts of Climatic Variations on Agriculture. Volume 1. Assessments in Cool Temperate and Cold Regions*, ed. M.L. Parry, T.R. Carter and N.T. Konijn, pp. 725–868. Dordrecht: Kluwer Academic Publishers.

Zheng, J.F. and Newman, J.E. (1986). The climatic resources for wheat production in China. *Agricultural and Forest Meteorology*, **38**, 205–16.

Concluding Comments

One of the strong motivations for better understanding the patterns and processes controlling vegetation pattern at the global scale has been an increased awareness of the Earth's systems (as a whole) and the realization that the Earth has changed in the past and may do so again. The climatic changes that are being predicted by computer models of the earth's "weather machine" (General Circulation Models, or GCMs) in response to a doubling of atmospheric carbon dioxide appear to be ecologically significant, at least at the higher northern latitudes (Shugart *et al.* 1986). Nevertheless, "uncertainty" is one of the key words that arises in any discussions evaluating the possibility and impact of anthropogenic global climate change. As reviewed in Bolin *et al.* (1986), the uncertainty of the magnitude of the increase in the Earth's CO_2 ambient and the temporal pattern of the projected increase is great, as is the regional pattern of climatec change (particularly with regard to the climate variables that involve water—precipitation, cloudiness, soil moisture, etc.)

From the point of view of an ecologist, a central question to be asked about the response of ecosystems to global climatic change is, "Will the rate and magnitude of global change be ecologically significant?" Will ecosystems simply adjust to change with no apparent effects or will changes occur too fast and be too large for such adjustment? If alterations occur, what will be the pattern and consequences of these responses? If the forests, for example, do change, what actions can be taken to adapt to the new climate regimes? We are just now beginning to ask the reciprocal question: "How may changes in the Earth's terrestrial ecosystems interact with and alter other major Earth systems, particularly atmospheric systems?"

These questions are more easily asked than answered. The answers are rendered complex by assumptions needed to develop the special cases to which our relatively limited knowledge of global ecology applies. A particularly important consideration is that although we have some spatial proxies for climate change (altitudinal gradients, geographic differences) to test our models, we have no obvious direct tests of a simultaneous climate change and change in CO_2 ambient atmospheric concentration. This puts a premium on the merging of models that can be tested to some degree for climate change and that can include what we feel are the fundamental mechanisms in the CO_2 ambient response of plants.

Projecting the effects of change on the vegetation of the earth is sufficiently daunting that to state the problem is to signify its magnitude. Yet, as one may gain in the reading of the preceding chapters, there is not a lack of approaches to the problem. Rather, there is a great richness in the possible techniques that one might take. Several options can be used to project large-scale and global vegetation pattern. These are based on different underlying models that, to one degree or another, contain fundamental biological and ecological mechanisms. A degree of agreement exists among scientists with differing points of view on what processes are essential to the understanding of global vegetation and, in particular, on what parts of the problem are particularly difficult.

These book chapters identify both the richness of methods and the areas of agreement. Several themes reoccur: The needs to understand the scaling up of processes that are important at one time- and space-scale to larger scales; the needs to optimize the conceptual benefits of representing processes in detail with the attendant logistical difficulties of estimating parameters and designing tests for elaborate models; the needs to identify the most appropriate system variables and their appropriate levels of aggregation or disaggregation. Assumptions in different models interweave to be parallel in some cases and in opposition in others. If global ecological studies had an infancy in the International Biological Program of the 1960s and 1970s, the current adolescence seems boisterous and eager to solve the larger-scale problems of the coming millennia.

It has been our hope to present as much as we could of the variety of the ideas in predicting large-scale vegetation dynamics. In doing so, we have also attempted to provide the reader with a sampling of the flavors of the class of research problems.

References

Bolin, B., Döös, B.R., Jäger, J.W., and Warrick, R.A. ed. (1986). *The Greenhouse Effect, Climatic Change,* and *Ecosystems*. (Scope 29) Chichester: John Wiley & Sons.

Shugart, H.H., Antonovsky, M.Y., Jarvis, P.G. and Sanford, A.P. (1986). CO_2, Climatic Change and Forest Ecosystems. In ed. Bolin, B., Döös, B.R., Jäger, J.W., and Warrick, R.A., The Greenhouse Effect, Climatic Change, and Ecosystems, pp. 475–521. Chichester: John Wiley & Sons.

Index